Like the Earth and planets, stars rotate. Understanding how stars rotate is central to modeling their structure, formation, and evolution and how they interact with their environment and companion stars. This authoritative volume provides a lucid introduction to stellar rotation and the definitive reference to the subject. It combines theory and observation in a comprehensive survey of how the rotation of stars affects the structure and evolution of the Sun, single stars, and close binaries.

This timely book will be of primary interest to graduate students and researchers studying solar and stellar rotation and close binary systems. It will also appeal to those with a more general interest in solar and stellar physics, star formation, binary stars, and the hydrodynamics of rotating fluids – including geophysicists, planetary scientists, and plasma physicists.

STELLAR ROTATION

Cambridge astrophysics series

Series editors

Andrew King, Douglas Lin, Stephen Maran, Jim Pringle and Martin Ward

Titles available in this series

7. Spectroscopy of Astrophysical Plasmas
   by A. Dalgarno and D. Layzer
10. Quasar Astronomy
   by D. W. Weedman
17. Molecular Collisions in the Interstellar Medium
   by D. Flower
18. Plasma Loops in the Solar Corona
   by R. J. Bray, L. E. Cram, C. J. Durrant and R. E. Loughhead
19. Beams and Jets in Astrophysics
   edited by P. A. Hughes
20. The Observation and Analysis of Stellar Photospheres
   by David F. Gray
21. Accretion Power in Astrophysics 2nd Edition
   by J. Frank, A. R. King and D. J. Raine
22. Gamma-ray Astronomy 2nd Edition
   by P. V. Ramana Murthy and A. W. Wolfendale
23. The Solar Transition Region
   by J. T. Mariska
24. Solar and Stellar Activity Cycles
   by Peter R. Wilson
25. 3K: The Cosmic Microwave Background Radiation
   by R. B. Partridge
26. X-ray Binaries
   by Walter H. G. Lewin, Jan van Paradijs and Edward P. J. van den Heuvel
27. RR Lyrae Stars
   by Horace A. Smith
28. Cataclysmic Variable Stars
   by Brian Warner
29. The Magellanic Clouds
   by Bengt E. Westerlund
30. Globular Cluster Systems
   by Keith M. Ashman and Stephen E. Zepf
31. Pulsar Astronomy 2nd Edition
   by Andrew G. Lyne and Francis Graham-Smith
32. Accretion Processes in Star Formation
   by Lee W. Hartmann
33. The Origin and Evolution of Planetary Nebulae
   by Sun Kwok
34. Solar and Stellar Magnetic Activity
   by Carolus J. Schrijver and Cornelis Zwaan
35. The Galaxies of the Local Group
   by Sidney van den Bergh
36. Stellar Rotation
   by Jean-Louis Tassoul

STELLAR ROTATION

JEAN-LOUIS TASSOUL
Université de Montréal

CAMBRIDGE
UNIVERSITY PRESS

CAMBRIDGE UNIVERSITY PRESS
Cambridge, New York, Melbourne, Madrid, Cape Town, Singapore, São Paulo

Cambridge University Press
The Edinburgh Building, Cambridge CB2 8RU, UK

Published in the United States of America by Cambridge University Press, New York

www.cambridge.org
Information on this title: www.cambridge.org/9780521772181

First published 2000
This digitally printed version 2007

A catalogue record for this publication is available from the British Library

Library of Congress Cataloguing in Publication data

Tassoul, Jean Louis.
Stellar rotation / Jean-Louis Tassoul.
p. cm. – (Cambridge astrophysics)
ISBN 0-521-77218-4
1. Stars – Rotation. I. Title. II. Series.
QB810.T36 2000
523.8 – dc21 99-16740
CIP

ISBN 978-0-521-77218-1 hardback
ISBN 978-0-521-03769-3 paperback

To my wife

Monique

There are epochs in the history of every great operation and in the course of every undertaking, to which the co-operations of successive generations of men have contributed (. . .), when it becomes desirable to pause for a while, and, as it were, to take stock; to review the progress made, and estimate the amount of work done: not so much for complacency, as for the purpose of forming a judgement of the efficiency of the methods resorted to, to do it; and to lead us to inquire how they may yet be improved, if such improvement be possible, to accelerate the furtherance of the object, or to ensure the ultimate perfection of its attainments. In scientific, no less than in material and social undertakings, such pauses and *résumés* are eminently useful, and are sometimes forced on our considerations by a conjuncture of circumstances which almost of necessity obliges us to take a *coup d'oeil* of the whole subject, and make up our minds, not only as to the validity of what is done, but of the manner in which it has been done, the methods employed, and the direction in which we are henceforth to proceed, and probability of further progress.*

Sir John Herschel (1792–1871)

* Quoted in Hatton Turnor, *Astra Castra – Experiments and Adventures in the Atmosphere*, p. v, London: Chapman and Hall, 1865.

Contents

Preface

When I wrote my first book – *Theory of Rotating Stars* (Princeton: Princeton University Press, 1978) – I was not aware of the fact that the 1970s were a period of transition and that major unexpected developments would take place in the field of stellar rotation during the 1980s.

In the mid-1970s, we had no direct information about the internal rotation of the Sun. Little was known about the rotation of main-sequence stars of spectral type G and later, although it was already well established that the surface rotation rate of these stars decayed as the inverse square root of their age. We certainly had much more information about axial rotation in the upper-main-sequence stars, but the actual distribution of specific angular momentum within these stars was still largely unknown. On the theoretical side, important progress in the study of rotating stars had been made by direct numerical integration of the partial differential equations of stellar structure. However, because there was no clear expectation for the actual rotation law in an early-type star, the angular momentum distribution always had to be specified in an ad hoc manner. The presence of large-scale meridional currents in a stellar radiative zone was also a serious problem: All solutions presented to date had unwanted mathematical singularities at the boundaries, and the back reaction of these currents on the rotational motion had never been properly taken into account. As far as I remember, there was only one bright spot that was emerging from this rather gloomy picture of stellar rotation: The observed degree of synchronism and orbital circularization in the short-period close binaries appeared to be in reasonable agreement with the (then current) theoretical views on tidal interaction in close binary systems. The year was 1977 and, as I said, we did not realize that the tide was turning fast.

Ten years later, helioseismology was already providing a wealth of detail about the internal rotation of the Sun through the inversion of p-mode frequency splittings. At the same time, spectroscopic rotational velocities for numerous lower-main-sequence stars and pre–main-sequence stars were derived on the basis of high signal-to-noise ratio data and Fourier analysis techniques. Modulation of starlight due to dark or bright areas on a rotating star was also currently used to obtain rotation periods for a number of low-mass main-sequence stars. Helioseismology has forced us to reconsider our views on the Sun's internal rotation. Similarly, the newly derived rotational velocities of stars belonging to open clusters have provided us with a general outline of the rotational history of solar-type stars. However, very little observational progress has been made in measuring the surface rotation rates of main-sequence stars more massive than the

Sun; and since asteroseismology is still in its infancy, we do not yet know their internal distribution of angular velocity. Unexpectedly, renewal of interest in the close binaries has led to the conclusion that synchronous rotators and circular orbits are observed in binaries with orbital periods substantially larger than previously thought possible. This is a most challenging result since it requires that we reconsider the currently held views on tidal interaction in close (and not so close) binaries.

Over the course of the past two decades, theoreticians have also made great progress in developing an understanding of the effects of rotation in stellar radiative zones. This progress has *not* resulted from the development of new observational techniques or faster supercomputers, however, but from the recognition that rotation generates meridional currents as well as a wide spectrum of small-scale, eddylike motions wherever radiative transfer prevails. The importance of these rotationally driven motions lies in the fact that, under certain conditions, they can produce chemical mixing in regions that remain unmixed in standard calculations of nonrotating stellar models. Meridional circulation and eddylike motions also explain in a natural way the correlation between slow rotation and abnormal spectrum in the Am and Ap stars. This new approach, which is based on the idea that eddylike motions are an ever-present feature of a stellar radiative zone, also resolves in a very simple manner the many contradictions and inconsistencies that have beset the theory of meridional streaming in rotating stars.

All these new developments provide sufficient justification for a new book on rotating stars that would summarize the basic concepts and present a concise picture of the recent important advances in the field. Unfortunately, because the subject has grown so much in breadth and in depth over the past twenty years, a complete coverage of all the topics discussed in my first book has become an almost impossible task for a single individual. This is the reason why I have tried to concentrate almost exclusively on topics dealing with main-sequence stars, making occasional incursions into their pre–main-sequence and post–main-sequence phases. Admittedly, although much attention has been paid in the book to the correspondence between theory and observation, the text is basically theoretical with greater emphasis on firm quantitative results rather than on quick heuristic arguments. The book's prime emphasis, therefore, is on problems of long standing rather than on more recent developments (such as rotationally induced mixing in stellar radiative zones) that are still in the process of rapid and diverse growth. The view adopted throughout the book is that the study of rotating stars is a multidisciplinary endeavor and that much can be learned from a parallel study of other rotating fluid systems, such as the Earth's atmosphere and the oceans.*

The contents of the various chapters are as follows: Chapter 1 presents the main observational data on which the subsequent discussion is based. Chapters 2 and 3 provide the theoretical background necessary for the understanding of the structure and evolution of a rotating star. In particular, Sections 2.5–2.7 describe some important geophysical concepts that will find their application in subsequent chapters. Even though the reader may not wish to go through these two chapters, I recommend reading the whole of Section 3.6,

* This is not the place to discuss the psychological impact that the new trends toward interdisciplinary modes of research may have on individual members of the scientific community. For pertinent comments, see Juan G. Roederer, "Tearing Down Disciplinary Barriers," *Astrophysics and Space Science*, **144**, 659, 1988.

however, because it summarizes several basic ideas and concepts that are recurring throughout the book. Chapter 4 describes the state of motion in a star that consists of a convective core surrounded by a radiative envelope, whereas Chapter 5 is concerned with the rotational deceleration of the Sun – a star that consists of a radiative core and an outer convection zone that is slowly but inexorably losing angular momentum to outer space. These twin chapters are purely theoretical in the sense that both of them attempt to develop a clear understanding of the many hydrodynamical phenomena that arise in the early-type and late-type stars as they slowly evolve on the main sequence. On the contrary, in the next two chapters I review the observational evidence for axial rotation in single stars and, as far as possible, I compare the theoretical models with observation. Chapter 6 is entirely devoted to stars more massive than the Sun, whereas Chapter 7 discusses the rotational history of solar-type stars. Finally, Chapter 8 is concerned with tidal interaction in close binary stars and contact binaries. Sections 8.4 and 8.5 present distinct applications of two well-known geophysical concepts, namely, Ekman pumping and geostrophy.

All chapters end with a short section entitled "Bibliographical notes," where references have been listed for elaboration of the material discussed in the corresponding sections. No attempt at completeness has been made, however, because that would have involved far too many entries. In each chapter, then, I have tried to include a useful selection of significant research papers and reviews from which further references may be obtained. Particular attention has been paid to original credits and priorities. For any inadvertent omission I offer a sincere apology in advance.

I am indebted to Paul Charbonneau and Georges Michaud who kindly provided valuable comments on portions of the manuscript. I appreciate also the untiring efforts of my wife, Monique, who typed and converted the original draft into LaTeX format, offered many helpful comments and corrections, and assisted with the proofreading and indexes. Their help is gratefully acknowledged, but of course they are in no way responsible for any errors of fact or judgment that the book may contain.

Montréal, Québec
December 1997

1

Observational basis

1.1 Historical development

The study of stellar rotation began at the turn of the seventeenth century, when sunspots were observed for the first time through a refracting telescope. Measurements of the westward motion of these spots across the solar disk were originally made by Johannes Fabricius, Galileo Galilei, Thomas Harriot, and Christopher Scheiner. The first public announcement of an observation came from Fabricius (1587–c. 1617), a 24-year old native of East Friesland, Germany. His pamphlet, *De maculis in Sole observatis et apparente earum cum Sole conversione*, bore the date of dedication June 13, 1611 and appeared in the *Narratio* in the fall of that year. Fabricius perceived that the changes in the motions of the spots across the solar disk might be the result of foreshortening, with the spots being situated on the surface of the rotating Sun. Unfortunately, from fear of adverse criticism, Fabricius expressed himself very timidly. His views opposed those of Scheiner, who suggested that the sunspots might be small planets revolving around an immaculate, nonrotating Sun. Galileo made public his own observations in *Istoria e Dimostrazioni intorno alle Macchie Solari e loro Accidenti*. In these three letters, written in 1612 and published in the following year, he presented a powerful case that sunspots must be dark markings on the surface of a rotating Sun. Foreshortening, he argued, caused these spots to appear to broaden and accelerate as they moved from the eastern side toward the disk center. The same effect made the sunspots seem to get thinner and slower as they moved toward the western side of the disk. Galileo also noticed that all spots moved across the solar disk at the same rate, making a crossing in about fourteen days, and that they all followed parallel paths. Obviously, these features would be highly improbable given the planetary hypothesis, which is also incompatible with the observed changes in the size and shape of sunspots.

The planetary hypothesis, championed by Scheiner among others, was thus convincingly refuted by Galileo. Eventually, Scheiner's own observations led him to realize that the Sun rotates with an apparent period of about 27 days. To him also belongs the credit of determining with considerably more accuracy than Galileo the position of the Sun's equatorial plane and the duration of its rotation. In particular, he showed that different sunspots gave different periods of rotation and, furthermore, that the spots farther from the Sun's equator moved with a slower velocity. Scheiner published his collected observations in 1630 in a volume entitled *Rosa Ursina sive Sol*, dedicated to the Duke of Orsini, who sponsored the work. (The title of the book derives from the badge of the Orsini family, which was a rose and a bear.) This was truly the first monograph on solar physics.

It is not until 1667 that any further significant discussion of stellar rotation was made. In that year the French astronomer Ismaël Boulliaud (1605–1694) suggested that the variability in light of some stars (such as Mira Ceti) might be a direct consequence of axial rotation, with the rotating star showing alternately its bright (unspotted) and dark (spotted) hemispheres to the observer. This idea was popularized in Fontenelle's *Entretiens sur la pluralité des mondes* – a highly successful introduction to astronomy that went through many revised editions during the period 1686–1742. To be specific, he noted "... that these fixed stars which have disappeared aren't extinguished, that these are really only half-suns. In other words they have one half dark and the other lighted, and since they turn on themselves, they sometimes show us the luminous half and then we see them sometimes half dark, and then we don't see them at all." * Although this explanation for the variable stars did not withstand the passage of time, it is nevertheless worth mentioning because it shows the interest that stellar rotation has aroused since its inception. As a matter of fact, nearly three centuries were to elapse before Boulliaud's original idea was fully recognized as a useful method of measuring the axial rotation of certain classes of stars, that is, stars that exhibit a detectable rotational modulation of their light output due to starspots or stellar plages.

For more than two centuries the problem of solar rotation was practically ignored, and it is not until the 1850s that any significant advance was made. Then, a long series of observations of the apparent motion of sunspots was undertaken by Richard Carrington and Gustav Spörer. They confirmed, independently, that the outer visible envelope of the Sun does not rotate like a solid body; rather, its period of rotation varies as a function of heliocentric latitude. From his own observations made during the period 1853–1861, Carrington derived the following expression for the Sun's rotation rate:

$$\Omega(\text{deg/day}) = 14^\circ42 - 2^\circ75 \sin^{7/4} \phi, \tag{1.1}$$

where ϕ is the heliocentric latitude. Somewhat later, Hervé Faye found that the formula

$$\Omega(\text{deg/day}) = 14^\circ37 - 3^\circ10 \sin^2 \phi \tag{1.2}$$

more satisfactorily represented the dependence of angular velocity on heliocentric latitude. Parenthetically, note that Carrington also found evidence for a mean meridional motion of sunspots. Convincing evidence was not found until 1942, however, when Jaakko Tuominen positively established the existence of an equatorward migration of sunspots at heliocentric latitudes lower that about 20° and a poleward migration at higher latitudes.

The spectroscope was the instrument that marked the beginning of the modern era of stellar studies. As early as 1871 Hermann Vogel showed that the Sun's rotation rate can be detected from the relative Doppler shift of the spectral lines at opposite edges of the solar disk, one of which is approaching and the other receding. Extensive measurements were made visually by Nils Dunér and Jakob Halm during the period 1887–1906. They showed a rotation rate and equatorial acceleration that were quite similar to those obtained from the apparent motion of sunspots. They concluded that Faye's empirical law

* Bernard le Bovier de Fontenelle, *Conversations on the Plurality of Worlds*, translation of the 1686 edition by H. A. Hargreaves, p. 70, Berkeley: University of California Press, 1990.

adequately represented the spectroscopic observations also, but their coverage of latitude was double that of the sunspot measurements. The first spectrographic determinations of solar rotation were undertaken at the turn of the twentieth century by Walter S. Adams at Mount Wilson Solar Observatory, California.

William de Wiveleslie Abney was the first scientist to express the idea that the axial rotation of single stars could be determined from measurements of the widths of spectral lines. In 1877, he suggested that the effect of a star's rotation on its spectrum would be to broaden all of the lines and that "... other conditions being known, the mean velocity of rotation might be calculated."* In 1893, while doubts were still being expressed with regard to measurable rotational motions in single stars, J. R. Holt suggested that axial rotation might be detected from small distortions in the radial velocity curve of an eclipsing binary. Thus, he argued,

> ... in the case of variable stars, like Algol, where the diminution of light is supposed to be due to the interposition of a dark companion, it seems to me that there ought to be a spectroscopic difference between the light at the commencement of the minimum phase, and that of the end, inasmuch as different portions of the edge would be obscured. In fact, during the progress of the partial eclipse, there should be a shift in position of the lines; and although this shift is probably very small, it ought to be detected by a powerful instrument.†

Confirmation of this effect was obtained by Frank Schlesinger in 1909, who presented convincing evidence of axial rotation in the brightest star of the system δ Librae. However, twenty more years were to elapse before Abney's original idea resulted in actual measurements of projected equatorial velocities in single stars. This notable achievement was due to the efforts of Otto Struve and his collaborators during the period 1929–1934 at Yerkes Observatory, Wisconsin.

A graphical method was originally developed by Grigori Shajn and Otto Struve. The measurements were made by fitting the observed contour of a spectral line to a computed contour obtained by applying different amounts of Doppler broadening to an intrinsically narrow line-contour having the same equivalent width as the observed line. Comparison with an observed line profile gave the projected equatorial velocity $v \sin i$ along the line of sight. These early measurements indicated that the values of $v \sin i$ fell into the range 0–250 km s^{-1} and may occasionally be as large as 400 km s^{-1} or even more. As early as 1930 it was found that the most obvious correlation between $v \sin i$ and other physical parameters is with spectral type, with rapid rotation being peculiar to the earliest spectral classes. This was originally recognized by Struve and later confirmed by statistical studies of line widths in early-type stars by Christian T. Elvey and Christine Westgate. The O-, B-, A-, and early F-type stars frequently have large rotational velocities, while in late F-type and later types rapid rotation occurs only in close spectroscopic binaries. A study of rotational line broadening in early-type close binaries was also made by Egbert Adriaan Kreiken. From his work it is apparent that the components of these binaries have their rotational velocities significantly diminished with respect to single, main-sequence stars of the same spectral type. The following year, 1936, Pol Swings

* *Mon. Not. R. Astron. Soc.*, **37** (1877), p. 278.

† *Astronomy and Astro-Physics*, **12** (1893), p. 646.

properly established that in close binaries of short periods axial rotation tends to be either perfectly or approximately synchronized with the orbital motion.

At this juncture the problem was quietly abandoned for almost fifteen years. Interest in the measurements of axial rotation in stars was revived in 1949 by Arne Slettebak. Extensive measurements of rotational velocities were made during the 1950s and 1960s by Helmut A. Abt, Robert P. Kraft, Slettebak, and others. However, because the only observational technique available was to determine line widths in stars from photographic spectra, these studies were limited almost entirely to stars more massive than the sun ($M \gtrsim 1.5 M_\odot$) and to main-sequence or post–main-sequence stars. Since appreciable rotation disappears in the middle F-type stars, higher-resolution spectra are therefore required to measure rotational broadening in the late-type stars. In 1967, Kraft pushed the photographic technique to its limit to measure $v \sin i$ as low as 6 km s^{-1} in solar-type stars. Now, as early as 1933, John A. Carroll had suggested the application of Fourier analysis to spectral line profiles for rotational velocity determinations. In 1973, the problem was reconsidered by David F. Gray, who showed that high-resolution data make it possible to distinguish between the Fourier transform profile arising from rotation versus those arising from other broadening mechanisms. Since the late 1970s systematic studies of very slow rotators have been made by Gray, Myron A. Smith, David R. Soderblom, and others. Current techniques limit the measurement accuracy of projected rotational velocities to 2 km s^{-1} in most stars.

Periodic variations in the light output due to dark or bright areas on some rotating stars have also been used to determine the rotation periods of these stars. Although the principle of rotational modulation was suggested as early as 1667 by Ismaël Boulliaud, convincing detection of this effect was not made until 1947, when Gerald E. Kron found evidence in the light curve of the eclipsing binary AR Lacertae for surface inhomogeneities in its G5 component. The principle was therefore well established when in 1949 Horace W. Babcock proposed the so-called oblique-rotator model for the magnetic and spectrum variations of the periodic Ap stars. Kron's result was forgotten till 1966, when interest in the principle of rotational modulation was independently revived by Pavel Chugainov. A large body of literature has developed since the late 1960s. This work generally divides according to the method used to estimate the rotation periods, with the two types being (i) photometric monitoring of light variations produced by large starspot groups or bright surface areas and (ii) measurements of the periodic variation in strength of some emission lines that are enhanced in localized active regions in the chromosphere. These techniques have the advantage that a rotation period can be determined to much higher precision than $v \sin i$ and are free of the $\sin i$ projection factor inherent to the spectrographic method. Moreover, very accurate rotation periods can be derived even for quite slowly rotating stars at rates that would be impossible to see as a Doppler broadening of their spectral lines.

A different line of inquiry was initiated by the discovery of the so-called five-minute oscillations in the solar photosphere. The first evidence for ubiquitous oscillatory motions was obtained in the early 1960s by Robert B. Leighton, Robert W. Noyes, and George W. Simon. However, it is not until 1968 that Edward N. Frazier suggested that "... the well known 5 min oscillations are primarily standing resonant acoustic waves."*

* *Zeit. Astrophys.*, **68** (1968), p. 345.

Two years later, Roger K. Ulrich presented a detailed theoretical description of the phenomenon, showing that standing acoustic waves may be trapped in a layer beneath the solar photosphere. This model was independently proposed in 1971 by John W. Leibacher and Robert F. Stein. In 1975, Franz-Ludwig Deubner obtained the first observational evidence for these trapped acoustic modes. Soon afterward, it was realized that a detailed analysis of the frequencies of these many oscillatory modes could provide a probe of the solar *internal* rotation. Indeed, because axial rotation breaks the Sun's spherical symmetry, it splits the degeneracy of the nonradial modes with respect to the azimuthal angular dependence. A technique for measuring the solar internal rotation from these frequency splittings was originally devised by Edward J. Rhodes, Jr., Deubner, and Ulrich in 1979. Since 1984, following the initial work of Thomas L. Duvall, John W. Harvey, and others,* diverse methods have been used to determine the Sun's internal angular velocity.

1.2 The Sun

In Section 1.1 we briefly discussed the early measurements of the axial rotation of the Sun. With the advent of more sensitive instruments, however, Doppler and tracer measurements have shown that the solar atmosphere exhibits motions on widely different scales. Besides the large-scale axisymmetric motions corresponding to differential rotation and meridional circulation, velocity fields associated with turbulent convection and also with oscillatory motions at about a five-minute period have been observed. Considerable attention has focused on analysis of these oscillations since, for the very first time, they make it possible to probe the Sun's *internal* rotation.

1.2.1 Large-scale motions in the atmosphere

The solar surface rotation rate may be obtained from measurements of the longitudinal motions of semipermanent features across the solar disk (such as sunspots, faculae, magnetic field patterns, dark filaments, or even coronal activity centers), or from spectrographic observations of Doppler displacements of selected spectral lines near the solar limb. Each of the two methods for deriving surface rotation rates has its own limitations, although few of these limitations are common to both. Actually, the determination of solar rotation from tracers requires that these semipermanent features be both randomly distributed throughout the fluid and undergo no appreciable proper motion with respect to the medium in which they are embedded. In practice, no tracers have been shown to possess both characteristics; moreover, most of them tend to occur in a limited range of heliocentric latitudes. By the spectrographic method, rotation rates can be found over a wider range of latitudes. But then, the accuracy is limited by the presence of inhomogeneities of the photospheric velocity field and by macroscopic motions within coronal and chromospheric features, so that the scatter between repeated measurements is large.

Figure 1.1 assembles sidereal rotation rates obtained from photospheric Doppler and tracer measurements. The observations refer to the sunspots and sunspot groups, magnetic field patterns, and Doppler shifts. In all cases the relationships shown in Figure 1.1 are

* *Nature*, **310** (1984), pp. 19 and 22.

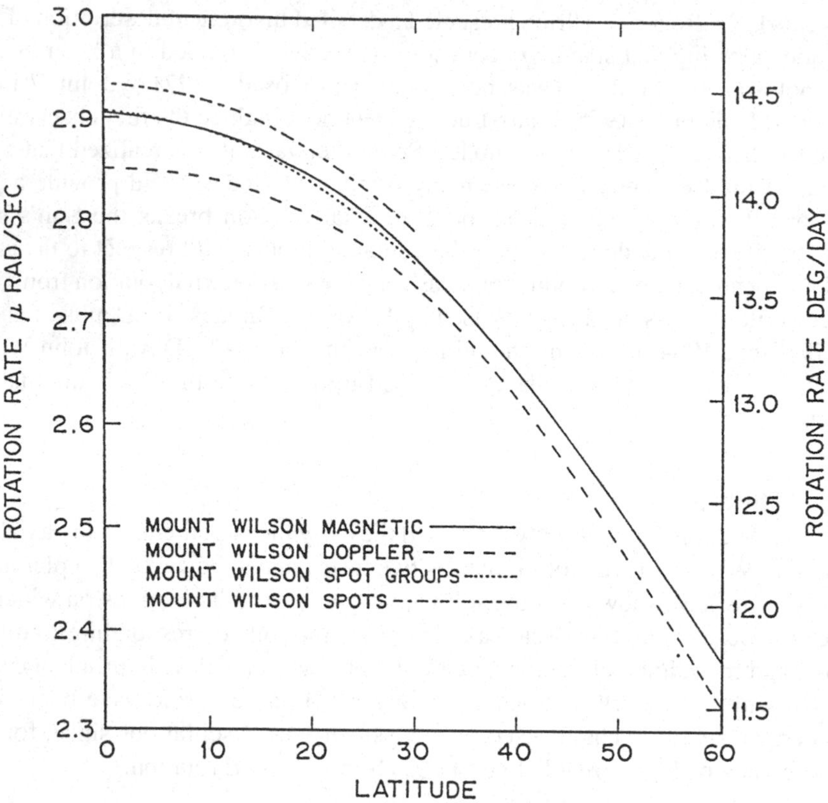

Fig. 1.1. Comparison of the solar differential rotation obtained by different methods. *Source:* Howard, R., *Annu. Rev. Astron. Astrophys.*, **22**, 131, 1984. (By permission. Copyright 1984 by Annual Reviews.)

smoothed curves obtained by fitting the data to expansions in the form

$$\Omega = A + B \sin^2 \phi + C \sin^4 \phi. \tag{1.3}$$

The decrease of angular velocity with increasing heliocentric latitude is clear. However, it is also apparent that different techniques for measuring the solar surface rotation rate yield significantly different results. In particular, the sunspot groups rotate more slowly in their latitudes than individual sunspots. Note also that the rotation rate for the magnetic tracers is intermediate between that for the individual spots and that for the photospheric plasma. It is not yet clear whether these different rotation rates represent real differences of rotation at various depths in the solar atmosphere or whether they reflect a characteristic behavior of the tracers themselves.

Chromospheric and coronal rotation measurements have also been reported in the literature. It seems clear from these results that the latitudinal gradient of angular velocity depends very much on the size and lifetime of the tracers located above the photosphere. To be specific, the long-lived structures exhibit smaller gradients than the short-lived ones, and the very long-lived coronal holes rotate almost uniformly. These noticeable differences remain poorly understood.

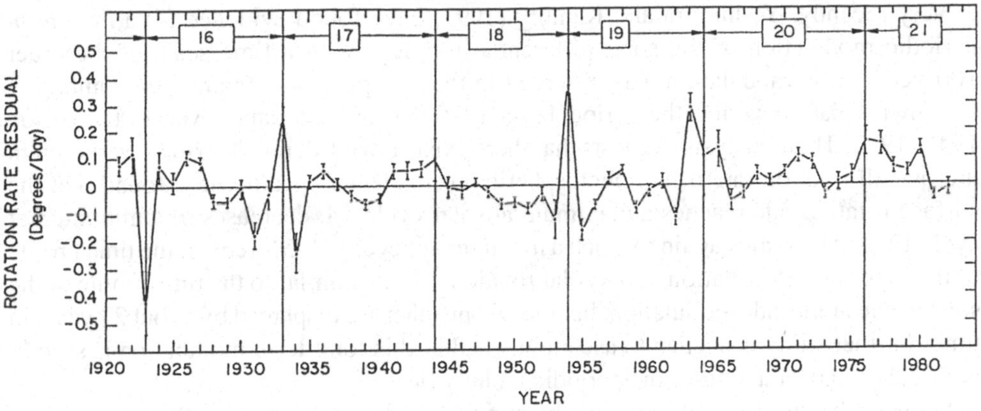

Fig. 1.2. Residuals of annual average sunspot rotation rates for the period 1921–1982. Solar cycle maxima timing and length are denoted by numbered boxes. Vertical lines denote year of sunspot minimum. *Source:* Gilman, P. A., and Howard, R., *Astrophys. J.*, **283**, 385, 1984.

Figure 1.1 merely illustrates the *mean* properties of the solar surface differential rotation. As was originally shown by Howard and LaBonte (1980), however, analysis of the residual motions in the daily Doppler measurements made at Mount Wilson suggests the presence of a *torsional oscillation* of very small amplitude in the photosphere. This oscillation is an apparently organized pattern of zonally averaged variations from a mean curve for the differential rotation, as defined in Eq. (1.3). The amplitude of the residuals constituting the torsional oscillation is of the order of 5 m s^{-1}. It is a traveling wave, with latitude zones of fast and slow rotation, that originates near the poles and moves equatorward over the course of a 22-year cycle. The latitude drift speed of the shear is of the order of 2 m s^{-1}. In the lower heliocentric latitudes, the torsional shear zone between the fast stream on the equator side and the slow stream on the pole side is the locus of solar activity. This coincidence strongly suggests that this torsional oscillation is somewhat related to the solar activity cycle.

Variations of the solar surface rotation rate over individual sunspot cycles have been reported by many investigators. Detailed analyses of the Mount Wilson sunspot data for the period 1921–1982 suggest that *on average* the Sun rotates more rapidly at sunspot minimum.* A similar frequency of rotation maxima is also seen in the Greenwich sunspot data for the years 1874–1976. The variability of the mean rotation rate is illustrated in Figure 1.2, which exhibits peaks of about 0.1 degree day^{-1} in the residuals near minima of solar activity. The Mount Wilson data also show variations from cycle to cycle, with the most rapid rotation found during cycles with fewer sunspots and less sunspot area.

* A similar result was obtained by Eddy, Gilman, and Trotter (1977) from their careful analysis of drawings of the Sun made by Christopher Scheiner (during 1625–1626) and Johannes Hevelius (during 1642–1644). During the earlier period, which occurred 20 years before the start of the Maunder sunspot minimum (1645–1715), solar rotation was very much like that of today. By contrast, in the later period, the equatorial velocity of the Sun was faster by 3 to 5% and the differential rotation was enhanced by a factor of 3. These results strongly suggest that the change in rotation of the solar surface between 1625 and 1645 was associated, as cause or effect, with the Maunder minimum anomaly.

Very recently, Yoshimura and Kambry (1993) have found evidence for a *long-term* periodic modulation of the solar differential rotation, with a time scale of the order 100 years. This modulation was observed in the sunspot data obtained by combining Greenwich data covering the period 1874–1976 and Mitaka data covering the period 1943–1992. Their analysis suggests that there exists a well-defined periodic variation in the overall rotation rate of the photospheric layers. To be specific, it is found that the surface rotation rate reaches a maximum at solar cycle 14, decreases to a minimum at cycle 17, and increases again to reach a maximum at cycle 21. Moreover, the time profile of the long-term modulation of the solar rotation is quite similar to the time profile of the solar-cycle amplitude modulation, but the two profiles are displaced by about 23 years in time. Further study is needed to ascertain whether this long-term modulation is strictly periodic or part of a long-term aperiodic undulation.

Several observational efforts have been made to detect a *mean* north–south motion on the Sun's surface. Unfortunately, whereas the latitudinal and temporal variations of the solar rotation are reasonably well established, the general features of the meridional flow are still poorly understood. Three different techniques have been used to measure these very slow motions: (i) the Doppler shift of selected spectral lines, (ii) the displacement of magnetic features on the solar disk, and (iii) the tracing of sunspots or plages. A majority of Doppler observations suggests a poleward motion of the order of 10 m s^{-1}, whereas others differ in magnitude and even in direction. Doppler data obtained with the Global Oscillation Network Group (GONG) instruments in Tucson from 1992 to 1995 indicate a poleward motion of the order of 20 m s^{-1}, but the results also suggest that the Sun may undergo episodes in which the meridional speeds increase dramatically. The analysis of magnetic features shows the existence of a meridional flow that is poleward in each hemisphere and is of the order of 10 m s^{-1}, which agrees with most of the Doppler measurements. On the contrary, sunspots or plages do not show a simple poleward meridional flow but a motion either toward or away from the mean latitude of solar activity, with a speed of a few meters per second. Analysis of sunspot positions generally shows equatorward motions at low heliocentric latitudes and poleward motions at high latitudes. Several authors have suggested that these discrepancies might be ascribed to the fact that different features are anchored at different depths in the solar convection zone. Accordingly, the meridional flow deep into this zone might be reflected by the sunspot motions, whereas the meridional flow in the upper part of this zone might be reflected by the other measurements. As we shall see in Section 5.2, these speculations have a direct bearing on the theoretical models of solar differential rotation.

1.2.2 Helioseismology: The internal rotation rate

The Sun is a very small amplitude variable star. Its oscillations are arising from a huge number of discrete modes with periods ranging from a few minutes to several hours. The so-called five-minute oscillations, which have frequencies between about 2 mHz and 4 mHz, have been extensively studied. They correspond to standing *acoustic* waves that are trapped beneath the solar surface, with each mode traveling within a well-defined shell in the solar interior. Since the properties of these modes are determined by the stratification of the Sun, accurate measurements of their frequencies thus provide a new window in the hitherto invisible solar interior.

To a first approximation, the Sun may be considered to be a spherically symmetric body. In that case, by making use of spherical polar coordinates (r, θ, φ), we can write the components of the Lagrangian displacement for each acoustic mode in the separable form

$$\boldsymbol{\xi} = \left(\xi_r P_l^m \, , \, \xi_h \frac{dP_l^m}{d\theta} \, , \, \xi_h \frac{P_l^m}{\cos\theta} \frac{\partial}{\partial\varphi} \right) \cos(m\varphi - \omega_{n,l} t), \tag{1.4}$$

where $P_l^m(\cos\theta)$ is the associated Legendre function of degree l and order m $(-l \leq m \leq +l)$. The eigenfunctions $\xi_r(r; n, l)$ and $\xi_h(r; n, l)$ define the radial and horizontal displacements of the mode. Both functions depend on the integer n, which is related to the number of zeros of the function ξ_r along the radius, and the integer l, which is the number of nodal lines on the solar surface. Because a spherical configuration has no preferred axis of symmetry, these eigenfunctions are independent of the azimuthal order m, so that to each value of the eigenfrequency $\omega_{n,l}$ correspond $2l + 1$ displacements. Rotation splits this degeneracy with respect to the azimuthal order m of the eigenfrequencies. Hence, we have

$$\omega_{n,l,m} = \omega_{n,l} + \Delta\omega_{n,l,m}. \tag{1.5}$$

Since the magnitude of the angular velocity Ω is much less than the acoustic frequencies $\omega_{n,l}$, perturbation theory can be applied to calculate these frequency splittings. One can show that

$$\Delta\omega_{n,l,m} = m \int_0^R \int_0^\pi K_{n,l,m}(r, \theta)\, \Omega(r, \theta)\, r\, dr\, d\theta, \tag{1.6}$$

where the rotational kernels $K_{n,l,m}(r, \theta)$ are functions that may be derived from a *non-rotating* solar model for which one has calculated the eigenfrequencies $\omega_{n,l}$ and their corresponding eigenfunctions. Given measurements of the rotational splittings $\Delta\omega_{n,l,m}$, it is therefore possible, in principle, to solve this integral equation for the angular velocity.

Measurement of the rotational splitting $\Delta\omega_{n,l,m}$ provides a measure of rotation *in a certain region* of the Sun. In fact, the acoustic modes of progressively lower l penetrate deeper into the Sun, so that the information on the angular velocity in the deeper layers is confined to splittings of low-l modes. Similarly, because only when an acoustic mode is quasi-zonal can it reach the polar regions, the information on the angular velocity at high heliocentric latitudes is confined to splittings of low-m modes. Since the measured splittings for the low-l and low-m modes have comparatively larger relative errors, determination of the function $\Omega(r, \theta)$ thus becomes increasingly difficult with increasing depth and increasing latitude.

Several groups of workers have observed the splittings of acoustic frequencies that arise from the Sun's differential rotation. Figures 1.3 and 1.4 illustrate the inverted solution of Eq. (1.6) based on frequency splitting determinations from the latest GONG data (1996). Note that the equatorial rotation rate presents a steep increase with radius near $r = 0.7R_\odot$, thus suggesting the possibility of a discontinuity near the base of the convection zone. Note also that the equatorial rotation rate peaks near $r = 0.95R_\odot$, before decreasing with radius in the outermost surface layers. Figure 1.4 illustrates the latitudinal dependence of the inverted profile. In the outer convection zone, for latitude $\phi < 30°$, the rotation rate is nearly constant on cylinders, owing to a rapidly rotating

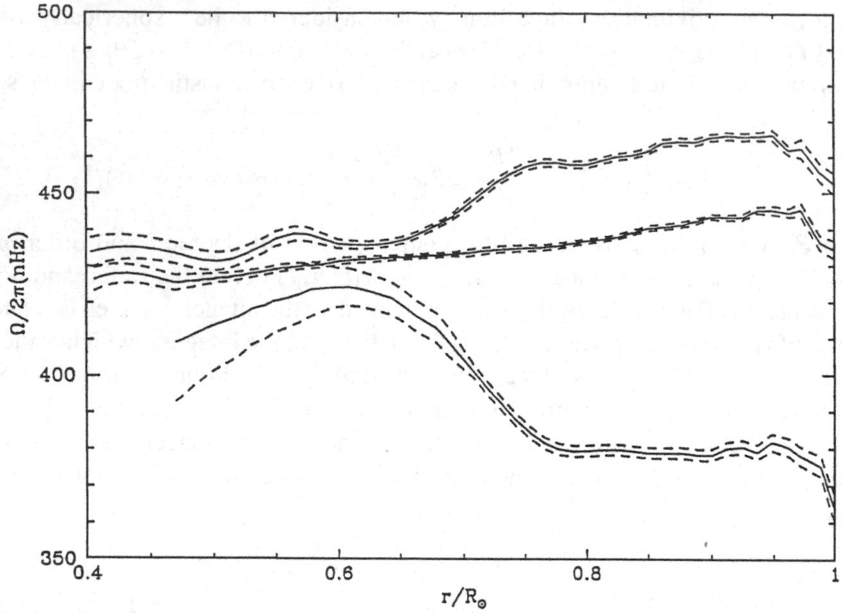

Fig. 1.3. Solar rotation rate inferred from the latest GONG data (1996). The curves are plotted as a function of radius at the latitudes of 0° (top), 30° (middle), and 60° (bottom). The dashed curves indicate error levels. *Source:* Sekii, T., in *Sounding Solar and Stellar Interiors* (Provost, J., and Schmider, F. X., eds.), I.A.U. Symposium No 181, p. 189, Dordrecht: Kluwer, 1997. (By permission. Copyright 1997 by Kluwer Academic Publishers.)

belt centered near $r = 0.95R_\odot$. At higher latitudes, however, the rotation rate becomes constant on cones. The differential character of the rotation disappears below a depth that corresponds to the base of the convection zone. This solution agrees qualitatively with the inverted profiles obtained by other groups. Perhaps the most interesting result of these inversions is that they show no sign of a tendency for rotation to occur at constant angular velocity on cylinders throughout the outer convection zone.

In summary, several inversion studies indicate that the rotation rate in the solar convection zone is similar to that at the surface, with the polar regions rotating more slowly than the equatorial belt. Near the base of the convection zone, one finds that there exists an abrupt unresolved transition to essentially uniform rotation at a rate corresponding to some average of the rate in the convection zone. This shear layer, which is known as the *solar tachocline*, is centered near $r = 0.7R_\odot$; recent studies indicate that it is quite thin, probably no more than $0.06R_\odot$. The actual rotation rate in the radiative core remains quite uncertain, however, because of a lack of accurately measured splittings for low-l acoustic modes. Several investigators have found that from the base of the convection zone down to $r \approx 0.1$–$0.2R_\odot$ their measurements are consistent with uniform rotation at a rate somewhat lower than the surface equatorial rate. Not unexpectedly, the rotation rate inside that radius is even more uncertain. Some studies suggest that the rotation rate of this inner core might be between 2 and 4 times larger than that at the surface. According to other investigators, however, it is more likely that this inner core rotates with approximately the same period as the outer parts of the radiative core. I shall not go into the disputes.

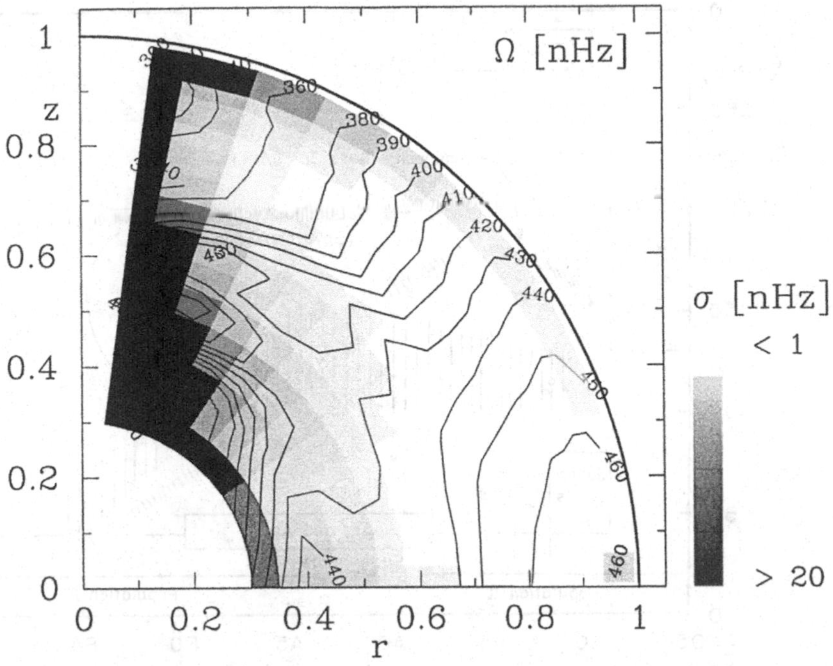

Fig. 1.4. Solar rotation rate as a function of normalized radius and latitude. Contours of isorotation are shown, superimposed on a gray-scale plot of the formal errors. A very dark background means a less reliable determination. *Source:* Korzennik, S. G., Thompson, M. J., Toomre, J., and the GONG Internal Rotation Team, in *Sounding Solar and Stellar Interiors* (Provost, J., and Schmider, F. X., eds.), I.A.U. Symposium No 181, p. 211, Dordrecht: Kluwer, 1997. (Courtesy of Dr. F. Pijpers. By permission; copyright 1997 by Kluwer Academic Publishers.)

1.3 Single stars

As was noted in Section 1.1, two basic methods have been used to measure rotational velocities of single stars. One of them consists of extracting rotational broadening from a spectral line profile, from which one infers the *projected* equatorial velocity $v \sin i$ along the line of sight. The other one consists of determining the modulation frequency of a star's light due to the rotation of surface inhomogeneities (such as spots or plages) across its surface. If observable, this modulation frequency is a direct estimate of the star's rotation period P_{rot}, which is free of projection effects. Hence, given a radius R for the star, this period can be transformed into a *true* equatorial velocity v $(= \Omega R = 2\pi R/P_{\text{rot}})$.

The spectrographic method has proven useful in determining the projected velocities for stars of spectral type O, B, A, and F. In fact, $v \sin i$ measurements can only be used in a statistical way because the inclination angle i is generally unknown. Evidence for random orientation of rotation axes is found in the lack of correlation between the measured values of $v \sin i$ and the galactic coordinates of the stars. For randomly oriented rotation axes, one can thus convert the average projected equatorial velocity $\langle v \sin i \rangle$ for a group of stars to an average equatorial velocity $\langle v \rangle$, taking into account that the average value $\langle \sin i \rangle$ is equal to $\pi/4$. Numerous statistical studies have been made over the period 1930–1970. The main results pertaining to stellar rotation have been assembled

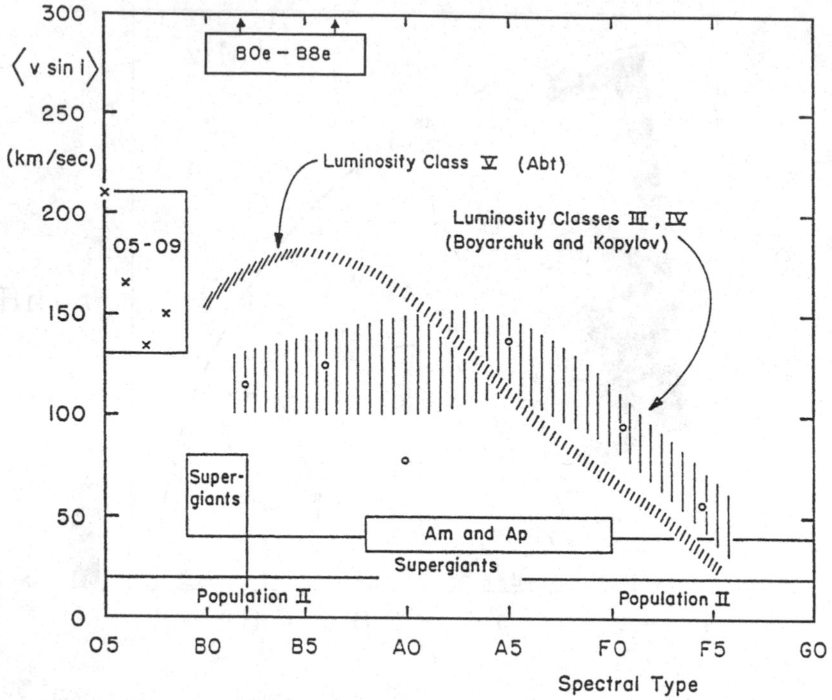

Fig. 1.5. Mean projected equatorial velocities for a number of different classes of stars as compared with normal main-sequence stars. *Source:* Slettebak, A., in *Stellar Rotation* (Slettebak, A., ed.), p. 5, New York: Gordon and Breach, 1970. (By permission. Copyright 1970 by Gordon and Breach Publishers.)

by Slettebak and are summarized in Figure 1.5. In this figure the mean observed rotational velocities for single, normal, main-sequence stars are compared with the mean observed $v \sin i$s for giant and supergiant stars, Be stars, peculiar A-type and metallic-line stars, and Population II objects.

The distribution of rotational velocities along the main sequence is quite remarkable: *Rotation increases from very low values in the F-type stars to some maximum in the B-type stars*. However, a different picture emerges when one considers the mean rotation periods rather than the mean equatorial velocities. This is illustrated in Table 1.1 which lists typical values of the masses, radii, equatorial velocities, angular velocities, and rotation periods. Note that the periods reach a minimum value of about 0.56 day near spectral type A5, and they increase rather steeply on both sides so that the G0- and O5-type stars have approximately the same rotation period. The large observed values $\langle v \rangle$ for the upper main-sequence stars are thus entirely due to the large radii of these stars.

The open circles in Figure 1.5 represent mean rotational velocities for stars belonging to the luminosity classes III and IV; they are connected by a broad cross-hatched band, thus suggesting uncertainties in the mean rotational velocities for the giant stars. According to Slettebak, the very low point at spectral type A0 can probably be interpreted in terms of selection effects. In any case, the broad band indicates that the early-type giants rotate more slowly than the main-sequence stars of corresponding spectral types, whereas for the late A- and F-types the giants rotate more rapidly than their main-sequence counterparts.

Table 1.1. *Average rotational velocities of main-sequence stars.*

Spectrum (class V)	M ($M_\odot$)	R ($R_\odot$)	v (km s^{-1})	Ω (10^{-5} s^{-1})	P_{rot} (days)
O5	39.5	17.2	190	1.5	4.85
B0	17.0	7.6	200	3.8	1.91
B5	7.0	4.0	210	7.6	0.96
A0	3.6	2.6	190	10.0	0.73
A5	2.2	1.7	160	13.0	0.56
F0	1.75	1.3	95	10.0	0.73
F5	1.4	1.2	25	3.0	2.42
G0	1.05	1.04	12	1.6	4.55

Source: McNally, D., *The Observatory*, **85**, 166, 1965.

This behavior can be interpreted as an evolutionary effect. As we know, the rapidly rotating B- and A-type main-sequence stars evolve to luminosity classes III and IV in later spectral types. But then, the drop in rotation as the star's radius increases is compensated by the steeper drop in rotation along the main sequence, so that the evolving star still has a larger equatorial velocity than its main-sequence counterpart. As we shall see in Section 6.5, the drop in rotation for the giants takes place between spectral types G0 III and G3 III; the drop for subgiants occurs a little earlier, at spectral types F6 IV to F8 IV.

Supergiants and Population II stars are shown schematically near the bottom of Figure 1.5. The supergiants of all spectral types do not show conspicuous rotations. They show no sudden decrease in rotation either, although rotational velocities up to 90 km s^{-1} are observed for spectral types earlier than F9. The apparent rotation velocities of Population II stars are also small, with $v \sin i$ values smaller than 30 km s^{-1}. Note also that the mean rotational velocities of the peculiar A-type stars and metallic-line stars are considerably smaller than the means for normal stars of corresponding spectral types. Finally, going to the other extreme, we note that the Be stars rotate most rapidly, and individual rotational velocities of 500 km s^{-1} have been observed by Slettebak. These stars are shown separately on Figure 1.5, with arrows indicating that their mean rotational velocities are in reality larger than shown. (As we shall see in Sections 6.3.2 and 6.3.4, however, there are no early-type stars with rotation rates anywhere near the critical rate at which centrifugal force balances gravity at the equator.) As a rule, the white dwarfs rotate rather slowly, with typical $v \sin i$ values of order 20 km s^{-1}, and none of them rotates faster than 60 km s^{-1}.

To put the relation between stellar age and axial rotation on a firm quantitative basis, several authors have obtained projected equatorial velocities for stars belonging to open clusters and associations. Detailed statistical analyses have been made by Bernacca and Perinotto (1974) and Fukuda (1982). In Figure 1.6, which is derived from data presented by Fukuda, we compare the average rotational velocity loci for field and cluster stars. As was done in Figure 1.5, the data have been grouped to smooth out irregularities in the distributions of $\langle v \sin i \rangle$ along the main sequence (see also Section 6.3). Figure 1.6 shows that field and cluster stars of spectral type O, B, and A have mean projected rotational

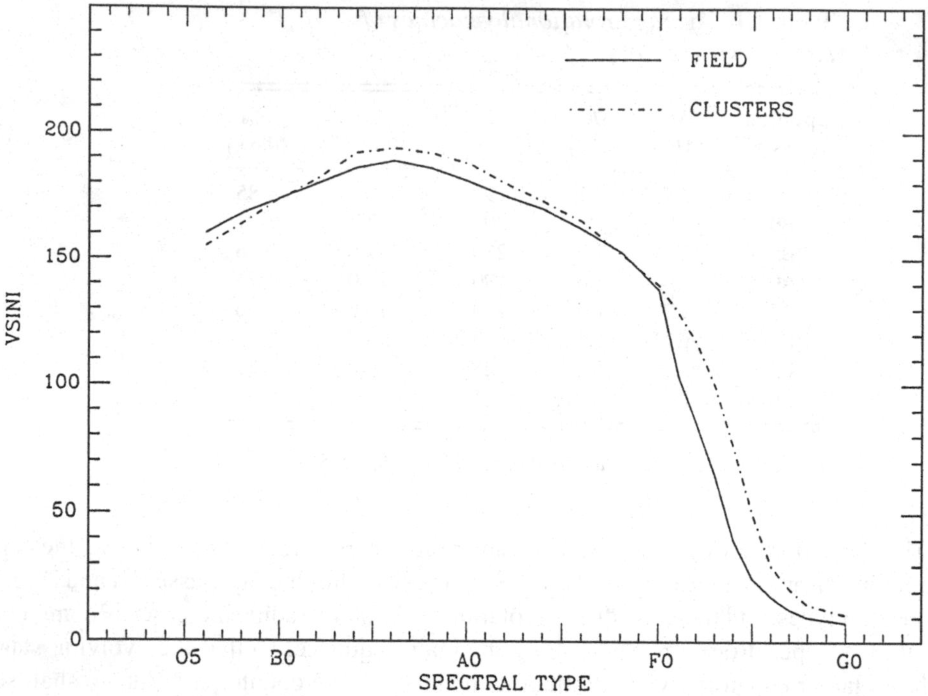

Fig. 1.6. Mean projected equatorial velocities for early-type field and cluster stars. Note that the open-cluster F dwarfs rotate more rapidly than their older, field counterparts. *Source:* Stauffer, J. R., and Hartmann, L. W., *Publ. Astron. Soc. Pacific*, **98**, 1233, 1986. (Courtesy of the Astronomical Society of the Pacific.)

velocities in the range 150–200 km s^{-1}. Within each spectral type, the mean rotational velocities of the field stars earlier than spectral type F0 are almost the same as those in clusters. Later than spectral type F0, however, the rotational velocities steeply decrease with increasing spectral type, dropping to below 20 km s^{-1} at spectral type G0. Note also that the F-type cluster stars, which are generally younger than the field stars, rotate more rapidly than their field counterparts. This result confirms Kraft's (1967) original finding that *the mean rotational velocities of late-F and early-G stars decline with advancing age.* This correlation between rotation and age was quantified shortly afterward by Skumanich (1972), who pointed out that the surface angular velocity of a solar-type star decays as the inverse square root of its age. To a good degree of approximation, we thus let

$$\Omega \propto t^{-1/2}, \tag{1.7}$$

which is known as *Skumanich's law.* (Other mathematical relations between rotation and age have been suggested, however.) As we shall see in Section 7.2, such a spin-down process is consistent with the idea that magnetically controlled stellar winds and/or episodic mass ejections from stars with outer convection layers continuously decelerate these stars as they slowly evolve on the main sequence.

An inspection of Figure 1.5 shows that appreciable rotational velocities are common among the *normal* O-, B-, and A-type stars along the main sequence, whereas they

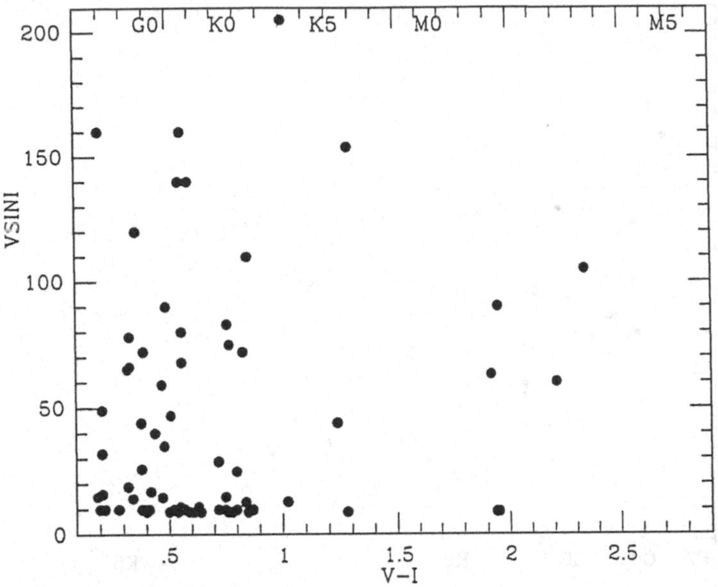

Fig. 1.7. Rotational velocity distribution for α Persei members. *Source:* Stauffer, J. R., Hartmann, L. W., and Jones, B. F., *Astrophys. J.*, **346**, 160, 1989.

virtually disappear near spectral type F5. Several photometric and spectroscopic studies made during the 1980s have confirmed that late-type, old field dwarfs with few exceptions are slow rotators, with true equatorial velocities less than 10 km s^{-1} in most stars. Fortunately, because continuous mass loss or discrete mass ejections cause spin-down of stars having convective envelopes, this sharp drop in rotational velocities along the main sequence is considerably reduced in younger stellar groups. Hence, clues to the rotational evolution of low-mass stars may be gained from the study of stars belonging to open clusters. This is illustrated in Figures 1.7 and 1.8, which depict, respectively, the rotational velocity distributions for lower main-sequence stars in the α Persei cluster (age $\sim$ 50 Myr) and in the Hyades (age $\sim$ 600 Myr). Figure 1.7 shows that the young α Persei cluster has a large number of very slowly rotating stars and a significant number of stars with projected equatorial velocities greater than 100 km s^{-1}. This is in contrast to the older Hyades, where G and K dwarfs are slow rotators, with the mean equatorial velocity appearing to decrease at least until spectral type K5. There is one prominent exception in Figure 1.8, however, a K8 dwarf that is the earliest known member of a population of relatively rapidly rotating late K- and M-type Hyades stars. These are genuine evolutionary effects that will be discussed in Section 7.4.2.

Other essential clues to the initial angular momentum distribution in solar-type stars can be obtained from the rotational velocity properties of low-mass, pre–main-sequence stars. These stars are commonly divided into two groups: the *classical* T Tauri stars, which have evidence of active accretion, and the *weak-line* T Tauri stars, which do not. Several photometric monitoring surveys have successfully determined rotation periods for a large number of these stars. It appears likely that most of the weak-line stars rotate faster than the classical T Tauri stars. Moreover, as was originally found by Attridge and Herbst

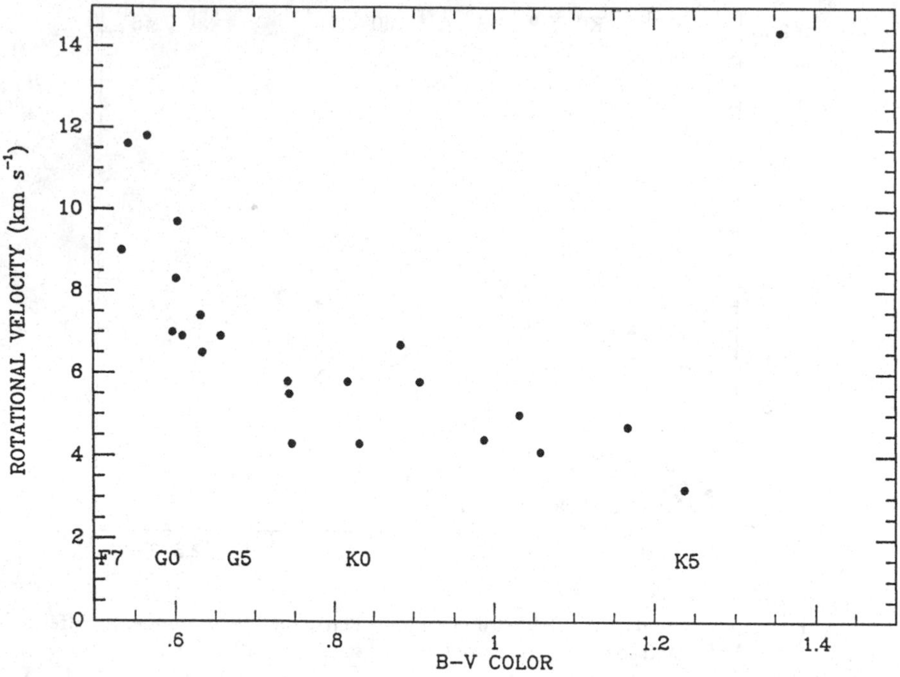

Fig. 1.8. Rotational velocity distribution for 23 Hyades stars. *Source:* Radick, R. R., Thompson, D. T., Lockwood, G. W., Duncan, D. K., and Baggett, W. E., *Astrophys. J.*, **321**, 459, 1987.

(1992), the frequency distribution of rotation periods for the T Tauri stars in the Orion Nebula cluster is distinctly bimodal. Figure 1.9 illustrates the frequency distribution of known rotation periods for these stars, combining the data for the Trapezium cluster, the Orion Nebula cluster, and other T associations. *This combined distribution is clearly bimodal, with a sparsely populated tail of extremely slow rotators*. The implications of this bimodality will be further discussed in Section 7.4.1.

1.4 Close binaries

In Section 1.1 we pointed out that the early-type components of close binaries rotate more slowly than the average of single stars of the same spectral type. In contrast, whereas the rotational velocities of single main-sequence stars of spectral type F5 and later are quite small (i.e., less than 10 km s^{-1}), appreciable rotations are common among the late-type components of close binaries. It has long been recognized that the distribution of rotational velocities in the close binaries is caused mostly by tidal interaction between the components, although some other processes – such as stellar winds, gravitational radiation, and large-scale magnetic fields – may also play a definite role in some binaries. To be specific, all types of tidal interaction involve an exchange of kinetic energy and angular momentum between the orbital and rotational motions. If we neglect stellar winds, the total angular momentum will be conserved in the tidal process. However, due to tidal dissipation of energy in the outer layers of the components, the total kinetic energy will decrease monotonically. Accordingly, as a result of

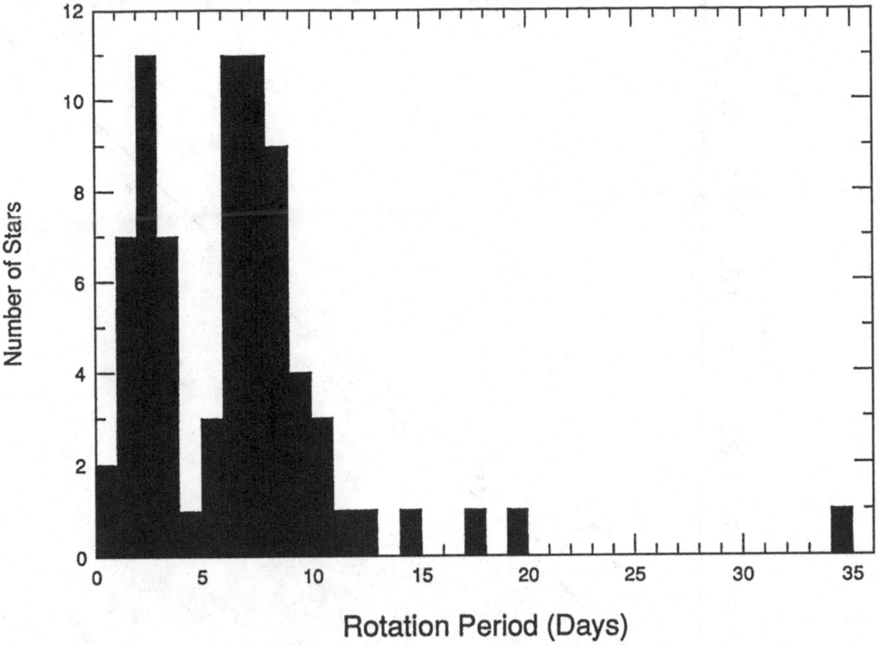

Fig. 1.9. Histogram showing the frequency distribution of rotation periods of T Tauri stars. This figure combines the data for the Trapezium cluster, the Orion Nebula cluster, and other T associations. *Source:* Eaton, N. L., Herbst, W., and Hillenbrand, L. A., *Astron. J.*, **110**, 1735, 1995.

various dissipative processes, a close binary starting from a wide range of initial spin and orbital parameters might eventually reach a state of minimum kinetic energy. This equilibrium state is characterized by a circular orbit, where the stellar spins are aligned and synchronized with the orbital spin.

As we shall see in Sections 8.2–8.4, however, in detached binaries the synchronization of the components proceeds at a much faster pace than the circularization of their orbits. Accordingly, the rotation of each component will quickly synchronize with the instantaneous orbital angular velocity *at periastron*,

$$\Omega_p = \frac{(1+e)^{1/2}}{(1-e)^{3/2}}\,\Omega_0, \tag{1.8}$$

where the tidal interaction is the most important during each orbital revolution. (As usual, e is the orbital eccentricity and Ω_0 is the mean orbital angular velocity.) Figure 1.10 illustrates this concept of *pseudo-synchronism* for a sample of selected eclipsing binaries with eccentric orbits for which we have accurate absolute dimensions. This figure compares the observed rotational velocities with the computed rotational velocities, assuming synchronization at periastron. We observe that most points scatter along the 45-degree line, indicating that pseudo-synchronization obtains in most close binaries of short orbital periods, either perfectly or approximately.

Observations show that an upper limit to the orbital period exists at which the observed rotational velocities begin to deviate very much from the synchronization (or

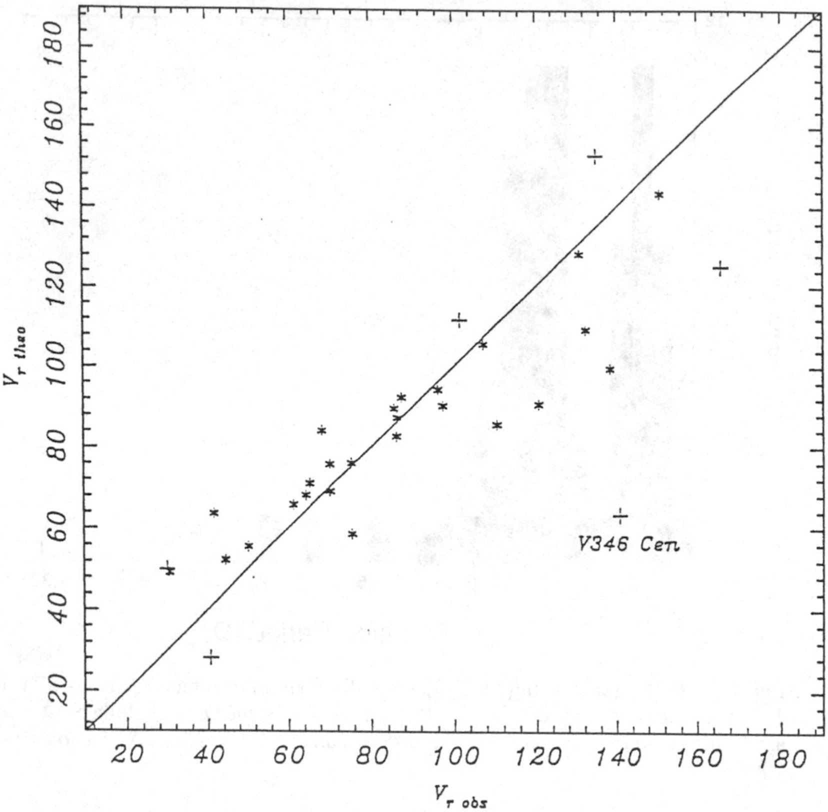

Fig. 1.10. Predicted versus observed rotational velocities assuming synchronization at periastron. The diagonal line is the locus of pseudo-synchronous rotation. *Source:* Claret, A., and Giménez, A., *Astron. Astrophys.*, **277**, 487, 1993.

pseudo-synchronization) period. As was originally noted by Levato (1976), the orbital period below which main-sequence binary components are still rotating in synchronism depends on spectral type. Specifically, he found that the largest orbital period for full synchronism is about 4–8 days in the early B spectral range, decreases to a minimum value of about 2 days at mid A-type, and increases up to 10–14 days at mid F-type. Subsequent investigations have confirmed that the tendency toward synchronization between the axial rotation and orbital revolution is indeed stronger in the F-type and later types than in the hotter ones. However, these studies have also demonstrated that *in the whole early spectral range synchronism (or pseudo-synchronism) extends up to binary separations substantially greater than previously held.* For example, the rotational properties of a large sample of early-type double-lined spectroscopic binaries have been investigated by Giuricin, Mardirossian, and Mezzetti (1985). Their statistical study indicates that a considerable tendency toward pseudo-synchronization extends up to a distance ratio $d/R \approx 20$ in the early-type (from O to F5) close binaries. (Here d is the mean distance between the components and R is the radius.) In fact, only for $d/R \gtrsim 20$ do pronounced deviations from synchronism at periastron become the rule in these binaries. In terms of orbital periods (for an easier comparison with Levato's underestimated upper limit

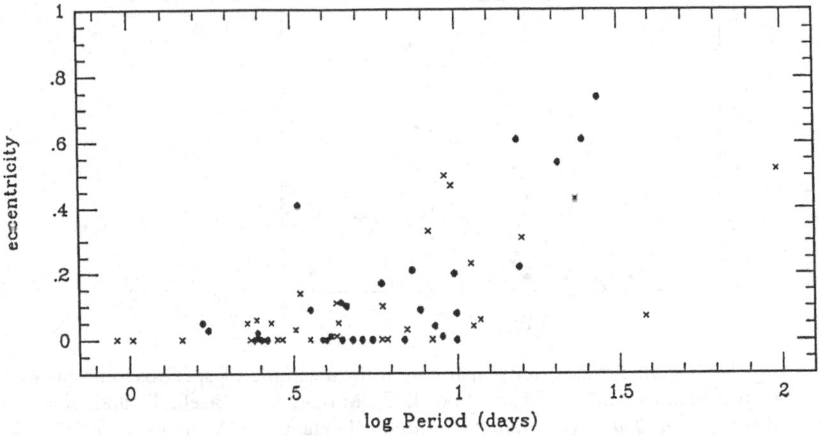

Fig. 1.11. Period–eccentricity distribution for a sample of spectroscopic binaries with A-type primaries. Single-lined binaries are shown as crosses; double-lined binaries are shown as filled circles. *Source:* Matthews, L. D., and Mathieu, R. D., in *Complementary Approaches to Double and Multiple Star Research* (McAlister, H., and Hartkopf, W. I., eds.), *A.S.P. Conference Series*, **32**, 244, 1992. (Courtesy of the Astronomical Society of the Pacific.)

periods), a limiting value of $d/R \approx 20$ corresponds to orbital periods of about 26, 18, and 13 days at spectral types B2, A0, and A5, respectively.

It is a well-known fact that circular (or nearly circular) orbits greatly predominate in short-period binaries. Since tidal interaction between the components of close binaries will tend to circularize their orbits, the precise determination of the cutoff period above which binaries display eccentric orbits appears to be a valuable test for the tidal theories. Giuricin, Mardirossian, and Mezzetti (1984) have studied the period–eccentricity distribution for a large sample of early-type detached binaries, excluding systems believed to have undergone (or to be undergoing) mass exchange between the components. They found that almost all binaries have circular or nearly circular orbits for orbital periods P smaller than 2 days. However, a mixed population of circular and eccentric orbits was found in the period range 2–10 days. Beyond $P = 10$ days all orbits are eccentric. A similar result was obtained by Matthews and Mathieu (1992), who investigated the period–eccentricity distribution of a sample of spectroscopic binaries with A-type primary stars. Figure 1.11 clearly shows that all binaries with orbital periods less than $P \approx 3$ days have circular or almost circular orbits (i.e., $e < 0.05$). Binaries with periods between 3 and 10 days are found with either circular or eccentric orbits, with the maximum eccentricity increasing with period. The longest-period circular orbit is at $P = 9.9$ days. This is exactly the kind of distribution one may expect to find for a sample of detached binaries with a *random* distribution of ages, where the population of circular and eccentric orbits becomes increasingly mixed as the Ps tend toward an upper limit period above which all orbits become eccentric.* For comparison, Figure 1.12 illustrates

* More recently, Mermilliod (1996, Fig. 2) has shown that this upper limit period was actually close to 25 days for a sample of 39 late-B and A-type binary stars belonging to open clusters. Note also that most of the O-type binaries with periods less than 30 days have circular orbits, whereas the long-period systems have eccentric orbits (Massey, 1982, p. 258).

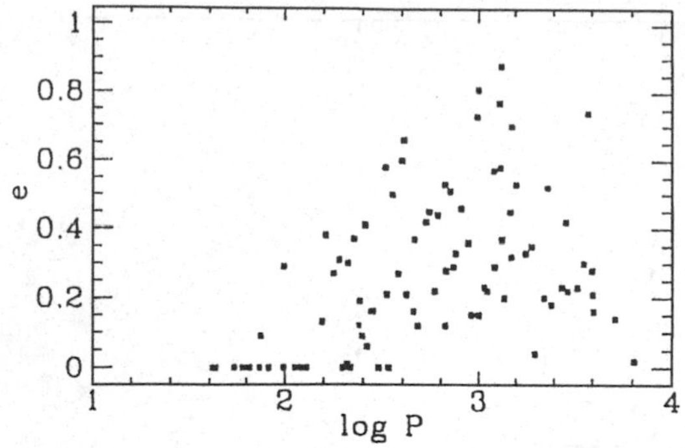

Fig. 1.12. Period–eccentricity distribution for a sample of spectroscopic binaries with red giant primaries. *Source:* Mermilliod, J. C., Mayor, M., Mazeh, T., and Mermilliod, J. C., in *Binaries as Tracers of Stellar Formation* (Duquennoy, A., and Mayor, M., eds.), p. 183, Cambridge: Cambridge University Press, 1992.

the period–eccentricity distribution of spectroscopic binaries with red giant primaries. Not unexpectedly, because red giants reach larger radii than main sequence stars, circular orbits are found for larger orbital periods. Note also the mixed population of circular and eccentric orbits in the period range 80–300 days. Again, this is caused by the mixing of all red giants, since the sample contains a range in age and mass.

It will be shown in Sections 8.2–8.4 that the degree of circularity of an orbit depends on how long the tidal forces have been acting on the components of a close binary. The study of binaries belonging to clusters is of particular interest, therefore, since these are the only stars for which one has some information about their ages. Mayor and Mermilliod (1984) were the first to study the orbital eccentricities for a *coeval* sample of late-type binaries in open clusters (33 red-dwarf binaries in the Hyades, Pleiades, Praesepe, and Coma Berenices open clusters). They found that all binaries with periods shorter than 5.7 days display circular orbits whereas all orbits with longer periods have significant eccentricities. More recently, it has been found that other coeval samples with different evolutionary ages exhibit transitions between circular and eccentric orbits at distinct cut-off periods. It is immediately apparent from Table 1.2 that *the transition period* P_{cut} *increases monotonically with the sample age* t_a. Accordingly, the observed t_a–P_{cut} relation strongly suggests that the circularization mechanism is operative *during* the main-sequence lifetime of the stars – pre–main-sequence tidal circularization is permitted but not required by present observations. This provides a very important test for the tidal mechanisms since the theoretical circularization time cannot exceed the sample age at cutoff period.

Tidal interaction in the RS CVn stars poses quite a challenging problem also. In fact, in these chromospherically active binaries there is still a tendency toward synchronization in the period range 30–70 days, up to $P = 100$ days. However, asynchronous rotators are present in all period groups, even among binaries with orbital periods of 30 days or less. In these systems one also finds that the rotation periods are either shorter or longer than the orbital periods, independent of the orbital eccentricities. As was shown by Tan, Wang, and Pan (1991), however, asynchronous RS CVn stars have orbital eccentricities that are larger, on the average, than the eccentricities of pseudo-synchronously rotating systems.

Table 1.2. *The observed t_a–$P_{\rm cut}$ relation.*

Binary Sample	Cutoff Period (day)	Age (Gyr)
Pre-main-sequence	4.3	0.003
Pleiades	7.05	0.1
Hyades/Praesepe	8.5	0.8
M67	12.4	4.0
Halo	18.7	17.6

Source: Mathieu, R. D., Duquennoy, A., Latham, D. W., Mayor, M., Mazeh, T., and Mermilliod, J. C., in *Binaries as Tracers of Stellar Formation* (Duquennoy, A., and Mayor, M., eds.), p. 278, Cambridge: Cambridge University Press, 1992.

These authors also found that the chromospheric activity in their sample of asynchronous binaries is lower, on the average, than in synchronous RS CVn stars. If so, then, other braking mechanisms (e.g., magnetically driven winds) must be interfering with tidal interaction in these giant binary stars. To make the problem even more complex, let us note that Stawikowski and Glebocki (1994) have found another basic difference between the synchronous and asynchronous long-period RS CVn stars, when their primary component is a late-type giant or subgiant: *Whereas for synchronously rotating stars the assumption about coplanarity of their equatorial and orbital planes is justified, in most asynchronous binaries the rotation axis of the primary is not perpendicular to the orbital plane.* A similar result was obtained by Glebocki and Stawikowski (1995, 1997) for late-type main-sequence binaries and short-period RS CVn stars with orbital periods shorter than about 10 days. Pseudo-synchronism and coplanarity will be further discussed in Section 8.2.1.

1.5 Bibliographical notes

Section 1.1. Historical accounts will be found in:

1. Mitchell, W. M., *Popular Astronomy*, **24**, 22, 1916; *ibid.*, p. 82; *ibid.*, p. 149; *ibid.*, p. 206; *ibid.*, p. 290; *ibid.*, p. 341; *ibid.*, p. 428; *ibid.*, p. 488; *ibid.*, p. 562.
2. Brunet, P., *L'introduction des théories de Newton en France au XVIIIe siècle*, pp. 223–228, Paris, 1931 (Genève: Slatkine Reprints, 1970).
3. Struve, O., *Popular Astronomy*, **53**, 201, 1945; *ibid.*, p. 259.
4. Bray, R. J., and Loughhead, R. E., *Sunspots*, London: Chapman and Hall, 1964.

Reference 1 contains facsimiles and English translations of all relevant papers by Fabricius, Galileo, and Scheiner; it also presents a brief account of Harriot's unpublished work. See also References 21 and 30, which contain detailed citations to many original papers on helioseismology and stellar rotation.

Section 1.2.1. The following review papers are particularly worth noting:

5. Howard, R., *Annu. Rev. Astron. Astrophys.*, **22**, 131, 1984.
6. Schröter, E. H., *Solar Phys.*, **100**, 141, 1985.
7. Bogart, R. S., *Solar Phys.*, **110**, 23, 1987.
8. Snodgrass, H. B., in *The Solar Cycle* (Harvey, K. L., ed.), *A.S.P. Conference Series*, **27**, 205, 1992.

Temporal variations of the solar rotation rate have been considered by:

9. Eddy, J. A., Gilman, P. A., and Trotter, D. E., *Science*, **198**, 824, 1977.
10. Howard, R., and LaBonte, B. J., *Astrophys. J. Letters*, **239**, L33, 1980.
11. Gilman, P. A., and Howard, R., *Astrophys. J.*, **283**, 385, 1984.
12. Balthazar, H., Vásquez, M., and Wöhl, H., *Astron. Astrophys.*, **155**, 87, 1986.
13. Hathaway, D. H., and Wilson, R. M., *Astrophys. J.*, **357**, 271, 1990.
14. Yoshimura, H., and Kambry, M. A., *Astron. Nachr.*, **314**, 9, 1993; *ibid.*, p. 21.

There is a wide literature on the vexing problem of meridional motions on the solar surface. The following papers may be noted:

15. Kambry, M. A., Nishikawa, J., Sakurai, T., Ichimoto, K., and Hiei, E., *Solar Phys.*, **132**, 41, 1991.
16. Cavallini, F., Ceppatelli, G., and Righini, A., *Astron. Astrophys.*, **254**, 381, 1992.
17. Komm, R. W., Howard, R. F., and Harvey, J. W., *Solar Phys.*, **147**, 207, 1993.
18. Hathaway, D. H., *Astrophys. J.*, **460**, 1027, 1996.
19. Snodgrass, H. B., and Dailey, S. B., *Solar Phys.*, **163**, 21, 1996.

Section 1.2.2. Among the many review papers on helioseismology and the Sun's internal rotation, my own preference goes to:

20. Christensen-Dalsgaard, J., in *Advances in Helio- and Asteroseismology* (Christensen-Dalsgaard, J., and Frandsen, S., eds.), I.A.U. Symposium No 123, p. 3, Dordrecht: Reidel, 1988.
21. Gough, D., and Toomre, J., *Annu. Rev. Astron. Astrophys.*, **29**, 627, 1991.
22. Gilliland, R. L., in *Astrophysical Applications of Stellar Pulsation* (Stobie, R. S., and Whitelock, P. A., eds.), *A.S.P. Conference Series*, **83**, 98, 1995.

There is also an interesting collective review in *Science*, **272**, pp. 1281–1309, 1996. The presentation in the text is largely based on:

23. Korzennik, S. G., Thompson, M. J., Toomre, J., and the GONG Internal Rotation Team, in *Sounding Solar and Stellar Interiors* (Provost, J., and Schmider, F. X., eds.), I.A.U. Symposium No 181, p. 211, Dordrecht: Kluwer, 1997.

Measurements of the rotation rate in the radiative core have been made by:

24. Brown, T. M., Christensen-Dalsgaard, J., Dziembowski, W. A., Goode, P., Gough, D. O., and Morrow, C. A., *Astrophys. J.*, **343**, 526, 1989.
25. Tomczyk, S., Schou, J., and Thompson, M. J., *Astrophys. J. Letters*, **448**, L57, 1995.

Rotation rates in the inner core are discussed in:

26. Jiménez, A., Pérez Hernández, F., Claret, A., Pallé, P. L., Régulo, C., and Roca Cortés, T., *Astrophys. J.*, **435**, 874, 1994.
27. Toutain, T., and Kosovichev, A. G., *Astron. Astrophys.*, **284**, 265, 1994.

See also Reference 39 of Chapter 5. Other relevant papers may be traced from the GONG publications.

Section 1.3. The following review papers may be noted:

28. Kraft, R. P., in *Spectroscopic Astrophysics* (Herbig, G. H., ed.), p. 385, Berkeley: University of California Press, 1970.
29. Slettebak, A., in *Stellar Rotation* (Slettebak, A., ed.), p. 3, New York: Gordon and Breach, 1970.
30. Slettebak, A., in *Calibration of Fundamental Stellar Quantities* (Hayes, D. S., Pasinetti, L. E., and Davis Philip, A. G., eds.), I.A.U. Symposium No 111, p. 163, Dordrecht: Reidel, 1985.
31. Stauffer, J. R., and Hartmann, L. W., *Publ. Astron. Soc. Pacific*, **98**, 1233, 1986.
32. Stauffer, J. R., in *Angular Momentum Evolution of Young Stars* (Catalano, S., and Stauffer, J. R., eds.), p. 117, Dordrecht: Kluwer, 1991.

An excellent introduction to these matters is given by:

33. Gray, D. F., *The Observation and Analysis of Stellar Photospheres*, 2nd Edition, pp. 368 – 400, Cambridge: Cambridge University Press, 1992.

Statistical studies of early-type stars will be found in:

34. Bernacca, P. L., and Perinotto, M., *Astron. Astrophys*, **33**, 443, 1974.
35. Fukuda, I., *Publ. Astron. Soc. Pacific*, **94**, 271, 1982.

The following key references are also quoted in the text:

36. Kraft, R. P., *Astrophys. J.*, **150**, 551, 1967.
37. Skumanich, A., *Astrophys. J.*, **171**, 565, 1972.

The rotational velocities of low-mass stars are discussed at length in References 30–32. More recent discussions of the T Tauri stars are due to:

38. Attridge, J. M., and Herbst, W., *Astrophys. J. Letters*, **398**, L61, 1992.
39. Bouvier, J., Cabrit, S., Fernández, M., Martín, E. L., and Matthews, J. M., *Astron. Astrophys.*, **272**, 176, 1993.
40 Choi, P. I., and Herbst, W., *Astron. J.*, **111**, 283, 1996.

Other references may be traced to:

41. Bouvier, J., Wichmann, R., Grankin, K., Allain, S., Covino, E., Fernández, M., Martín, E. L., Terranegra, L., Catalano, S., and Marilli, E., *Astron. Astrophys.*, **318**, 495, 1997.
42. Stauffer, J. R., Hartmann, L. W., Prosser, C. F., Randich, F., Balachandran, S., Patten, B. M., Simon, T., and Giampapa, M., *Astrophys. J.*, **479**, 776, 1997.

Section 1.4. Early-type binaries have been considered by:

43. Levato, H., *Astrophys. J.*, **203**, 680, 1976.
44. Rajamohan, R., and Venkatakrishnan, P., *Bull. Astron. Soc. India*, **9**, 309, 1981.
45. Giuricin, G., Mardirossian, F., and Mezzetti, M., *Astron. Astrophys.*, **131**, 152, 1984; *ibid.*, **134**, 365, 1984; *ibid.*, **135**, 393, 1984.
46. Giuricin, G., Mardirossian, F., and Mezzetti, M., *Astron. Astrophys. Suppl. Ser.*, **59**, 37, 1985.
47. Hall, D. S., *Astrophys. J. Letters*, **309**, L83, 1986.

See also:

48. Massey, P., in *Wolf-Rayet Stars: Observations, Physics, Evolution* (de Loore, C. W. H., and Willis, A. J., eds.), I.A.U. Symposium No 99, p. 251, Dordrecht: Reidel, 1982.
49. Mermilliod, J. C., in *The Origins, Evolution, and Destinies of Binary Stars in Clusters* (Milone, E. F., and Mermilliod, J. C., eds.), *A.S.P. Conference Series*, **90**, 95, 1996.

The eccentricity distribution of low-mass binaries in open clusters was originally discussed by:

50. Mayor, M., and Mermilliod, J. C., in *Observational Tests of the Stellar Evolution Theory* (Maeder, A., and Renzini, A., eds.), p. 411, Dordrecht: Reidel, 1984.

Detailed surveys are summarized in:

51. Mathieu, R. D., Duquennoy, A., Latham, D. W., Mayor, M., Mazeh, T., and Mermilliod, J. C., in *Binaries as Tracers of Stellar Formation* (Duquennoy, A., and Mayor, M., eds.), p. 278, Cambridge: Cambridge University Press, 1992.

Statistical studies of the RS CVn stars will be found in:

52. Tan, H. S., and Liu, X. F., *Chinese Astron. Astrophys.*, **11**, 15, 1987.
53. Fekel, F. C., and Eitter, J. J., *Astron. J.*, **97**, 1139, 1989.
54. Tan, H. S., Wang, X. H., and Pan, K. K., *Chinese Astron. Astrophys.*, **15**, 461, 1991.

See also:

55. de Medeiros, J. R., and Mayor, M., *Astron. Astrophys.*, **302**, 745, 1995.

The problem of coplanarity has been considered by:

56. Merezhin, V. P., *Astrophys. Space Sci.*, **218**, 223, 1994.
57. Stawikowski, A., and Glebocki, R., *Acta Astronomica*, **44**, 33, 1994; *ibid.*, p. 393.
58. Glebocki, R., and Stawikowski, A., *Acta Astronomica*, **45**, 725, 1995.
59. Glebocki, R., and Stawikowski, A., *Astron. Astrophys.*, **328**, 579, 1997.

2

Rotating fluids

2.1 Introduction

As we may infer from the observations, most stars remain in a state of mechanical equilibrium, with the pressure-gradient force balancing their own gravitation corrected for the centrifugal force of axial rotation. Accordingly, theoretical work has tended to focus on the figures of equilibrium of a rotating star, assuming the motion to be wholly one of pure rotation. However, detailed study of the Sun has demonstrated the existence of large-scale motions in its convective envelope, both around the rotation axis and in meridian planes passing through the axis. Theoretical work has shown that large-scale meridional currents also exist in the radiative regions of a rotating star. Moreover, as new results become available, it is becoming increasingly apparent that these regions contain a wide spectrum of turbulent motions embedded in the large-scale flow. All these problems are the domain of *astrophysical fluid dynamics* – a field that has developed quite slowly until recently.

Over the course of the past fifty years, however, meteorologists and oceanographers have made important advances in our knowledge of the behavior of rotating fluids. I thus find it appropriate to review some dynamical concepts that are applicable to both the Earth's atmosphere and the oceans. As we shall see, all of them play a key role in providing useful ideas for quantitative analysis of large-scale motions in a rotating star. Accordingly, unless the reader is already familiar with *geophysical fluid dynamics*, I recommend reading this introductory chapter, which is essential for understanding the many hydrodynamical problems treated in the book.

2.2 The equations of fluid motion

Fluid dynamics proceeds on the hypothesis that the length scale of the flow is always taken to be large compared with the mean free path of the constitutive particles, so that the fluid may be treated as a *continuum*. This model makes it possible to treat fluid properties (such as velocity, pressure, density, etc.) at a *point* in space, with the physical variables being continuous functions of space and time. In other words, we assume that the macroscopic behavior of our systems is the same as if their distribution of matter were perfectly continuous in structure. Accordingly, whenever we speak of the velocity of a "mass element" (or "fluid particle") we always mean the average velocity of a large number of constitutive particles contained within a small volume of finite extent, although this volume must be regarded as a point.

The mathematical description of a fluid motion from the continuum point of view allows two distinct methods of approach. The first one, called the *Lagrangian description*, identifies each mass element and describes what happens to it over time. Mathematically, we represent the motion by the function

$$\mathbf{r} = \mathbf{r}(\mathbf{R}, t), \tag{2.1}$$

where $\mathbf{R} = (X_1, X_2, X_3)$ is the original position of a fluid particle, at time $t = 0$ (say), and $\mathbf{r} = (x_1, x_2, x_3)$ is the location of the same mass element at the subsequent instant t. The dependent vector $\mathbf{r}$ is thus determined as a function of the independent variables $\mathbf{R}$ and t. The velocity and acceleration of a fluid particle are

$$\mathbf{v}(\mathbf{R}, t) = \frac{\partial \mathbf{r}}{\partial t} \quad \text{and} \quad \mathbf{a}(\mathbf{R}, t) = \frac{\partial^2 \mathbf{r}}{\partial t^2}, \tag{2.2}$$

where the partial derivative indicates that the differentiation must be carried out for a given mass element (i.e., holding $\mathbf{R}$ constant).

The second approach, called the *Eulerian description*, focuses attention on a particular point in space and describes the flow at that point over time. Mathematically, the state of motion is described by the velocity field

$$\mathbf{v} = \mathbf{v}(\mathbf{r}, t), \tag{2.3}$$

where the independent variables are location in space, represented by the vector $\mathbf{r} = (x_1, x_2, x_3)$, and time. The acceleration of a fluid particle is the material derivative of the velocity. Hence, we let

$$\mathbf{a}(\mathbf{r}, t) = \frac{D\mathbf{v}}{Dt} = \frac{\partial \mathbf{v}}{\partial t} + (\mathbf{v} \cdot \mathrm{grad})\mathbf{v}. \tag{2.4}$$

Similarly, one can define the material derivative

$$\frac{DQ}{Dt} = \frac{\partial Q}{\partial t} + (\mathbf{v} \cdot \mathrm{grad})Q, \tag{2.5}$$

which measures the rate of change of the quantity Q as one follows a fluid particle along its path.

2.2.1 Conservation principles

It is not my purpose to demonstrate the basic equations of fluid dynamics, since their derivation can be found in numerous textbooks. In this section I shall list these equations in an inertial frame of reference, making use of the Eulerian specification.

In the absence of sources or sinks of matter within the fluid, the condition of mass conservation is expressed by the *continuity equation*,

$$\frac{1}{\rho}\frac{D\rho}{Dt} + \mathrm{div}\,\mathbf{v} = 0. \tag{2.6}$$

This equation states that the fractional rates of change of density and volume of a mass element are equal in magnitude and opposite in sign.

Newton's second law of action can be written in the form

$$\frac{D\mathbf{v}}{Dt} = \mathbf{g} - \frac{1}{\rho}\,\mathrm{grad}\,p + \frac{1}{\rho}\,\mathbf{f}(\mathbf{v}), \tag{2.7}$$

where $\mathbf{g}$ is the acceleration due to gravity, ρ is the density, and p is the pressure. The vector $\mathbf{f}$ is the viscous force per unit volume, which can be written as the vectorial divergence of the viscous stress tensor $\boldsymbol{\tau}$. For Newtonian fluids, the six components of this symmetric tensor are

$$\tau_{ij} = \mu \left(\frac{\partial v_i}{\partial x_j} + \frac{\partial v_j}{\partial x_i} - \frac{2}{3}\,\delta_{ij}\,\frac{\partial v_k}{\partial x_k} \right) + \mu_\vartheta\,\delta_{ij}\frac{\partial v_k}{\partial x_k}, \tag{2.8}$$

where the coefficients of *shear viscosity* μ and *bulk viscosity* μ_ϑ both depend on local thermodynamic properties only ($\delta_{ij} = 1$ if $i = j$, $\delta_{ij} = 0$ if $i \neq j$; summation over repeated indices). Thus, we have

$$\mathbf{f}(\mathbf{v}) = \mu\nabla^2\mathbf{v} + \left(\mu + \frac{1}{3}\,\mu_\vartheta\right) \operatorname{grad}(\operatorname{div}\mathbf{v}), \tag{2.9}$$

where, for the sake of simplicity, we have assumed that the viscosity coefficients remain constant over the field of motion. Equation (2.7) is often known as the *Navier–Stokes equation*.

If the flow is such that the pressure variations do not produce any significant density variations, the compressibility of the fluid may be neglected. (This occurs in most liquid flows; it also occurs in many of the gas flows for which the speed is everywhere much smaller than the speed of sound.) In compressible flows, however, it is always necessary to augment Eqs. (2.6) and (2.7) with an equation based on the principles of thermodynamics. To be specific, the *conservation of thermal energy* implies that

$$\rho\,\frac{DU}{Dt} + p\ \operatorname{div}\mathbf{v} = \operatorname{div}(k_c \operatorname{grad} T) + \Phi_v + \rho Q, \tag{2.10}$$

where U is the thermal energy per unit mass, T is the temperature, k_c is the coefficient of thermal conductivity, Φ_v is the rate (per unit volume and unit time) at which heat is generated by viscous friction, and Q is the net heat addition per unit mass by internal heat sources. For all situations to be discussed in this book, the dissipation function Φ_v is utterly negligible. Since the function Q is of particular relevance to stellar interiors, it will be discussed further in Section 3.2.

Now, assuming quasi-static changes at every point of the fluid, we can write

$$T\frac{DS}{Dt} = \frac{DU}{Dt} + p\,\frac{D}{Dt}\left(\frac{1}{\rho}\right), \tag{2.11}$$

where S is the entropy per unit mass. By virtue of Eq. (2.6), a comparison between Eqs. (2.10) and (2.11) leads to the result

$$\rho T\,\frac{DS}{Dt} = \operatorname{div}(k_c \operatorname{grad} T) + \Phi_v + \rho Q, \tag{2.12}$$

expressing the change of specific entropy as we follow a mass element along its motion.

To complete the system of equations, further thermodynamic relations are required. For example, for an ideal gas, one has

$$U = c_V T \tag{2.13}$$

and

$$p = \frac{\mathcal{R}}{\bar{\mu}} \rho T, \tag{2.14}$$

where $\bar{\mu}$ is the mean molecular weight. One also has $\mathcal{R}/\bar{\mu} = c_p - c_V$, where c_p and c_V are the specific heats, at constant pressure and constant volume. Inserting these relations into Eq. (2.11), one obtains

$$S = c_p \log \Theta + \textit{constant}. \tag{2.15}$$

The quantity

$$\Theta = T \left(\frac{p_0}{p} \right)^{(\gamma-1)/\gamma}, \tag{2.16}$$

where p_0 is a constant reference pressure and γ is the adiabatic exponent, is known as the *potential temperature*. For isentropic motions (i.e., motions for which the right-hand side of Eq. [2.12] identically vanishes) the potential temperature of each fluid particle remains a constant along its path.

2.2.2 *Boundary conditions*

In order to solve the partial differential equations that govern the fluid motion, it is necessary to prescribe initial conditions specified over all space and boundary conditions specified over all time. Whereas initial conditions are always peculiar to the problem at hand, the appropriate boundary conditions are of a rather general nature.

On a fixed *solid boundary*, there can be no fluid motion across the boundary. This condition implies that

$$\mathbf{n} \cdot \mathbf{v} = 0, \tag{2.17}$$

where **n** is the outer normal to the surface. A second condition is provided by the no-slip condition that there should be no relative tangential velocity between a rigid wall and the viscous fluid next to it. Hence, we must also prescribe that

$$\mathbf{n} \times \mathbf{v} = \mathbf{0}, \tag{2.18}$$

on a fixed solid wall.

At an *interfacial boundary* (such as the top of an ocean or the outer surface of a star), one must prescribe that no mass element cross this boundary so that fluid particles on the boundary will remain on the boundary. Thus, if $\xi(\mathbf{r}, t)$ defines the surface elevation above an equilibrium level, this kinematic boundary condition on the velocity is

$$\frac{D\xi}{Dt} = \mathbf{n} \cdot \mathbf{v} \tag{2.19}$$

at the material boundary. If this boundary is fixed (i.e., $\xi \equiv 0$), condition (2.19) reduces to

$$\mathbf{n} \cdot \mathbf{v} = 0, \tag{2.20}$$

expressing that matter is always flowing *along* the prescribed material boundary.

Beyond this kinematic boundary condition, it is also clear that we must ensure the balance of forces at any nonsolid boundary. For example, the gravitational attraction must be continuous across the free surface of a star. Similarly, the components of the stress vector acting on a nonsolid boundary must be continuous across that boundary (see Eq. [2.8]). Thus, we let

$$[n_k(-p\delta_{ik} + \tau_{ik})] = 0, \tag{2.21}$$

where brackets designate the jump that the quantity experiences on a nonsolid boundary ($i = 1, 2, 3$). In particular, at the free surface of a stellar model embedded into a vacuum, these three components must identically vanish.

2.2.3 *Rotating frame of reference*

In some applications, it is convenient to describe the motions as they appear to an observer at rest in a frame rotating with the constant angular velocity $\mathbf{\Omega}$. We can write

$$\mathbf{v}(\mathbf{r}, t) = \mathbf{u}(\mathbf{r}, t) + \mathbf{\Omega} \times \mathbf{r}, \tag{2.22}$$

where the velocity $\mathbf{u}$ refers to the moving axes. Similarly, the material acceleration (2.4) has the form

$$\mathbf{a}(\mathbf{r}, t) = \frac{D\mathbf{u}}{Dt} + 2\mathbf{\Omega} \times \mathbf{u} + \mathbf{\Omega} \times (\mathbf{\Omega} \times \mathbf{r}), \tag{2.23}$$

where

$$\frac{D\mathbf{u}}{Dt} = \frac{\partial \mathbf{u}}{\partial t} + (\mathbf{u} \cdot \operatorname{grad})\mathbf{u} \tag{2.24}$$

is the acceleration relative to the rotating frame. The quantities $2\mathbf{\Omega} \times \mathbf{u}$ and $\mathbf{\Omega} \times (\mathbf{\Omega} \times \mathbf{r})$ represent, respectively, the Coriolis acceleration and the centrifugal acceleration. Since the tensor (2.8) is invariant with respect to a uniform rotation, Eq. (2.7) then becomes

$$\frac{D\mathbf{u}}{Dt} + 2\mathbf{\Omega} \times \mathbf{u} = \mathbf{g} - \mathbf{\Omega} \times (\mathbf{\Omega} \times \mathbf{r}) - \frac{1}{\rho} \operatorname{grad} p + \frac{1}{\rho} \mathbf{f}(\mathbf{u}). \tag{2.25}$$

It is a simple matter to show that

$$\mathbf{\Omega} \times (\mathbf{\Omega} \times \mathbf{r}) = -\operatorname{grad}\left(\frac{1}{2}|\mathbf{\Omega} \times \mathbf{r}|^2\right). \tag{2.26}$$

Because the vector $\mathbf{g}$ is derivable also from a scalar potential, V (say), we can thus rewrite the momentum equation (2.25) in the form

$$\frac{D\mathbf{u}}{Dt} + 2\mathbf{\Omega} \times \mathbf{u} = \mathbf{g}_e - \frac{1}{\rho} \operatorname{grad} p + \frac{1}{\rho} \mathbf{f}(\mathbf{u}), \tag{2.27}$$

where

$$\mathbf{g}_e = -\operatorname{grad}\left(V - \frac{1}{2}|\mathbf{\Omega} \times \mathbf{r}|^2\right) \tag{2.28}$$

is the *effective gravity*. Comparing Eq. (2.27) with Eq. (2.7), one readily sees that the Coriolis acceleration is the only structural change of Newton's second law for motion relative to a rotating frame of reference.*

* As far back as 1835, the French engineer Gaspard Coriolis (1792–1843) made a detailed mathematical study of the absolute acceleration of moving solids in a rotating frame of reference. His work had little

For steady flows, the relative importance of the acceleration measured in the rotating frame and the Coriolis acceleration can be estimated as

$$\frac{|\mathbf{u} \cdot \operatorname{grad} \mathbf{u}|}{|\boldsymbol{\Omega} \times \mathbf{u}|} \approx \frac{U^2/L}{\Omega U} = \frac{U}{\Omega L}, \tag{2.29}$$

where U and L are the characteristic velocity and length of the flow. This ratio is a nondimensional number called the *Rossby number*, and it is designated by

$$Ro = \frac{U}{\Omega L}. \tag{2.30}$$

Similarly, by making use of Eq. (2.9), one can easily estimate the ratio of the viscous force to the Coriolis force. One obtains

$$\frac{|\nu \nabla^2 \mathbf{u}|}{|\boldsymbol{\Omega} \times \mathbf{u}|} \approx \frac{\nu U/L^2}{\Omega U} = \frac{\nu}{\Omega L^2}, \tag{2.31}$$

where $\nu = \mu/\rho$ is the coefficient of kinematic viscosity. The nondimensional number

$$E = \frac{\nu}{\Omega L^2} \tag{2.32}$$

is known as the *Ekman number*.

2.3 The vorticity equation

To visualize a fluid motion, it is often convenient to construct the streamlines of the flow. Since a streamline is an imaginary line that is everywhere tangent to the fluid velocity $\mathbf{v}(\mathbf{r}, t)$, the family of such lines is given by the integration of

$$\frac{dx_1}{v_1} = \frac{dx_2}{v_2} = \frac{dx_3}{v_3}. \tag{2.33}$$

In steady flows, streamlines and particle paths are identical.

In many instances, however, it is also instructive to describe the flow in terms of the *absolute vorticity*

$$\boldsymbol{\omega} = \operatorname{curl} \mathbf{v}, \tag{2.34}$$

which represents the local and instantaneous rate of rotation of the fluid measured in an inertial frame of reference. By definition, a continuous line that is everywhere tangent to the vector $\boldsymbol{\omega}(\mathbf{r}, t)$ is called an absolute-vorticity line. The family of such lines is defined

impact on the meteorological studies of that time, however, so that few advances in our knowledge of the behavior of rotating fluids were made during the nineteenth century. As a matter of fact, it is not until the late 1850s that the American meteorologist William Ferrel (1817–1891) gave the first mathematical formulation of atmospheric motions on a rotating Earth. Moreover, as we shall see in Section 2.6.1, the importance of the deflective force of the Earth's rotation on wind-driven currents in the oceans was not recognized until the turn of the twentieth century. For comparison, Sir Arthur Eddington (1882–1944) in 1925 noticed that large-scale meridional currents in the radiative regions of a star would be deflected east and west by the star's rotation, but it is not until 1941 that Gunnar Randers (1914–1992) made the first detailed analysis of the steady motion exhibiting a balance between the viscous and deflective forces in a rotating star (see Eq. [4.49]).

by the pair of differential equations

$$\frac{dx_1}{\omega_1} = \frac{dx_2}{\omega_2} = \frac{dx_3}{\omega_3}. \tag{2.35}$$

By virtue of Eq. (2.34), one always has

$$\operatorname{div} \boldsymbol{\omega} = 0. \tag{2.36}$$

Hence, *absolute-vorticity lines cannot begin or end in the fluid; they are either closed curves or terminate on the boundary*. By making use of Eq. (2.22), one can also write

$$\boldsymbol{\omega} = \operatorname{curl}(\mathbf{u} + \boldsymbol{\Omega} \times \mathbf{r}) = \boldsymbol{\zeta} + 2\boldsymbol{\Omega}, \tag{2.37}$$

where $\boldsymbol{\zeta}$ is the *relative vorticity*, that is, the curl of the velocity measured in the rotating frame of reference.

Let us now derive the equation expressing the rate of change of vorticity in a continuous motion. Using a formula well known in vector analysis,

$$\frac{1}{2} \operatorname{grad} |\mathbf{u}|^2 = \mathbf{u} \times \operatorname{curl} \mathbf{u} + \mathbf{u} \cdot \operatorname{grad} \mathbf{u}, \tag{2.38}$$

we can take the curl of Eq. (2.27) to obtain

$$\frac{\partial \boldsymbol{\zeta}}{\partial t} + \operatorname{curl}(\boldsymbol{\omega} \times \mathbf{u}) = \frac{1}{\rho^2} \operatorname{grad} \rho \times \operatorname{grad} p + \operatorname{curl}\left(\frac{1}{\rho}\mathbf{f}\right). \tag{2.39}$$

By making use of Eq. (2.36), one finds that

$$\operatorname{curl}(\boldsymbol{\omega} \times \mathbf{u}) = \mathbf{u} \cdot \operatorname{grad} \boldsymbol{\omega} - \boldsymbol{\omega} \cdot \operatorname{grad} \mathbf{u} + \boldsymbol{\omega} \operatorname{div} \mathbf{u}. \tag{2.40}$$

Since $\boldsymbol{\Omega}$ is a constant vector, one also has $\partial \boldsymbol{\zeta}/\partial t = \partial \boldsymbol{\omega}/\partial t$. Hence, Eq. (2.39) becomes

$$\frac{D\boldsymbol{\omega}}{Dt} = \boldsymbol{\omega} \cdot \operatorname{grad} \mathbf{u} - \boldsymbol{\omega} \operatorname{div} \mathbf{u} + \frac{1}{\rho^2} \operatorname{grad} \rho \times \operatorname{grad} p + \operatorname{curl}\left(\frac{1}{\rho}\mathbf{f}\right). \tag{2.41}$$

Combining Eqs. (2.6) and (2.41), one obtains the *vorticity equation*

$$\frac{D}{Dt}\left(\frac{\boldsymbol{\omega}}{\rho}\right) = \frac{\boldsymbol{\omega}}{\rho} \cdot \operatorname{grad} \mathbf{u} + \frac{1}{\rho^3} \operatorname{grad} \rho \times \operatorname{grad} p + \frac{1}{\rho} \operatorname{curl}\left(\frac{1}{\rho}\mathbf{f}\right). \tag{2.42}$$

The first term on the right-hand side of this equation represents the action of velocity variations on the ratio $\boldsymbol{\omega}/\rho$. The second term, the so-called baroclinic vector, modifies this ratio whenever the surfaces of constant pressure and constant density do not coincide in the fluid. The third term represents the rate of change of the ratio $\boldsymbol{\omega}/\rho$ due to diffusion of vorticity by viscous friction.

Since the vector $(\boldsymbol{\omega}/\rho) \cdot \operatorname{grad} \mathbf{u}$ has no counterpart in the momentum equation, it warrants further discussion. Thus neglecting the baroclinic vector and the curl of the frictional force, we obtain

$$\frac{D}{Dt}\left(\frac{\boldsymbol{\omega}}{\rho}\right) = \frac{\boldsymbol{\omega}}{\rho} \cdot \operatorname{grad} \mathbf{u}. \tag{2.43}$$

Now, by making use of the Lagrangian variables **R** and t (see Eqs. [2.1] and [2.2]), one can integrate this equation at once to obtain

$$\frac{\omega_i}{\rho} = \frac{\omega_{0k}}{\rho_0} \frac{\partial x_i}{\partial X_k}, \tag{2.44}$$

where $\boldsymbol{\omega}_0(\mathbf{R}, 0)$ and $\rho_0(\mathbf{R}, 0)$ are the initial values of $\boldsymbol{\omega}(\mathbf{R}, t)$ and $\rho(\mathbf{R}, t)$. As was shown by Helmholtz, this solution simply means that *the particles that compose an absolute-vorticity line at one instant will continue to form an absolute-vorticity line at any subsequent instant*. The proof lies in the fact that a tangent to such a line is carried by the fluid so that it always remains tangent to an absolute-vorticity line. Let dX_i be the components of the vector representing a line element, at the instant $t = 0$, of an absolute vortex line. As we follow its motion, we have

$$dx_i = \frac{\partial x_i}{\partial X_k} dX_k, \tag{2.45}$$

where the dx_is are the new components, at time t, of this line element. Now, by hypothesis, we can always write

$$dX_i = \epsilon \frac{\omega_{0i}}{\rho_0}, \tag{2.46}$$

where ϵ is some constant. From Eqs. (2.44)–(2.46), it follows that

$$dx_i = \epsilon \frac{\omega_{0k}}{\rho_0} \frac{\partial x_i}{\partial X_k} = \epsilon \frac{\omega_i}{\rho}, \tag{2.47}$$

thus implying that the vector with components dx_i is also tangent to an absolute-vorticity line. This concludes the proof. By virtue of Eqs. (2.46) and (2.47), we also note that the ratio $\boldsymbol{\omega}/\rho$ is proportional to the length of a line element along an absolute-vorticity line. This is known as *vortex line stretching* or *shrinking*.

In summary, we have shown that the absolute-vorticity lines move with the fluid in the absence of baroclinicity and friction. However, although one can also construct lines of relative vorticity, it is only the absolute-vorticity lines that may remain coincident with material lines. Moreover, when the last two terms in Eq. (2.42) do not identically vanish, viscous friction allows the absolute-vorticity lines to diffuse across the fluid, with the baroclinicity also being able to create new vortices.

2.3.1 The Taylor–Proudman theorem

Let us consider steady motions in a rotating fluid. Then, if both the Rossby number and the Ekman number of the flow are small, and if the baroclinic vector is identically zero, Eq. (2.42) becomes

$$\boldsymbol{\Omega} \cdot \operatorname{grad} \mathbf{u} = \mathbf{0}, \tag{2.48}$$

since $|\boldsymbol{\zeta}| \ll |\boldsymbol{\Omega}|$ when $Ro \ll 1$. This condition implies that the velocity relative to the moving axes must be independent of the coordinate parallel to $\boldsymbol{\Omega}$. If this vector is along the x_3 axis, we can thus write

$$\frac{\partial u_1}{\partial x_3} = \frac{\partial u_2}{\partial x_3} = \frac{\partial u_3}{\partial x_3} = 0. \tag{2.49}$$

In particular, if we consider a system with solid boundaries perpendicular to the rotation axis so that one has $u_3 = 0$ at some specified value of x_3, it follows at once that

$$\frac{\partial u_1}{\partial x_3} = \frac{\partial u_2}{\partial x_3} = 0 \quad \text{and} \quad u_3 = 0, \tag{2.50}$$

everywhere in the fluid. The flow is then entirely two dimensional in planes perpendicular to the rotation axis.

Motions that satisfy the Taylor–Proudman constraint can be observed in laboratory experiments (e.g., Greenspan 1968, Fig. 1.2, and Tritton 1988, Sec. 16.4). For example, let us consider a case in which the relative motion between the fluid and an obstacle is perpendicular to the rotation axis. Obviously, the fluid is deflected past the obstacle. However, because the flow must be two dimensional, this deflection also occurs above and below the obstacle. Accordingly, one observes the formation of a column of fluid, extending parallel to the rotating axis from the obstacle, around which the fluid is deflected *as if it too were solid*. Since the neglected terms never exactly vanish, especially at the edge of the column, there is in fact some interchange between the exterior and the interior of the column. Yet, Eq. (2.48) clearly demonstrates the tendency for coupling of the relative motion in the direction parallel to the vector $\mathbf{\Omega}$.

2.4 Reynolds stresses and eddy viscosities

Laboratory experiments show that the transition from laminar to turbulent motions in an incompressible fluid depends on the *Reynolds number*

$$Re = \frac{LU}{\nu}, \tag{2.51}$$

which is a measure of the relative magnitude of the inertial to viscous forces occurring in the flow (see Eq. [2.7]). Here U is the characteristic velocity of the flow, L is a characteristic length for the problem on hand, and $\nu = \mu/\rho$ is the coefficient of kinematic viscosity. Turbulent flows always occur when the nondimensional number Re exceeds some critical value Re_c (say). This critical number is not a universal constant but takes different values for each type of flow. (A laminar flow in a pipe normally becomes turbulent when $Re > Re_c \approx 2{,}200$.) This explains why the majority of fluid motions in systems with large dimensions and low viscosity are turbulent.

Damping due to molecular viscosity is very small and its effects on the large-scale motions encountered in geophysics and astrophysics is utterly negligible. However, for the very reason that one can make direct measurements in the Earth's atmosphere and in the oceans, it has long been recognized that these systems contain a wide spectrum of eddylike motions that coexist with the largest-scale motions. (As we shall see in Section 3.6, similar small-scale motions exist in stellar interiors, but their existence can be inferred by reasoning only.) Since there is as yet no practical and accurate theory that describes all scales of motion, from the largest to the smallest scales, it is convenient to restrict consideration to the large-scale motions only. Because Eq. (2.7) contains the nonlinear terms $\mathbf{v} \cdot \text{grad}\, \mathbf{v}$, this isolation can never in fact be complete, with motions on one spatial scale necessarily interacting with motions on other spatial scales. These interactions are often modeled by the inclusion of a large *anisotropic eddy viscosity* in the momentum equation, of much larger magnitude than the molecular viscosity; the functional form of this frictional force is analogous to that of Eq. (2.9). Unfortunately, because turbulence is not a feature of fluids but of fluid flows, the momentum exchange by eddylike motions only superficially resembles molecular exchange of momentum. Yet, although the empirical concept of eddy viscosity cannot be derived rigorously from first principles alone, it has proven to be both useful and effective in many dynamical problems that demand some frictional forces to be present.

At any given point and time, the physical variables of a system may be expressed in terms of mean values (denoted by overbars) and fluctuating values (denoted by primes). For such a decomposition to make sense, a suitable averaging period has to be found so that the mean values are substantially independent of this averaging period. Here we shall assume that it is possible. Hence, we let

$$\mathbf{v} = \overline{\mathbf{v}} + \mathbf{v}', \tag{2.52}$$

and we write similar expressions for the other physical variables. By definition, the components of the mean velocity are given by

$$\overline{v}_k = \frac{\overline{\rho v_k}}{\bar{\rho}}, \tag{2.53}$$

so that

$$\overline{\rho v_k'} = 0. \tag{2.54}$$

Note that we have also

$$\overline{v_k'} = -\frac{\overline{\rho' v_k'}}{\bar{\rho}}, \tag{2.55}$$

which vanishes only in the case of an incompressible fluid. Equation (2.54) ensures that, on the average, there is no transfer of mass due to turbulence and that Eq. (2.6) remains valid for the mean flow. It follows at once that

$$\frac{1}{\bar{\rho}}\frac{D\bar{\rho}}{Dt} + \operatorname{div}\overline{\mathbf{v}} = 0. \tag{2.56}$$

Combining next Eqs. (2.6) and (2.7), we can recast the momentum equation in the form

$$\frac{\partial}{\partial t}(\rho v_i) + \frac{\partial}{\partial x_k}(\rho v_i v_k) = \rho g_i - \frac{\partial p}{\partial x_i} + \frac{\partial \tau_{ik}}{\partial x_k}, \tag{2.57}$$

where the viscous stress tensor is defined in Eq. (2.8). If we suppose the body force to be unaffected by turbulence, the average of Eq. (2.57) is

$$\frac{\partial}{\partial t}(\bar{\rho}\bar{v}_i) + \frac{\partial}{\partial x_k}(\bar{\rho}\bar{v}_i\bar{v}_k) = \bar{\rho} g_i - \frac{\partial \bar{p}}{\partial x_i} + \frac{\partial}{\partial x_k}(\bar{\tau}_{ik} + \sigma_{ik}), \tag{2.58}$$

since the operations of averaging and differentiation commute. The tensor $\overline{\boldsymbol{\tau}}$ is the average of the tensor $\boldsymbol{\tau}$. The new tensor $\boldsymbol{\sigma}$ has the components

$$\sigma_{ik} = -\overline{\rho v_i' v_k'}. \tag{2.59}$$

These six quantities define the *Reynolds stresses*. Equation (2.58) is identical to Eq. (2.7) with all quantities replaced by their mean values, except for the additional Reynolds stresses. This symmetric tensor represents the flux of momentum due to the eddylike motions. The term div σ in Eq. (2.58) thus exchanges momentum between these small-scale motions and the mean flow, even though the three components $\overline{\rho v_k'}$ of the mean momentum of the turbulent velocity fluctuations are zero. Whenever eddylike motions prevail, the average viscous stresses $\overline{\tau}$ are usually negligible compared to the Reynolds stresses σ.

The central problem in this representation of small-scale motions lies in the fact that Eq. (2.58) introduces six unknown quantities, namely, the six components of the tensor σ. The simplest approach is to draw an analogy with molecular viscosity. Following Boussinesq, we shall assume that the turbulent motions act on the large-scale flow in a manner that mimics the microscopic transfer of momentum between the constitutive particles, when a macroscopic velocity gradient prevails. In order to apply this method to geophysical problems, we shall make use of Cartesian coordinates. The relevant equations for a rotating star will be discussed further in Section 3.6.

In the Earth's atmosphere and in the oceans, the horizontal dimensions of the large-scale motions are much greater than the vertical ones. This anisotropy of the large-scale flows strongly suggests that the turbulent transport of momentum in these two directions cannot be expected to be the same. If the axes are chosen so that the x_3 axis is in the vertical direction, a particularly simple expression for the Reynolds stresses is

$$\sigma_{11} = 2A_H \frac{\partial \bar{v}_1}{\partial x_1}, \qquad \sigma_{22} = 2A_H \frac{\partial \bar{v}_2}{\partial x_2}, \qquad \sigma_{33} = 2A_V \frac{\partial \bar{v}_3}{\partial x_3}, \tag{2.60}$$

$$\sigma_{12} = \sigma_{21} = A_H \frac{\partial \bar{v}_1}{\partial x_2} + A_H \frac{\partial \bar{v}_2}{\partial x_1}, \tag{2.61}$$

$$\sigma_{13} = \sigma_{31} = A_V \frac{\partial \bar{v}_1}{\partial x_3} + A_H \frac{\partial \bar{v}_3}{\partial x_1}, \tag{2.62}$$

$$\sigma_{23} = \sigma_{32} = A_V \frac{\partial \bar{v}_2}{\partial x_3} + A_H \frac{\partial \bar{v}_3}{\partial x_2}, \tag{2.63}$$

where A_H and A_V are the horizontal and vertical coefficients of eddy viscosity. Neglecting molecular viscosity and omitting the overbars, one can thus rewrite Eq. (2.58) in the form

$$\frac{D\mathbf{v}}{Dt} = \mathbf{g} - \frac{1}{\rho} \operatorname{grad} p + \frac{1}{\rho} \mathbf{F}(\mathbf{v}), \tag{2.64}$$

where $\mathbf{F}$ is the turbulent viscous force per unit volume, which is the vectorial divergence of the tensor σ. Neglecting compressibility effects, one obtains

$$\mathbf{F}(\mathbf{v}) = A_H \left(\frac{\partial^2 \mathbf{v}}{\partial x_1^2} + \frac{\partial^2 \mathbf{v}}{\partial x_2^2} \right) + A_V \frac{\partial^2 \mathbf{v}}{\partial x_3^2}, \tag{2.65}$$

where we have assumed that A_H and A_V are constant quantities. The preferred vertical direction is thus properly taken into account. (Compare with Eqs. [2.7] and [2.9].)

Because the eddy viscosities cannot be calculated from first principles alone, crude measurements of their values in the Earth's atmosphere and in the oceans have been made. Typical atmospheric values of K_V $(= A_V/\rho)$ lie in the range 10^4–10^6 cm^2 s^{-1}, whereas one has $\nu \approx 10^{-1}$ cm^2 s^{-1} for air. It follows that

$$\frac{K_V}{\nu} \approx 10^5\text{–}10^7, \tag{2.66}$$

in the atmosphere (Houghton 1986). For the oceans, estimates of K_V range from 1 cm^2 s^{-1} to 10^2 cm^2 s^{-1}. This implies that

$$\frac{K_V}{\nu} \approx 10^2\text{–}10^4, \tag{2.67}$$

in the oceans, since one has $\nu = 10^{-2}$ cm^2 s^{-1} for water. The smaller values go with smaller-scale motions, and conversely (Apel 1987). It is also worth noting that in the Earth's lower atmosphere one has $A_H/A_V \lesssim 10^2$, whereas this ratio may be as large as 10^5 in the surface layer of the ocean where large-scale currents are observed.

2.5 Applications to the Earth's atmosphere

Since the atmosphere is essentially a thin layer of fluid on a sphere, a convenient set of axes at any point on the Earth's surface has x directed toward the east, y to the north, and z vertically upward (i.e., along the effective gravity $\mathbf{g}_e$, which combines the effects of the gravitational force and centrifugal force). If $\mathbf{i}$, $\mathbf{j}$, and $\mathbf{k}$ are unit vectors directed along these rotating axes, the relative velocity of the mean flow may be expressed as

$$\mathbf{u} = u\mathbf{i} + v\mathbf{j} + w\mathbf{k}. \tag{2.68}$$

Letting $\mathbf{g}_e = -g\mathbf{k}$, one can rewrite the components of the momentum equation in the form:

$$\frac{Du}{Dt} - \frac{uv}{R}\tan\varphi + \frac{uw}{R} = -\frac{1}{\rho}\frac{\partial p}{\partial x} + 2\Omega v\sin\varphi - 2\Omega w\cos\varphi + \frac{1}{\rho}F_x, \tag{2.69}$$

$$\frac{Dv}{Dt} + \frac{u^2}{R}\tan\varphi + \frac{wv}{R} = -\frac{1}{\rho}\frac{\partial p}{\partial y} - 2\Omega u\sin\varphi + \frac{1}{\rho}F_y, \tag{2.70}$$

$$\frac{Dw}{Dt} - \frac{u^2+v^2}{R} = -\frac{1}{\rho}\frac{\partial p}{\partial z} - g + 2\Omega u\cos\varphi + \frac{1}{\rho}F_z, \tag{2.71}$$

where R is the radius of the Earth, Ω is its angular velocity of rotation, and φ is the geographical latitude. By virtue of Eq. (2.65), the turbulent viscous force is given by

$$\mathbf{F}(\mathbf{u}) = A_H\left(\frac{\partial^2\mathbf{u}}{\partial x^2} + \frac{\partial^2\mathbf{u}}{\partial y^2}\right) + A_V\frac{\partial^2\mathbf{u}}{\partial z^2}. \tag{2.72}$$

If one further assumes that the fluid is incompressible, Eq. (2.6) becomes

$$\frac{\partial u}{\partial x} + \frac{\partial v}{\partial y} + \frac{\partial w}{\partial z} = 0, \tag{2.73}$$

which closes the system of equations.

In this section we shall be concerned with midlatitude synoptic scale motions, that is, systems of typically 10^3 km in the horizontal dimension and 10 km in vertical extent. For this scale, the vertical velocity (typically less than 1 cm s^{-1}) is much smaller than the horizontal velocity (typically 10^3 cm s^{-1}). Hence, to a first approximation, terms involving w can be neglected in Eqs. (2.69)–(2.71). Similarly, because the curvature terms are also much smaller than the other terms, they too can be neglected. The resulting approximate horizontal momentum equations are

$$\frac{Du}{Dt} - fv = -\frac{1}{\rho}\frac{\partial p}{\partial x} + \frac{1}{\rho}F_x \tag{2.74}$$

and

$$\frac{Dv}{Dt} + fu = -\frac{1}{\rho}\frac{\partial p}{\partial y} + \frac{1}{\rho}F_y, \tag{2.75}$$

where

$$f = 2\Omega \sin\varphi \tag{2.76}$$

is the *Coriolis parameter*. To this order of approximation, Eq. (2.71) becomes

$$\frac{\partial p}{\partial z} = -\rho g, \tag{2.77}$$

which is the hydrostatic approximation.

The Coriolis parameter f is the local component of the planetary vorticity normal to the Earth's surface. If the north–south particle motions are extensive enough in latitude, the values of this parameter also change. For small changes about a mean latitude φ_0 where $f = f_0$, one can write

$$f = f_0 + \frac{df}{dy}y + \cdots = f_0 + \beta y + \cdots. \tag{2.78}$$

At midlatitudes, $\varphi = 45°$ (say), one has $f_0 = 10^{-4}$ s^{-1} and $\beta = 1.619 \times 10^{-13}$ cm^{-1} s^{-1}. The tangent-plane approximation with f constant is called an f plane; if we assume a linear relation between f and y, it is known as the β-plane approximation.

2.5.1 *The geostrophic approximation*

For synoptic motions far from the Earth's surface, turbulent friction and convective accelerations can be neglected altogether in Eqs. (2.74) and (2.75). Accordingly, if the response of the atmosphere to gravity leads ultimately to a steady state, that state will be given by the time-independent solution $u = u_g$ and $v = v_g$, say, where

$$u_g = -\frac{1}{\rho f}\frac{\partial p}{\partial y} \quad \text{and} \quad v_g = +\frac{1}{\rho f}\frac{\partial p}{\partial x}. \tag{2.79}$$

This is known as the *geostrophic balance* and describes the familiar situation in which the flow is along contours of constant pressure.* If we define the geopotential

$$\Phi = \int_0^z g\,dz, \tag{2.80}$$

which is the work required to raise a unit mass from the Earth's surface to height z, this approximate solution becomes

$$u_g = -\frac{1}{f}\left(\frac{\partial \Phi}{\partial y}\right)_p \quad \text{and} \quad v_g = +\frac{1}{f}\left(\frac{\partial \Phi}{\partial x}\right)_p, \tag{2.81}$$

where the subscript "p" refers to differentiation at constant pressure. As we shall see in Section 8.5, such a motion is also relevant to the theory of contact binaries.

Now, if f is regarded as a constant, it is a simple matter to differentiate Eq. (2.81) with respect to pressure and to make use of the fact that $\partial\Phi/\partial p = -1/\rho$ to obtain

$$\frac{\partial u_g}{\partial p} = -\frac{1}{\rho^2 f}\left(\frac{\partial \rho}{\partial y}\right)_p \quad \text{and} \quad \frac{\partial v_g}{\partial p} = +\frac{1}{\rho^2 f}\left(\frac{\partial \rho}{\partial x}\right)_p. \tag{2.82}$$

Thence, in combining Eqs. (2.77) and (2.82), one finds that

$$\frac{\partial u_g}{\partial z} = +\frac{g}{\rho f}\left(\frac{\partial \rho}{\partial y}\right)_p \quad \text{and} \quad \frac{\partial v_g}{\partial z} = -\frac{g}{\rho f}\left(\frac{\partial \rho}{\partial x}\right)_p. \tag{2.83}$$

For the atmosphere, Eq. (2.14) implies that one can rewrite these relations in the forms:

$$\frac{\partial u_g}{\partial z} = -\frac{g}{T f}\left(\frac{\partial T}{\partial y}\right)_p \quad \text{and} \quad \frac{\partial v_g}{\partial z} = +\frac{g}{T f}\left(\frac{\partial T}{\partial x}\right)_p. \tag{2.84}$$

This is the *thermal wind* equation, which relates the increase of the horizontal geostrophic velocity with height to the horizontal temperature gradient within a surface of constant pressure. In other words, if the surfaces of constant pressure and constant temperature do not coincide, the geostrophic wind generally has vertical shear. On the contrary, if these two families of surfaces are coincident, its velocity must be independent of height. This result implies that the Taylor–Proudman theorem is a direct consequence of the geostrophic approximation (see Eq. [2.48]).

2.5.2 *Ekman layer at a rigid plane boundary*

In the lowest kilometer of the atmosphere, the geostrophic solution (2.79) does not apply because the vertical viscous force generally is comparable in magnitude to the

* An empirical law that describes the approximate agreement between the geostrophic wind (Eq. [2.79]) and the actual wind was originally derived in 1857 by the Dutch meteorologist Christoph Buys Ballot (1817–1890). This rule of thumb states that in the northern hemisphere a person standing with his back to the wind has the higher pressure to his right and the lower pressure to his left; in the southern hemisphere, the lower pressure is to the right of the observer and the higher pressure to the left. Buys Ballot also noticed that the wind blows in general perpendicular to the pressure gradient and that the wind speed increases with increasing pressure gradient (see Eq. [2.79]). As we shall see in Section 2.5.2, however, in both hemispheres the wind near the ground does not flow exactly parallel to the isobars but has a component toward lower pressure because of surface friction. *Buys Ballot's law*, which is also known as *Ferrel's law* or *the baric wind law*, is not applicable in the equatorial regions.

pressure-gradient and Coriolis forces. In this boundary layer, the acceleration terms are still small compared to the remaining terms in Eqs. (2.74) and (2.75). For a situation in which there is a shear in the vertical direction only, we can thus write

$$-fv = -\frac{1}{\rho}\frac{\partial p}{\partial x} + \frac{1}{\rho} A_V \frac{\partial^2 u}{\partial z^2} \tag{2.85}$$

and

$$+fu = -\frac{1}{\rho}\frac{\partial p}{\partial y} + \frac{1}{\rho} A_V \frac{\partial^2 v}{\partial z^2}. \tag{2.86}$$

Let us further assume that the fluid is homogeneous. Then, taking the first-order derivatives of Eq. (2.77) with respect to x and y, one readily sees that the horizontal pressure gradient does not depend on height. Hence, by making use of Eq. (2.79), one obtains

$$K_V \frac{d^2 u}{dz^2} + f(v - v_g) = 0 \tag{2.87}$$

and

$$K_V \frac{d^2 v}{dz^2} - f(u - u_g) = 0, \tag{2.88}$$

where $K_V = A_V/\rho$ is regarded as a constant. At ground level, in close analogy with molecular viscosity, we shall assume that eddy viscosity inhibits the tangential fluid motion. Hence, we let

$$u = v = 0 \quad \text{at} \quad z = 0. \tag{2.89}$$

(see Eqs. [2.17] and [2.18]). Since the flow must also match the geostrophic solution at high levels, it is also required that

$$u \to u_g \quad \text{and} \quad v \to v_g \quad \text{as} \quad z \to \infty, \tag{2.90}$$

where u_g and v_g are defined in Eq. (2.79).

The appropriate boundary-layer solution for the horizontal velocity is

$$u = u_g - e^{(-z/\Delta)}[u_g \cos(z/\Delta) + v_g \sin(z/\Delta)] \tag{2.91}$$

and

$$v = v_g - e^{(-z/\Delta)}[v_g \cos(z/\Delta) - u_g \sin(z/\Delta)], \tag{2.92}$$

where

$$\Delta = \left(\frac{2K_V}{f}\right)^{1/2}. \tag{2.93}$$

This steady solution was originally obtained by Ekman (1905). Figure 2.1 illustrates the wind velocity vector as a function of the nondimensional height z/Δ. Owing to the combined effects of the Coriolis force and turbulent friction, the tip of the velocity vector traces a spiral as z/Δ decreases to zero. As the solid boundary is approached, this vector is at 45° to the left of the geostrophic velocity. As $z/\Delta = \pi$, the wind is parallel to the geostrophic flow but slightly greater than geostrophic in magnitude. The level $z = \pi\Delta$ may be considered as the top of the viscous boundary layer. Measurements indicate that

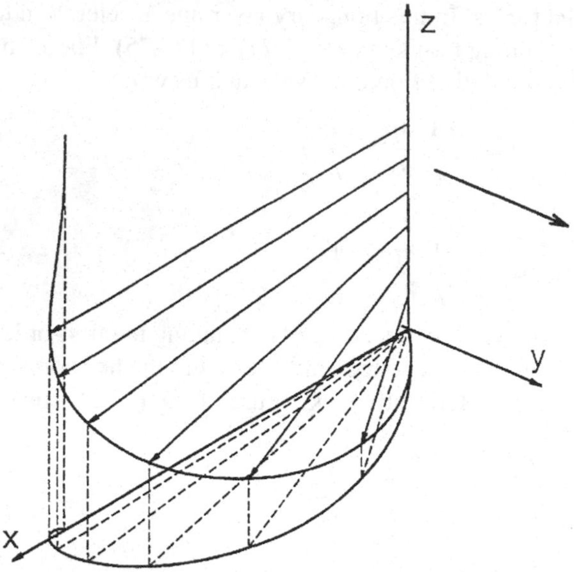

Fig. 2.1. The velocity vector within the Ekman layer, at various heights above a solid boundary. The values of the nondimensional height z/Δ are, respectively, π, $5\pi/6$, $4\pi/6$, $3\pi/6$, $2\pi/6$, and $\pi/6$ (see Eq. [2.93]). The large arrow indicates the direction of the applied pressure gradient.

the wind approaches its geostrophic value at about one kilometer above the ground. Letting $f = 10^{-4}\ \mathrm{s}^{-1}$ and $\nu = 10^{-1}\ \mathrm{cm}^2\ \mathrm{s}^{-1}$, one finds that $K_V = 5 \times 10^4\ \mathrm{cm}^2\ \mathrm{s}^{-1}$ and $K_V/\nu \approx 5 \times 10^5$ (see Eq. [2.66]). Note that this ideal solution is rarely observed because the coefficient K_V must vary rapidly with height near the ground. On qualitative grounds, however, it gives an adequate picture of the frictional coupling between the geostrophic flow in the free atmosphere and the Earth's surface.

2.5.3 Ekman pumping and secondary circulation

Away from the ground, the atmosphere adjusts to a geostrophic equilibrium in which the pressure-gradient force balances the Coriolis force associated with a steady flow along the surfaces of constant pressure. If this motion extends to the ground, the effect of turbulent friction is to disrupt this geostrophic balance, thus producing a flow across these surfaces from high to low pressure. Hence, work is being done on the fluid within the surface boundary layer by the pressure-gradient force. This work supplies the necessary energy to maintain this layer against the dissipative forces within it. Accordingly, unless the geostrophic flow is forcibly maintained, it will decay under the action of the bottom friction.

To calculate the typical spin-down time of a geostrophic flow, let us again consider quasi-horizontal motions in a cyclonic vortex. Since the motion is geostrophically balanced away from the ground, the center of the vortex is at low pressure compared to its outer edge. In the surface boundary layer, then, turbulent friction produces a radial flow of matter toward the vortex center. By continuity, this requires upward motion and a compensating outward radial flow *above* the friction layer. Figure 2.2 presents a

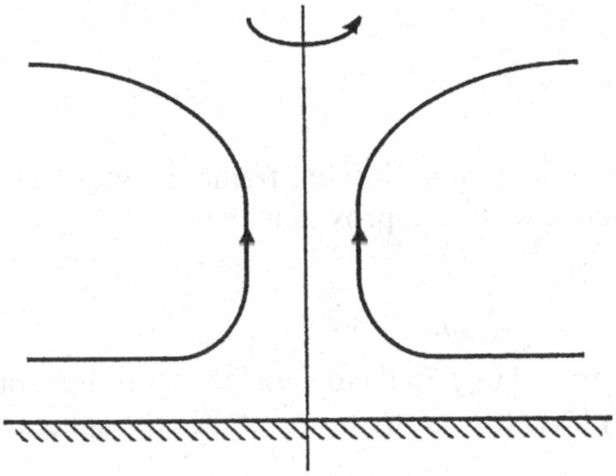

Fig. 2.2. Qualitative sketch of the streamlines of meridional circulation in a mid-latitude cyclonic vortex. The rotation axis is vertical.

qualitative sketch of the streamlines of this secondary flow. Since this slow but inexorable motion approximately conserves angular momentum, high angular velocity fluid is thus progressively replaced by low angular velocity fluid in the atmosphere. As we shall see, this axially symmetric circulation driven by turbulent friction in the surface layers serves to spin down the azimuthal motion of the cyclonic vortex far more rapidly than could turbulent diffusion of momentum. This mechanism, which exchanges mass between the surface boundary layer and the free atmosphere above it, is known as *Ekman pumping*.

For the sake of simplicity, let us assume that the atmosphere, of height H, is of uniform density. Assume further that in the surface boundary layer, of depth d, the radial inflow of matter is adequately described by Eqs. (2.91) and (2.92). Above the boundary layer, in the free atmosphere, the azimuthal motion of the cyclonic vortex has its relative vorticity – $\zeta_g(x, y)$ – along the z axis (see Eq. [2.37]). By virtue of Eq. (2.79), we thus have

$$\zeta_g = \frac{\partial v_g}{\partial x} - \frac{\partial u_g}{\partial y}, \tag{2.94}$$

which is called the *geostrophic vorticity*. To calculate the upward velocity w_E at the top of the boundary layer, let us integrate Eq. (2.73) through the depth of the layer. Because $w = 0$ at $z = 0$, it follows that

$$w_E = -\int_0^d \left(\frac{\partial u}{\partial x} + \frac{\partial v}{\partial y} \right) dz. \tag{2.95}$$

Substituting for u and v from Eqs. (2.91) and (2.92), one obtains

$$w_E = \left(\frac{K_V}{2f} \right)^{1/2} \zeta_g = \frac{\Delta}{2} \zeta_g. \tag{2.96}$$

This relation merely states that the vertical velocity of the matter that is pumped into the free atmosphere is proportional to the geostrophic vorticity.

For synoptic scale motions, the vorticity equation can be derived from Eqs. (2.74) and (2.75) by cross differentiation with respect to x and y. Neglecting turbulent friction, we thus have

$$\frac{D}{Dt}(f+\zeta) = -(f+\zeta)\left(\frac{\partial u}{\partial x}+\frac{\partial v}{\partial y}\right) = (f+\zeta)\frac{\partial w}{\partial z}. \tag{2.97}$$

(Compare with Eq. [2.43], which is written in an inertial frame of reference.) If f is regarded as a constant, Eq. (2.97) can be written approximately as

$$\frac{\partial \zeta_g}{\partial t} = f\,\frac{\partial w}{\partial z}, \tag{2.98}$$

where we have also neglected ζ_g compared to f in the divergence term. Integrating this equation from the top of the boundary layer ($z = d$) to the top of the atmosphere ($z = H$), we obtain

$$H\,\frac{\partial \zeta_g}{\partial t} = -f\,w_E, \tag{2.99}$$

since $d \ll H$ and $w = 0$ at $z = H$. Substituting for w_E from Eq. (2.96) and integrating this equation with respect to time, one finds that

$$\zeta_g = \zeta_g(0)\exp[-(fK_V/2H^2)^{1/2}t\,], \tag{2.100}$$

where $\zeta_g(0)$ is the geostrophic vorticity at $t = 0$. By virtue of Eq. (2.100), the spin-down time of a cyclonic vortex, t_{sd}, is

$$t_{\mathrm{sd}} = \left(\frac{2H^2}{fK_V}\right)^{1/2}. \tag{2.101}$$

This result was originally derived by Charney and Eliassen (1949).

To illustrate the problem, we shall let $H = 10\,\mathrm{km}$, $f = 10^{-4}\,\mathrm{s}^{-1}$, and $K_V = 10^5\,\mathrm{cm}^2\,\mathrm{s}^{-1}$. By making use of Eq. (2.101), one finds that $t_{\mathrm{sd}} \approx 4$ days. In contrast, the characteristic time t_υ for turbulent diffusion to penetrate a depth H is of the order H^2/K_V, which, for the above values of H and K_V, gives $t_\upsilon \approx 100$ days, which is much longer than 4 days. We conclude that *Ekman pumping is a far more effective mechanism for destroying a cyclonic vortex in the Earth's atmosphere than is turbulent diffusion of momentum*. Yet, letting $\zeta_g = f$ in Eq. (2.96), which means an intense cyclonic vortex, one finds that the vertical speed w_E does not exceed 2.3 cm s^{-1} at the top of the boundary layer.

As shown in Figure 2.3, it is an analogous meridional circulation that is responsible for the decay of the azimuthal motion created when a cup of tea is stirred. Physically, the spin-down mechanism is essentially that described for the cyclonic vortex, except that in the cup of tea it is the centrifugal force that balances the pressure-gradient force, not the Coriolis force. Visualization of the transient meridional flow is provided by the tea leaves, which are always observed to cluster near the center at the bottom of the spinning fluid. As was shown by Greenspan and Howard (1963), the spin-down time t_{sd} is, roughly, of the order of $(L^2/\nu\Omega)^{1/2}$, where L is a characteristic dimension, parallel to the rotation axis, ν is the kinematic viscosity, and Ω is the initial angular velocity. Letting $L = 4$ cm, $\nu = 10^{-2}$ cm^2 s^{-1}, and $\Omega = 2\pi$ s^{-1}, one obtains $t_{\mathrm{sd}} = 16$ s, in agreement with casual observation. One also finds that $t_\upsilon = L^2/\nu = 1{,}600$ s! Obviously, *the azimuthal motion in a cup of tea decays much more rapidly through Ekman pumping than by mere viscous diffusion of momentum*.

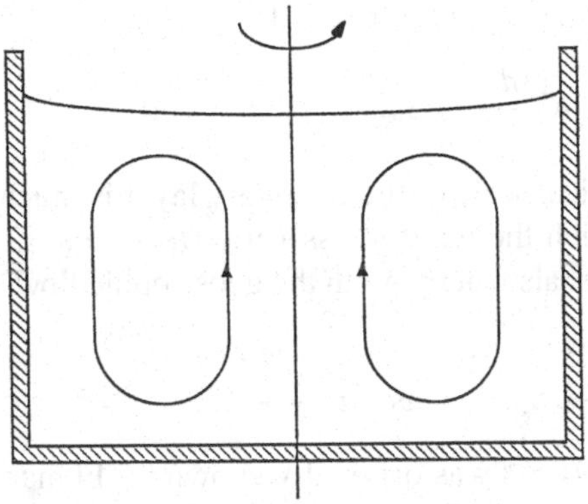

Fig. 2.3. Qualitative sketch of the streamlines of meridional circulation in a cup of tea. The rotation axis is vertical.

As we shall see in Section 8.4, a similar mechanism may be invoked to explain the high degree of synchronism that is observed in the close binary stars, although the physical cause of Ekman pumping in a nonsynchronous binary component is completely different.

2.6 The wind-driven oceanic circulation

When the wind blows over an ocean, a stress is exerted on its surface. The applied wind stress produces a horizontal mass transport in a thin surface layer, which is mostly between 10 and 100 meters deep. If the applied stress were spatially uniform, the ocean below this layer would be little affected by the wind. However, spatial variations of the wind also cause spatial variations in the horizontal mass transport near the surface. These horizontal divergences or convergences of matter result in vertical motions, with water flowing upward or downward to replace the displaced surface water. Deep in the ocean, however, this vertical mass flux must move so as to preserve the specific angular momentum of each fluid particle. As we shall see in Section 2.6.2, this may be accomplished by large-scale horizontal motions in the deep interior, where the basic momentum balance for this flow is given by the geostrophic approximation. Such a simple solution is clearly not complete, however, since those terms in the vorticity equation that are negligible in the open ocean become important near the ocean's lateral boundaries. In Section 2.6.3 we shall thus present a geostrophic flow that is frictionally eroded by horizontal eddy viscosity acting near these boundaries but is maintained by wind stresses acting over the ocean's surface.

2.6.1 Ekman layer at the ocean–atmosphere interface

We suppose that the ocean is bounded by the horizontal surface $z = 0$ at which the wind stress

$$\mathbf{S} = S_x \mathbf{i} + S_y \mathbf{j} \tag{2.102}$$

is applied. Since we are making allowance for a large-scale geostrophic motion in the inviscid interior, the basic flow in the surface layer is described by Eqs. (2.87) and (2.88).

However, the presence of an applied wind stress requires that

$$\rho K_V \frac{du}{dz} = S_x \quad \text{and} \quad \rho K_V \frac{dv}{dz} = S_y, \qquad \text{at} \quad z = 0, \tag{2.103}$$

where we took into account that the vertical scale of the boundary layer is much smaller than that of the horizontal scale on which the wind stress varies (see Eqs. [2.62] and [2.63]). The flow in the surface layer must also merge with the geostrophic flow at depth. Hence, we let

$$u \to u_g \quad \text{and} \quad v \to v_g, \qquad \text{as} \quad z \to -\infty, \tag{2.104}$$

where u_g and v_g are defined in Eq. (2.79). As was originally shown by Ekman (1905), the boundary-layer solution satisfying these conditions is

$$u = u_g + \frac{1}{2}\frac{\Delta}{\rho K_V} e^{+z/\Delta}[(S_y + S_x)\cos(z/\Delta) - (S_y - S_x)\sin(z/\Delta)] \tag{2.105}$$

and

$$v = v_g + \frac{1}{2}\frac{\Delta}{\rho K_V} e^{+z/\Delta}[(S_y - S_x)\cos(z/\Delta) + (S_y + S_x)\sin(z/\Delta)], \tag{2.106}$$

where Δ is defined in Eq. (2.93).*

Let us define

$$\mathbf{u}_E = (u - u_g)\mathbf{i} + (v - v_g)\mathbf{j}, \tag{2.107}$$

which is the friction velocity in the surface boundary layer. Figure 2.4 illustrates the spiral distribution of this vector. At the surface, the velocity $\mathbf{u}_E$ is 45° to the right of the applied wind stress. As the depth below the free surface increases, the direction of this vector rotates uniformly in a clockwise sense and its magnitude falls off exponentially. The horizontal mass flux associated with the velocity $\mathbf{u}_E$ is

$$\mathbf{M}_E = \int_{-\infty}^{0} \mathbf{u}_E \, dz = \frac{1}{\rho f}\mathbf{S} \times \mathbf{k}, \tag{2.108}$$

which does not depend on the eddy viscosity. Note also that the vector $\mathbf{M}_E$ is orthogonal to the applied stress. This is a consequence of the fact that a net mass flux in that direction would give rise to a net Coriolis force that would remain unbalanced.

For further reference, we can also integrate the continuity equation for the velocity $\mathbf{u}_E$ over the entire depth of the Ekman layer to obtain the vertical velocity w_E flowing into that layer. Taking into account that the surface wind usually varies much more rapidly

* This solution was motivated by observations made by Fridtjof Nansen (1861–1930) during the Norwegian North Polar Expedition of 1893–1896. Looking at observations of the wind and of ice drift taken from his ship *Fram* (i.e., "Forward") while she drifted in the arctic ice, he saw that the direction of the ice drift showed a systematic deviation to the right relative to the wind direction. Nansen correctly guessed that this deviation was in some way related to the Earth's rotation. The problem was given to the young Swedish scientist Vagn Walfrid Ekman (1874–1954), who came out in 1905 with a full scale theory of the so-called Ekman spiral. For a detailed account of these and related matters, see Arnt Eliassen, "Vilhelm Bjerknes and his Students," *Annual Review of Fluid Mechanics*, **14**, 1, 1982.

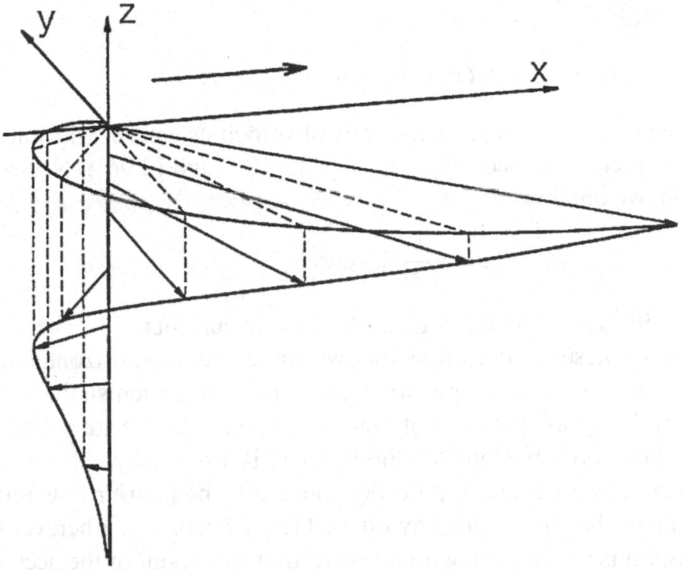

Fig. 2.4. The velocity vector within the Ekman layer, at various depths below a free surface. The values of the nondimensional depths $|z/\Delta|$ are, respectively, 0, $\pi/12$, $2\pi/12$, $3\pi/12$, $5\pi/12$, $7\pi/12$, $9\pi/12$, and π (see Eq. [2.93]). The large arrow indicates the direction of the applied wind stress.

than f, one finds that

$$w_E = \frac{1}{\rho f}\left(\frac{\partial S_y}{\partial x} - \frac{\partial S_x}{\partial y}\right) = \frac{1}{\rho f}\mathbf{k}\cdot \text{curl}\,\mathbf{S}, \tag{2.109}$$

which is the relation between the vertical velocity at the lower edge of the surface boundary layer and the z component of the *curl* of the surface wind stress.

2.6.2 The Sverdrup relation

To illustrate the basic features of wind-driven currents in the oceans, we shall assume that our model, of constant depth H, is of uniform density. The rotating fluid thus consists of three regions: a thin surface Ekman layer, the inviscid interior, and a thin bottom frictional layer. For low Rossby numbers, the basic momentum balance is given by the geostrophic approximation. By making use of Eqs. (2.78) and (2.97), we can also write

$$\frac{D\zeta}{Dt} + \beta v = f\,\frac{\partial w}{\partial z}, \tag{2.110}$$

where ζ is the z component of the relative vorticity of the interior flow and βv is the change with latitude of the planetary vorticity f. For steady motions and low Rossby numbers, this equation becomes

$$\beta v = f\,\frac{\partial w}{\partial z}. \tag{2.111}$$

Since the interior flow is homogeneous and geostrophic, the functions u, v, and ζ must be independent of z (see Eq. [2.83]). Hence, Eq. (2.111) may be integrated over the whole depth of the ocean to give

$$\beta v = \frac{f}{H}\left[w_E(x, y, 0) - w_E(x, y, -H)\right], \tag{2.112}$$

where $w_E(x, y, 0)$ and $w_E(x, y, -H)$ are the vertical velocities entering the interior flow on its upper and lower edge, respectively (see Eqs. [2.109] and [2.96]). Neglecting the bottom contribution, we obtain

$$\beta v = \frac{1}{\rho H}\mathbf{k} \cdot \text{curl}\,\mathbf{S}. \tag{2.113}$$

This is Sverdrup's (1947) solution for large-scale oceanic currents.

The interpretation of these relations is as follows. Wherever the horizontal mass flow in the surface layer is divergent, with upwelling taking place to replenish the surface water transported away, the right-hand sides of Eqs. (2.109) and (2.113) are positive so that the change of planetary vorticity along the motion, βv, is also positive. Since the planetary vorticity f increases poleward, it follows at once that the interior flow must move poleward to maintain the balance defined by Eq. (2.113). Conversely, wherever the horizontal surface transport is convergent, with downwelling as a result of the accumulation of mass, the interior flow must move in the direction along which the planetary vorticity f decreases, that is to say, equatorward. Such a mass transport is observed over much of the circulation pattern in the North Atlantic Ocean, with equatorward flowing currents on the eastern side of the ocean basin, north of approximately 20°N (e.g., Pedlosky [1987], Fig. 5.1.1, or Apel [1987], Fig. 2.21). The southward Sverdrup flow cannot apply to the whole ocean basin, however, since it must necessarily return to the north so as to ensure continuity and conservation of specific angular momentum of each fluid particle. As we shall see below, this circumstance results in the formation of the intense and narrow Gulf Stream current, which flows along the western edge of the North Atlantic Ocean, from Florida to Cape Hatteras, where it rejoins the generally clockwise oceanic circulation. Although the Gulf Stream is the best-known example of a western boundary current, such a vorticity-balancing and mass-balancing flow is present off the east coasts of continents everywhere in the world.

2.6.3 *Western boundary currents: The Munk layer*

Although the southward Sverdrup flow occurs over a very large portion of an ocean basin, Eq. (2.113) alone cannot satisfy all lateral boundary conditions, nor can it satisfy mass conservation for the basin as a whole. The inadequacy of this solution strongly suggests the existence of turbulent boundary layers adjacent to the perimeter of the basin. Making allowance for horizontal friction in Eqs. (2.74) and (2.75), we can rewrite the vorticity equation in the form

$$\frac{D\zeta}{Dt} + \beta v - K_H\left(\frac{\partial^2\zeta}{\partial x^2} + \frac{\partial^2\zeta}{\partial y^2}\right) = f\frac{\partial w}{\partial z}, \tag{2.114}$$

where $K_H = A_H/\rho$ (see Eq. [2.72]). Again for steady motions and low Rossby numbers, this equation may be integrated over the thickness of the ocean to yield

$$\beta v - K_H\left(\frac{\partial^2\zeta}{\partial x^2} + \frac{\partial^2\zeta}{\partial y^2}\right) = \frac{1}{\rho H}\mathbf{k} \cdot \text{curl}\,\mathbf{S}. \tag{2.115}$$

(Compare with Eq. [2.113].) Since the horizontal motion is geostrophic, the functions u and v may be written in terms of a stream function Ψ. We can thus write

$$u = -\frac{\partial \Psi}{\partial y} \quad \text{and} \quad v = +\frac{\partial \Psi}{\partial x}, \tag{2.116}$$

so that

$$\zeta = \frac{\partial^2 \Psi}{\partial x^2} + \frac{\partial^2 \Psi}{\partial y^2}. \tag{2.117}$$

To be specific, we shall consider the case where the meridional boundaries $x = X_W$ and $x = X_E$ are independent of latitude. In the interior of the ocean, away from these boundaries, we have

$$\beta \frac{\partial \Psi_I}{\partial x} = \frac{1}{\rho H} \mathbf{k} \cdot \operatorname{curl} \mathbf{S}, \tag{2.118}$$

where Ψ_I is the interior stream function, which depends on the applied wind stress. In the boundary layers, however, Eq. (2.115) can be written as

$$\beta \frac{\partial \Psi}{\partial x} - K_H \left(\frac{\partial^4 \Psi}{\partial x^4} + \cdots \right) = \frac{1}{\rho H} \mathbf{k} \cdot \operatorname{curl} \mathbf{S}, \tag{2.119}$$

since only the highest derivative with respect to x will be retained in the boundary-layer analysis.

Near the eastern boundary, it is convenient to make use of the stretched variable

$$\xi = \frac{X_E - x}{\delta}, \tag{2.120}$$

where

$$\delta = \left(\frac{K_H}{\beta} \right)^{1/3}. \tag{2.121}$$

In order to solve Eq. (2.119), we shall also let

$$\Psi = \Psi_I(x, y) + \Psi_E(\xi, y), \tag{2.122}$$

where Ψ_E must go to zero as $\xi \to \infty$. Inserting this relation into Eq. (2.119), one obtains

$$\frac{\partial^4 \Psi_E}{\partial \xi^4} + \frac{\partial \Psi_E}{\partial \xi} = 0. \tag{2.123}$$

As usual, the complete solution must satisfy the conditions of no normal flow and no slip at the boundary $x = X_E$ (see Eqs. [2.17] and [2.18]). Thus, we also have

$$\frac{\partial \Psi}{\partial x} = \frac{\partial \Psi}{\partial y} = 0, \quad \text{at} \quad \xi = 0. \tag{2.124}$$

The solution that satisfies these conditions is

$$\Psi = \Psi_I(x, y) - \delta \left(\frac{\partial \Psi_I}{\partial x} \right)_{x = X_E} e^{-\xi}, \tag{2.125}$$

where

$$\Psi_I = \frac{1}{\rho\beta H}\int_{X_E}^{x} \mathbf{k}\cdot\text{curl}\,\mathbf{S}\,dx'. \tag{2.126}$$

It is immediately apparent from Eq. (2.125) that this solution acts only to satisfy the no-slip condition on the eastern boundary, having little effect on the large-scale mass transport in the ocean.

The situation is quite different on the western side of the ocean. Here we shall define the stretched variable

$$\eta = \frac{x - X_W}{\delta}, \tag{2.127}$$

where δ is still defined in Eq. (2.121). We shall also let

$$\Psi = \Psi_I(x, y) + \Psi_W(\eta, y), \tag{2.128}$$

where Ψ_W must go to zero as $\eta \to \infty$. By virtue of Eq. (2.119), the function Ψ_W satisfies

$$\frac{\partial^4 \Psi_W}{\partial \eta^4} - \frac{\partial \Psi_W}{\partial \eta} = 0. \tag{2.129}$$

The boundary conditions at the western boundary $x = X_W$ are

$$\frac{\partial \Psi}{\partial x} = \frac{\partial \Psi}{\partial y} = 0, \quad \text{at} \quad \eta = 0. \tag{2.130}$$

The appropriate solution is

$$\Psi = \Psi_I(x, y)\left[1 - e^{-\eta/2}\left(\cos\frac{\sqrt{3}}{2}\eta + \frac{1}{\sqrt{3}}\sin\frac{\sqrt{3}}{2}\eta\right)\right], \tag{2.131}$$

which also ensures that the net mass flux in the meridional direction exactly vanishes. Accordingly, this western boundary current returns northward a mass flux that precisely balances the southward Sverdrup mass flux. By virtue of the second equation (2.116), the northward velocity in this western boundary current is given by

$$v = \Psi_I(X_W, y)\frac{2}{\delta\sqrt{3}}e^{-\eta/2}\sin\frac{\sqrt{3}}{2}\eta. \tag{2.132}$$

Both solutions were originally derived by Munk (1950). Figure 2.5 illustrates the zonal variation of the transport stream function, as given by Ψ/Ψ_I, and the north–south velocity v across the western boundary layer. Note the intense northward flow and the small but significant counterflow just to the east of this boundary flow. This counterflow is actually found in observations of the Gulf Stream, and thus Munk's frictional model is *qualitatively* similar to the general oceanic circulation. (This large-scale flow possesses speeds of the order of 1–10 cm s^{-1}, whereas the northward velocity in the Gulf Stream is typically 100 cm s^{-1}, with a maximum speed of 200 cm s^{-1}.) Unfortunately, one readily sees from Eq. (2.121) that the lateral scale of the flow is set by the horizontal eddy viscosity K_H, which is an *adjustable* parameter of the theory. Letting $\delta \approx$ 50–100 km, which is the lateral dimension of the Gulf Stream, one finds that K_V should be of the order of 10^7–10^8 cm^2 s^{-1}. Because such a value requires a very sizable frictional dissipation,

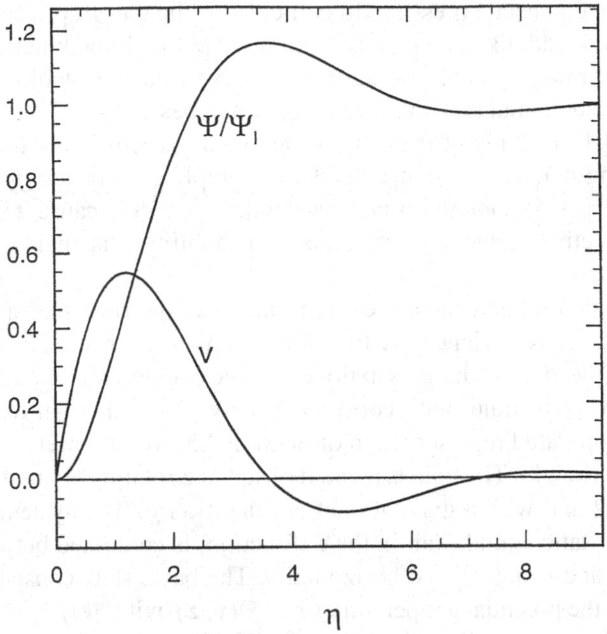

Fig. 2.5. The transport stream function, Ψ/Ψ_I, and the north–south velocity v across the western boundary layer. The quantity v is measured in units of Ψ_I/δ (see Eqs. [2.131] and [2.132]).

Bryan (1963) has made numerical calculations that retain both the frictional terms and the nonlinear terms (i.e., $D\zeta/Dt$) in Eq. (2.114). His nonlinear solutions, which require a smaller amount of lateral dissipation, converge to the purely frictional solution in the limit of large viscosities (e.g., Pedlosky [1987], pp. 309–311). As we shall see in Section 4.3, a similar problem arises in the discussion of large-scale meridional currents in stellar radiative zones.

2.7 Barotropic and baroclinic instabilities

Although a steady solution may prove useful for interpreting the general features of a large-scale motion, observations reveal that small-scale disturbances are inevitably present in any natural system. As a matter of fact, geophysicists have long recognized that the presence of fully developed disturbances – such as midlatitude cyclones – can be attributed to hydrodynamical instabilities. One form of instability that is of particular interest is called *baroclinic instability*. As was originally shown by Charney (1947), this mechanism depends on the presence of a vertical velocity shear in the mean zonal flow. By making use of Eq. (2.84), one readily sees that this shear requires the presence of horizontal temperature gradients. Since the surfaces of constant temperature and constant pressure do not coincide in such a system, it follows that the main source of energy for the baroclinic instability is the available potential energy that may be released and transferred to the small-scale disturbances. Another form of instability of geophysical interest depends on the presence of a horizontal velocity shear in the mean zonal flow. It is called *barotropic instability*, because it can occur in a system for which the surfaces of constant temperature and constant pressure do coincide. In the barotropic case, thus, the source of energy for the eddylike disturbances is associated with the kinetic energy stored in the mean flow. In more complex situations, however, these instabilities draw their energy from both the potential and kinetic energy of the basic state.

Deferring to Section 3.4 the study of these barotropic and baroclinic instabilities as they may occur in a rotating star, here we shall consider the simple geophysical model first introduced by Eady (1949). This configuration is particularly useful because it provides us with an overall perspective of the various kinds of instability that may arise in a rotating fluid.

The model neglects dissipative and curvature effects and uses the Boussinesq approximation for compressibility effects. Hence, we treat the density as a constant in all terms in the equations, except the one in the gravitational acceleration. Thus we assume a Boussinesq, isentropic, inviscid fluid, with constant density ρ and thermal expansion coefficient α. In the plane-parallel representation of Section 2.5, we interpret x as longitude, y as latitude, and z as height. The fluid is located on a plane rotating about the z axis with angular velocity $f/2$ and with a gravitational acceleration g. The system, which we assume to be in hydrostatic equilibrium in the z direction, is contained between the planes $z = 0$ and $z = H$ and is unbounded horizontally. The basic state consists of the zonal wind $u = U(z)$ and the potential temperature $\Theta = \Theta(y, z)$, with $\partial\Theta/\partial z > 0$ since we do not want to consider convective motions (see Eq. [2.16]). Hence, by making use of Eqs. (2.75) and (2.77), one can write

$$fU = -\frac{1}{\rho}\frac{\partial p}{\partial y} \tag{2.133}$$

and

$$-\alpha g\Theta = -\frac{1}{\rho}\frac{\partial p}{\partial z}, \tag{2.134}$$

where f is regarded as a constant. Eliminating the deviation from hydrostatic pressure by cross-differentiation, we obtain the thermal wind relation

$$f\,\frac{dU}{dz} = -\alpha g\,\frac{\partial\Theta}{\partial y}, \tag{2.135}$$

which relates the vertical velocity shear to the latitudinal potential-temperature gradient (see Eq. [2.84]).

At this juncture, it is convenient to define the squared buoyancy frequency

$$N^2 = \alpha g\,\frac{\partial\Theta}{\partial z}, \tag{2.136}$$

which is a measure of the stability of the fluid layer against vertical disturbances. (When $N^2 > 0$, the frequency N merely corresponds to buoyancy oscillations, i.e., stable gravity modes; when $N^2 < 0$, it corresponds to rising or sinking motions.) N is usually called the *Brunt–Väisälä frequency*. We shall also define the nondimensional number

$$Ri = \frac{N^2}{(dU/dz)^2}, \tag{2.137}$$

which is known as the *Richardson number*. A positive value of this number corresponds to a convectively stable stratification in the vertical direction.

Equations (2.133)–(2.135) define the unperturbed state of the flow. We now assume small deviations from this basic state. Hence, we linearize Eqs. (2.73)–(2.75), (2.77), and (2.12). Thus neglecting nonhydrostatic effects in the z component of the momentum equation, one obtains the following set of equations:

$$\frac{\partial u_1}{\partial x} + \frac{\partial v_1}{\partial y} + \frac{\partial w_1}{\partial z} = 0, \tag{2.138}$$

$$\left(\frac{\partial}{\partial t} + U\frac{\partial}{\partial x}\right) u_1 + \frac{dU}{dz} w_1 - f v_1 = -\frac{1}{\rho}\frac{\partial p_1}{\partial x}, \tag{2.139}$$

$$\left(\frac{\partial}{\partial t} + U\frac{\partial}{\partial x}\right) v_1 + f u_1 = -\frac{1}{\rho}\frac{\partial p_1}{\partial y}, \tag{2.140}$$

$$-\alpha g \Theta_1 = -\frac{1}{\rho}\frac{\partial p_1}{\partial z}, \tag{2.141}$$

$$\left(\frac{\partial}{\partial t} + U\frac{\partial}{\partial x}\right) \Theta_1 + \frac{\partial \Theta}{\partial y} v_1 + \frac{\partial \Theta}{\partial z} w_1 = 0, \tag{2.142}$$

where u_1, v_1, w_1, p_1, and Θ_1 designate, respectively, the small Eulerian changes in the velocity components, pressure, and potential temperature.

We now consider a zonal flow of magnitude u_0 and constant vertical shear u_0/H and a potential temperature with constant stratification $\partial\Theta/\partial z$ and constant horizontal gradient $\partial\Theta/\partial y$. For convenience, we shall make use of the following basic units: $x \approx u_0/f$, $y \approx u_0/f, z \approx H, t \approx 1/f, U \approx u_0, \Theta \approx H(\partial\Theta/\partial z), u_1 \approx u_0, v_1 \approx u_0$, and $w_1 \approx fH$. In these units, our basic steady state becomes

$$U = z \tag{2.143}$$

and

$$\Theta = z - \frac{y}{Ri} + constant, \tag{2.144}$$

where $Ri = \alpha g(\partial\Theta/\partial z)/(u_0/H)^2$ is a constant. Since the coefficients of the linearized equations depend on z only, we can thus let

$$w_1 = W(z)\exp[i(\sigma t + kx + ly)], \tag{2.145}$$

and we can write similar expressions for the other Eulerian variations in Eqs. (2.138)–(2.142). Given our choice of units, we have $k \approx f/u_0, l \approx f/u_0$, and $\sigma \approx f$. With a little algebra, these equations can be reduced to the following equation for the function W:

$$[1 - (\sigma + kz)^2]\frac{d^2W}{dz^2} - 2\left[\frac{k}{\sigma + kz} - il\right]\frac{dW}{dz} - \left[Ri(k^2 + l^2) + \frac{2ikl}{\sigma + kz}\right] W = 0. \tag{2.146}$$

We shall also prescribe that

$$W = 0, \quad \text{at} \quad z = 0 \quad \text{and} \quad z = 1. \tag{2.147}$$

Equations (2.146) and (2.147) constitute an eigenvalue problem for the complex frequency $\sigma = \sigma_r + i\sigma_i$. Since the unit of horizontal scale is u_0/f, the nondimensional wavenumbers k and l are simply the zonal and latitudinal *Rossby numbers* of the perturbations (see Eq. [2.30]).

A detailed study of this eigenvalue problem has been made by Stone (1966, 1970, 1972), who integrated Eqs. (2.146) and (2.147) for a wide range of values for the nondimensional parameters k, l, and Ri. His analysis shows that three basically different instabilities can occur for strictly positive Richardson numbers: a symmetric instability of the kind discussed by Solberg and Høiland (see Section 3.4.2), a baroclinic instability of the kind first discussed by Charney (1947) and Eady (1949), and a shear-flow instability analogous to the Kelvin–Helmholtz instability of two superposed fluids with different velocities and densities.* Not unexpectedly, because there is no latitudinal shear in the mean zonal flow $U = z$, barotropic instability does not occur in this simple model (see, however, Section 3.4.3).

2.7.1 *The symmetric instability*

The maximum growth rates for this instability are associated with perturbations that correspond to $k = 0$, $l \gg 1$, and large growth rates ($|\sigma_i| \approx 1$). Letting $k = 0$ in Eq. (2.146), one obtains

$$(1 - \sigma^2)\frac{d^2W}{dz^2} + 2il\frac{dW}{dz} - l^2 Ri W = 0. \tag{2.148}$$

Its solution, subject to boundary conditions (2.147), is

$$W = \exp\left(\frac{-ilz}{1 - \sigma^2}\right) \sin m\pi z \tag{2.149}$$

and

$$\sigma^2 = 1 + \frac{Ri}{2}\left(\frac{l}{m\pi}\right)^2 \pm \frac{l}{m\pi}\left[1 + \left(\frac{Ri}{2}\right)^2 \left(\frac{l}{m\pi}\right)^2\right]^{1/2}, \tag{2.150}$$

with $m = 1, 2, 3, \ldots$. Only the eigenvalues corresponding to the minus sign in front of the square root may lead to unstable motions. By virtue of Eq. (2.150), this instability occurs if and only if

$$Ri < 1. \tag{2.151}$$

The growth rate $|\sigma_i|$ is maximum for $l = \infty$ and has the value

$$|\sigma_i| = \left(\frac{1}{Ri} - 1\right)^{1/2}. \tag{2.152}$$

* See, e.g., Chandrasekhar, S., *Hydrodynamic and Hydromagnetic Stability*, Sections 100–104, Oxford: Clarendon Press, 1961 (New York: Dover Publications, 1981); Drazin, P. G., and Reid, W. H., *Hydrodynamic Stability*, Section 44, Cambridge: Cambridge University Press, 1981.

By making use of Eqs. (2.135) and (2.137), one readily sees that this instability may be visualized as the response to a large horizontal potential-temperature gradient in the form of an axisymmetric motion ($k = 0$) with small latitudinal wavelengths ($l \gg 1$), that is, a series of rolls parallel to the mean zonal flow. As was shown by Stone (1972), these motions draw their energy from both the kinetic and potential energy of the basic flow. For the most unstable modes ($l \to \infty$), however, the potential energy release is negligible. Accordingly, this instability may also be viewed as a form of barotropic instability. The link between condition (2.151) and the Solberg–Høiland conditions will be established at the end of Section 3.4.2.

2.7.2 The baroclinic instability

The maximum growth rates for this instability are associated with perturbations that are independent of latitude ($l = 0$), have large zonal scales ($k \ll 1$), and have small growth rates ($|\sigma_i| \approx k$). Thus we let $l = 0$ and $\sigma = kc$ in Eq. (2.146) to obtain

$$[1 - k^2(c+z)^2]\frac{d^2W}{dz^2} - \frac{2}{c+z}\frac{dW}{dz} - k^2 Ri\, W = 0. \tag{2.153}$$

Since the largest growth rates are found in the range $0 < k \ll 1$, we shall expand the solutions of the eigenvalue problem in powers of k^2. Letting

$$W = W_0 + k^2 W_1 + \cdots \quad \text{and} \quad c = c_0 + k^2 c_1 + \cdots \tag{2.154}$$

in Eq. (2.153), one can show that

$$W_0 = (c_0 + z)^3 - c_0^3, \tag{2.155}$$

$$W_1 = 3c_1(c_0 + z)^2 + \frac{Ri}{2}c_0^3(c_0 + z)^2 + \frac{6 + Ri}{10}(c_0 + z)^5, \tag{2.156}$$

etc. Applying boundary conditions (2.147) to these solutions, we obtain*

$$c = -\frac{1}{2} - \frac{i}{2\sqrt{3}}\left[1 - \frac{2}{15}k^2(1 + Ri)\right] + \cdots. \tag{2.157}$$

Hence, the growth rates of this instability can be written approximately as

$$|\sigma_i| = \frac{k}{2\sqrt{3}}\left[1 - \frac{2}{15}k^2(1 + Ri)\right]. \tag{2.158}$$

Ignoring terms of order k^5 or higher, one finds that the most rapidly growing perturbation is the one with the wavenumber

$$|k| = \left(\frac{5/2}{1 + Ri}\right)^{1/2} \tag{2.159}$$

* In this simple mathematical model, which does not include the latitudinal variation of the Coriolis parameter (i.e., the β term in Eq. [2.78]), one thus finds a cutoff wavelength, below which all disturbances are stable, and above which those of larger scale are unstable. As was originally shown by Green (1960), however, when Eady's (1949) problem is modified by taking $\beta > 0$, the flow becomes unstable to disturbances of *all* wavelengths, even for small values of β. In a more realistic formulation of baroclinic instability, there is thus no short wave limit for the instability region of the wave spectrum (e.g., Pedlosky [1987], Fig. 7.8.4).

and growth rate

$$|\sigma_i| = \left(\frac{5/54}{1+Ri}\right)^{1/2}. \tag{2.160}$$

One can show that this instability dominates over the symmetric instability whenever $Ri \gtrsim 1$. Stone's (1972) analysis also shows that the kinetic energy of the growing perturbations is drawn from both the potential and kinetic energy of the basic state. When $Ri \gg 1$, however, the kinetic energy release is negligible compared to the potential energy release. This is the reason why this instability is called a baroclinic instability. It will be discussed further in Section 3.4.3.

2.7.3 The shear-flow instability

This instability is associated with relatively small-scale perturbations ($k \gg 1$) and has small growth rates ($|\sigma_i| \approx k$). Again, letting $\sigma = kc$ in Eq. (2.146) and assuming that at most $l \approx k$ in the limit $k \to \infty$, one finds that Eq. (2.146) reduces to

$$(c+z)^2 \frac{d^2W}{dz^2} + Ri\left(1+\frac{l^2}{k^2}\right)W = 0. \tag{2.161}$$

Following Stone (1966), we obtain the solution for this equation:

$$W = (c+z)^{1/2}[A(c+z)^q + B(c+z)^{-q}], \tag{2.162}$$

where

$$q = \left[\frac{1}{4} - Ri\left(1+\frac{l^2}{k^2}\right)\right]^{1/2}. \tag{2.163}$$

Applying boundary conditions (2.147), one finds that a nontrivial solution exists if $c = 0$, $c = 1$, or

$$c = -\frac{1}{2} \pm \frac{i}{2}\,\mathrm{ctn}\,\frac{m\pi}{2q}, \tag{2.164}$$

with $m = 1, 2, 3, \ldots$. This equation shows that unstable solutions exist if q is real. In particular, q will be real inside the region

$$l^2 < k^2\left(\frac{1}{4Ri} - 1\right), \qquad k \gg 1, \tag{2.165}$$

and such a region exists as long as

$$Ri < \frac{1}{4}. \tag{2.166}$$

Equation (2.165) shows that this instability is greatest for small latitudinal wavenumbers, with the perturbations consisting of a series of rolls perpendicular to the mean zonal flow. Like the symmetric instability, it is also a form of barotropic instability, because it draws its energy mainly from the kinetic energy of the basic flow, although it may also store up potential energy. It will be considered further in Section 3.4.3.

2.8 Self-gravitating fluid masses

In this section we shall consider some general properties of self-gravitating bodies rotating freely in space. As we shall see in Section 6.2.1, although the interest of these models may be of a rather formal character, some of them have a direct bearing on the internal structure of rotating stars.

2.8.1 *The virial equations*

If we neglect friction altogether, Eq. (2.7) can be rewritten in the form

$$\frac{Dv_k}{Dt} = -\frac{\partial V}{\partial x_k} - \frac{1}{\rho}\frac{\partial p}{\partial x_k} \tag{2.167}$$

($k = 1, 2, 3$). Here we have

$$V = -G\int_{\mathcal{V}} \frac{\rho(\mathbf{r}', t)}{|\mathbf{r} - \mathbf{r}'|}\, dv', \tag{2.168}$$

where G is the constant of gravitation, $\mathcal{V}$ is the total volume of the configuration, and dv is the volume element.

Multiply now the left-hand side of Eq. (2.167) by ρx_i and integrate over the entire volume. By virtue of mass conservation, one has

$$\int_{\mathcal{V}} \rho x_i \frac{Dv_k}{Dt}\, dv = \frac{d}{dt}\int_{\mathcal{V}} \rho x_i v_k\, dv - 2K_{ik}, \tag{2.169}$$

where

$$K_{ik} = \frac{1}{2}\int_{\mathcal{V}} \rho v_i v_k\, dv \tag{2.170}$$

(see Eq. [3.53] below). Similarly, by making use of Eq. (2.168), we can write

$$\int_{\mathcal{V}} \rho x_i \frac{\partial V}{\partial x_k}\, dv = G\int_{\mathcal{V}}\int_{\mathcal{V}} \rho(\mathbf{r}, t)\,\rho(\mathbf{r}', t)\,\frac{x_i(x_k - x_k')}{|\mathbf{r} - \mathbf{r}'|^3}\, dv\, dv'. \tag{2.171}$$

Thus, if we let

$$W_{ik} = -\frac{1}{2}G\int_{\mathcal{V}}\int_{\mathcal{V}} \rho(\mathbf{r}, t)\,\rho(\mathbf{r}', t)\,\frac{(x_i - x_i')(x_k - x_k')}{|\mathbf{r} - \mathbf{r}'|^3}\, dv\, dv', \tag{2.172}$$

Eq. (2.171) reduces to

$$\int_{\mathcal{V}} \rho x_i \frac{\partial V}{\partial x_k}\, dv = -W_{ik}. \tag{2.173}$$

Finally, the last term in Eq. (2.167) can be integrated by parts to give

$$\int_{\mathcal{V}} x_i \frac{\partial p}{\partial x_k}\, dv = -\delta_{ik}\int_{\mathcal{V}} p\, dv, \tag{2.174}$$

since the pressure must vanish on the free surface.

Combining Eqs. (2.169), (2.173), and (2.174), we obtain

$$\frac{d}{dt}\int_{\mathcal{V}} \rho x_i v_k\, dv = 2K_{ik} + W_{ik} + \delta_{ik}\int_{\mathcal{V}} p\, dv. \tag{2.175}$$

Since all tensors on the right-hand side are symmetric, it follows that the left-hand side must also be symmetric. Hence, we can write

$$\frac{d}{dt} \int_{\mathcal{V}} \rho x_i v_k \, dv = \frac{d}{dt} \int_{\mathcal{V}} \rho x_k v_i \, dv. \tag{2.176}$$

This equation, which embodies the conservation of the total angular momentum J, implies that

$$\frac{d}{dt} \int_{\mathcal{V}} \rho x_i v_k \, dv = \frac{1}{2} \frac{d^2}{dt^2} \int_{\mathcal{V}} \rho x_i x_k \, dv. \tag{2.177}$$

Equation (2.175) thus becomes

$$\frac{1}{2} \frac{d^2 I_{ik}}{dt^2} = 2K_{ik} + W_{ik} + \delta_{ik} \int_{\mathcal{V}} p \, dv, \tag{2.178}$$

where

$$I_{ik} = \int_{\mathcal{V}} \rho x_i x_k \, dv. \tag{2.179}$$

These are the second-order virial equations in their usual form.

By contracting the indices in Eq. (2.178), we obtain the scalar virial equation,

$$\frac{1}{2} \frac{d^2 I}{dt^2} = 2K + W + 3 \int_{\mathcal{V}} p \, dv, \tag{2.180}$$

where I is the moment of inertia with respect to the center of mass and K is the total kinetic energy. By virtue of Eqs. (2.168) and (2.172), one also has

$$W = \frac{1}{2} \int_{\mathcal{V}} \rho V \, dv, \tag{2.181}$$

which is the gravitational potential energy.

For a self-gravitating fluid that rotates steadily about the x_3 axis with some assigned angular velocity Ω, Eq. (2.180) becomes

$$2K - |W| + 3 \int_{\mathcal{V}} p \, dv = 0, \tag{2.182}$$

where

$$K = \frac{1}{2} \int_{\mathcal{V}} \rho \Omega^2 \left(x_1^2 + x_2^2 \right) dv, \tag{2.183}$$

which is the rotational kinetic energy. Since the volume integral over the pressure always remains a nonnegative quantity, it follows at once that the ratio

$$\tau = K/|W| \tag{2.184}$$

is limited by equilibrium requirements to range from $\tau = 0$ (a spherical body) to $\tau = 0.5$. Of course, at this stage we do not know a priori whether the entire domain of values for τ (or J) can be covered by suitable models.

2.8.2 The Maclaurin and Jacobi ellipsoids

From a purely mechanical point of view, the specification of a particular model in a state of permanent rotation depends on three quantities: (i) the total mass M, (ii) the total angular momentum J, and (iii) an assigned distribution for the angular momentum per unit mass $\Omega(x_1^2+x_2^2)$. To clarify some aspects of the general problem, let us consider uniformly rotating, homogeneous ellipsoids. For that purpose, consider a system that is at rest with respect to a frame of reference rotating with the constant angular velocity Ω. The equations of relative equilibrium referred to rectangular axes rotating around the x_3 axis are

$$\frac{1}{\rho}\frac{\partial p}{\partial x_i} = -\frac{\partial V}{\partial x_i} + (1-\delta_{i3})\,\Omega^2 x_i \tag{2.185}$$

($i = 1, 2, 3$; no summation over repeated indices). Now, the components of the gravitational attraction in a homogeneous ellipsoid (with semi-axes a_1, a_2, and a_3) have the form

$$\frac{\partial V}{\partial x_i} = 2\pi G \rho A_i x_i, \tag{2.186}$$

where

$$A_i = a_1 a_2 a_3 \int_0^\infty \frac{du}{(a_i^2+u)\Delta} \tag{2.187}$$

and

$$\Delta^2 = (a_1^2+u)(a_2^2+u)(a_3^2+u). \tag{2.188}$$

By virtue of Eq. (2.186), the three components of Eq. (2.185) can be readily integrated to give

$$\frac{p}{\rho} = \frac{1}{2}\Omega^2\left(x_1^2+x_2^2\right) - \pi G\rho\left(A_1x_1^2 + A_2x_2^2 + A_3x_3^2\right) + \mathit{constant}, \tag{2.189}$$

so that the surfaces of constant pressure take the form

$$\left(A_1 - \frac{\Omega^2}{2\pi G\rho}\right)x_1^2 + \left(A_2 - \frac{\Omega^2}{2\pi G\rho}\right)x_2^2 + A_3x_3^2 = \mathit{constant}. \tag{2.190}$$

In expressing that the boundary of the ellipsoid,

$$\frac{x_1^2}{a_1^2} + \frac{x_2^2}{a_2^2} + \frac{x_3^2}{a_3^2} = 1, \tag{2.191}$$

coincides with one of the surfaces defined in Eq. (2.190), one finds that

$$a_1^2\left(A_1 - \frac{\Omega^2}{2\pi G\rho}\right) = a_2^2\left(A_2 - \frac{\Omega^2}{2\pi G\rho}\right) = a_3^2 A_3. \tag{2.192}$$

Hence, we must have

$$a_1^2a_2^2(A_1 - A_2) + (a_1^2 - a_2^2)a_3^2A_3 = 0 \tag{2.193}$$

and

$$\frac{\Omega^2}{2\pi G\rho} = \frac{a_1^2A_1 - a_2^2A_2}{a_1^2 - a_2^2} = \frac{a_1^2A_1 - a_3^2A_3}{a_1^2} = \frac{a_2^2A_2 - a_3^2A_3}{a_2^2}. \tag{2.194}$$

Obviously, the first equality (2.194) obtains only if $a_1 \neq a_2 \neq a_3$. If we next make use of Eq. (2.187), Eq. (2.193) becomes

$$\left(a_1^2 - a_2^2\right) \int_0^\infty \left[\frac{a_1^2 a_2^2}{(a_1^2+u)(a_2^2+u)} - \frac{a_3^2}{a_3^2+u} \right] \frac{du}{\Delta} = 0. \tag{2.195}$$

Finally, the three equalities (2.194) lead to the following relations:

$$\frac{\Omega^2}{2\pi G\rho} = a_1 a_2 a_3 \int_0^\infty \frac{u}{(a_1^2+u)(a_2^2+u)} \frac{du}{\Delta}, \tag{2.196}$$

when $a_1 \neq a_2 \neq a_3$; without any restriction, we also find

$$\frac{\Omega^2}{2\pi G\rho} = \frac{a_2 a_3}{a_1} \left(a_1^2 - a_3^2\right) \int_0^\infty \frac{u}{(a_1^2+u)(a_3^2+u)} \frac{du}{\Delta} \tag{2.197}$$

and a similar expression in which the index 1 replaces the index 2, and conversely.

From Eq. (2.197) and its unwritten companion, we first observe that $a_1 \geq a_3$ and $a_2 \geq a_3$. Thus, the rotation must always take place about the least axis. However, we may have either $a_1 \geq a_2$ or $a_1 \leq a_2$, since there is no physical difference between any two configurations for which we exchange the indices 1 and 2. Finally, we perceive at once that Eq. (2.195) can be satisfied in two different ways. Either we let $a_1 = a_2$ *or*, whenever possible, we let $a_1 > a_2$ (say) and make the integral factor vanish in Eq. (2.195). The former solution defines the *Maclaurin spheroids* while the latter corresponds to the *Jacobi ellipsoids*.* The Maclaurin spheroids range from a sphere ($\tau = 0$) to an infinitely thin disk that is at rest ($\tau = 0.5$). A numerical integration of Eq. (2.195) reveals that the Jacobi ellipsoids exist only in the domain $\tau_b \leq \tau \leq 0.5$, where $\tau_b = 0.1375$; they range from the bifurcation spheroid ($\tau = \tau_b$, where $a_3/a_1 = 0.5827$) to an infinitely long needle that is devoid of rotational motion ($\tau = 0.5$). Figure 2.6 illustrates the behavior of Ω^2 as a function of τ. Thus, when $0 \leq \tau \leq \tau_b$, the Maclaurin spheroids are the only possible figures of equilibrium; in contrast, in the range $\tau_b \leq \tau \leq 0.5$, to each value of τ correspond two ellipsoidal configurations in relative equilibrium: one Maclaurin spheroid and one Jacobi ellipsoid.

For fixed values of J, M, and $\mathcal{V}$, one can show that the total mechanical energy $K + W$ is smaller in the body with triplanar symmetry than in the corresponding axisymmetric

* The Scottish mathematician Colin Maclaurin (1698–1746) was the first to show in 1740 that *any* oblate homogeneous spheroid is a possible figure of equilibrium for uniformly rotating bodies. The next important discovery was not made until 1834, however, when the German mathematician Carl Jacobi (1804–1851) pointed out that homogeneous ellipsoids with three *unequal* axes can very well be figures of equilibrium (*Poggendorff Ann.*, **33**, 229, Oct. 4, 1834). Competition was fierce then, as it is today. Indeed, about three weeks later Joseph Liouville (1809–1882) published the detailed analytical proof of that theorem; however, noting that Jacobi had merely reported Eqs. (2.195) and (2.196) in his paper, Liouville could not refrain from saying that "this theorem, simple as it is, seems to have been enunciated as a challenge to the French mathematicians." And then Liouville added: "Mr. Jacobi was promising more indeed, when he announced that he was going to take over celestial mechanics from the pitiful state in which, so he said, Laplace had left it" (*J. Ecole Polytech. Paris*, **14**, Cahier 23, p. 291n, Oct. 27, 1834). Thus, Liouville perceived at once – but was reluctant to admit – that Jacobi had made an important discovery; yet, none of them could have foreseen that they were discussing the first known case of *broken symmetry* in physics. For the interested reader, the above quotations from Liouville's paper should clarify Chandrasekhar's (1969, p. 7) presentation of the Jacobi ellipsoids.

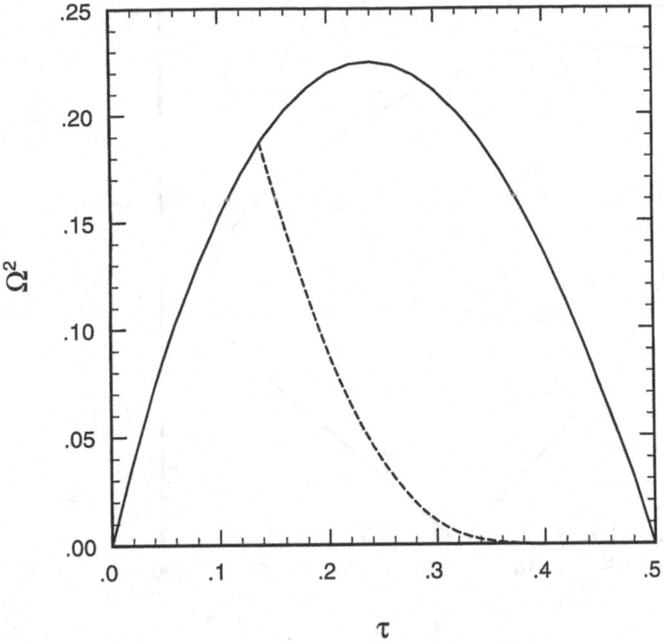

Fig. 2.6. The squared angular velocity Ω^2 along the Maclaurin (*solid line*) and the Jacobi (*dashed line*) sequences, as a function of the ratio $\tau = K/|W|$. The unit of Ω^2 is $2\pi G\rho$.

configuration. Accordingly, *if some dissipative mechanism is operative*, we may expect that beyond the point of bifurcation, $\tau = \tau_b$, an incompressible Maclaurin spheroid will evolve gradually to the Jacobi ellipsoid having the same total angular momentum. A detailed study of the global oscillations that transform a Maclaurin spheroid into a genuine triaxial body while preserving its plane of symmetry is thus in order.

Such a study was made by Lebovitz (1961). In particular, by making use of the perturbation forms of the six components of Eq. (2.178), he was able to calculate the five oscillation frequencies that reduce to the quintuple Kelvin frequency belonging to the spherical harmonics Y_2^m in the limit $\Omega = 0$ ($|m| \leq 2$). As we shall see, it is the toroidal (or barlike) modes that are of direct relevance to our discussion. When dissipative effects may be neglected, the corresponding frequencies are

$$\sigma_{-2} = \Omega - (2\omega - \Omega^2)^{1/2} \quad \text{and} \quad \sigma_{+2} = \Omega + (2\omega - \Omega^2)^{1/2}, \tag{2.198}$$

and two similar frequencies in which $-\Omega$ replaces Ω. Here we have let

$$\omega = 2\pi G\rho a_1^2 a_3 \int_0^\infty \frac{u}{(a_1^2 + u)^2} \frac{du}{\Delta}. \tag{2.199}$$

This is an exact analytical result. Figure 2.7 illustrates the behavior of the frequencies σ_{-2} and σ_{+2} along the Maclaurin sequence. We observe that σ_{-2} vanishes when $\Omega^2 = \omega$, that is, at the point $\tau = \tau_b$ where the Jacobi sequence branches off the Maclaurin sequence. In addition, both frequencies become complex when $\Omega^2 > 2\omega$, that is, beyond the point $\tau = \tau_i = 0.2738$, where $a_3/a_1 = 0.3033$; clearly, this implies instability by an overstable oscillation of frequency Ω.

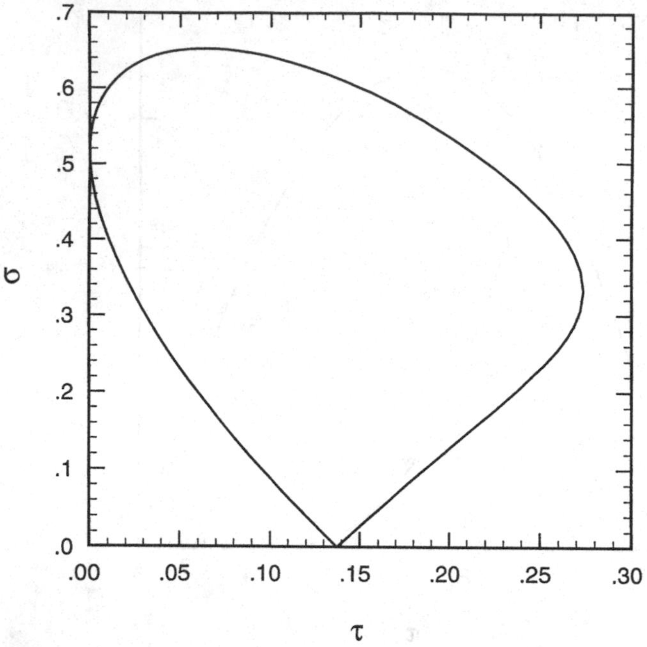

Fig. 2.7. Frequencies of the barlike modes ($|\sigma_{-2}|$ and $|\sigma_{+2}|$) along the Maclaurin sequence, as functions of the ratio $\tau = K/|W|$. The frequencies are given in units of $(4\pi G\rho)^{1/2}$; they are not represented beyond the point $\tau = \tau_i$ where they become complex. After Lebovitz (1961). *Source:* Ostriker, J. P., and Bodenheimer, P., *Astrophys. J.*, **180**, 171, 1973.

The situation is somewhat different when dissipation is properly taken into account. In that case, one can show that the barlike modes of oscillation are damped prior to the neutral point $\tau = \tau_b$. In the range $\tau_b < \tau < \tau_i$, however, the slightest amount of dissipation will carry slowly the Maclaurin spheroid into another configuration having a genuine triplanar symmetry. The system is then said to be secularly unstable. Beyond the point $\tau = \tau_i$, the Maclaurin spheroid becomes dynamically unstable as in the nondissipative case.

2.8.3 Rotating polytropes

In Section 2.8.2 we have summarized the main properties of the Maclaurin-Jacobi ellipsoids. To what extent can we extrapolate these results to centrally condensed bodies that we force to rotate with some prescribed angular momentum distribution? In particular, is it always possible to build a model in a state of permanent rotation for all values that we may assign to the total angular momentum J? To illustrate these problems, let us briefly consider polytropes, that is, barotropic structures for which the pressure p and the density ρ are related by the relation

$$p = K_0 \rho^{1+1/n}, \tag{2.200}$$

where K_0 and n are constants ($0 \leq n \leq 5$). Because the value $n = 0$ corresponds to a configuration of uniform density, it is therefore convenient to think of the incompressible Maclaurin spheroids as a sequence of uniformly rotating polytropes of index $n = 0$.

It has been known since the pioneering work of Jeans (1919), which was subsequently refined by James (1964), that every sequence of axially symmetric, uniformly rotating polytropes terminates with a configuration in which the gravitational and centrifugal accelerations exactly balance at points on the equator. As a rule, each sequence terminates at a point $\tau = \tau_{\max}$ (say), and the values of $\tau_{\max}$ decrease sharply with polytropic index. (While $\tau_{\max} = 0.5$ along the Maclaurin sequence, $\tau_{\max} \approx 0.12$ when $n = 1$, and $\tau_{\max}$ is already reduced to less than one percent when $n = 3$.) To be specific, for low polytropic index (i.e., $n < 0.8$) sequences of axisymmetric models reach points of bifurcation and, where further models become secularly unstable (as for $n = 0$), the sequences may bifurcate into analogs of the Jacobi ellipsoids. When $n > 0.8$, however, sequences of axially symmetric, uniformly rotating polytropes always terminate at models in which the effective gravity vanishes at the equator.

In the early 1980s, different groups have actually constructed complete sequences of *non*axisymmetric, uniformly rotating polytropes for $n < 0.8$. Their independent calculations clearly show that all these Jacobi-like sequences bifurcate from their corresponding axisymmetric counterparts at about the same value of τ (≈ 0.137). However, as was shown by Eriguchi and Hachisu (1982), even for an index as low as $n = 0.1$, they terminate after only a small increase in J away from the axisymmetric models. Accordingly, until limited by the onset of equatorial breakup, the equilibrium figures of uniformly rotating polytropes with low polytropic index resemble, in all essential respects, the Maclaurin–Jacobi ellipsoids. This is in sharp contrast to the more centrally condensed polytropes for which the rotational kinetic energy does not exceed a small fraction of the gravitational potential energy ($\tau \ll 0.5$). Indeed, as was shown by Tassoul and Ostriker (1970), because uniformly rotating configurations having a genuine triplanar symmetry always branch off at the point $\tau \approx 0.137$, which is almost independent of n, bifurcation – and the ensuing secular instability – does not occur along polytropic sequences with $n > 0.8$. In other words, uniformly rotating, centrally condensed polytropes remain unaffected by the global instabilities described in Section 2.8.2 because they cannot store a large amount of angular momentum!

As was originally shown by Bodenheimer and Ostriker in 1973, a completely different picture emerges from the study of frictionless, differentially rotating polytropes for which we prescribe a given angular momentum distribution. In that case, the centrally condensed models closely simulate incompressible Maclaurin spheroids, except that they do not maintain uniform rotation. In particular, it was found that the polytropic sequences do not terminate and that bifurcation may occur when τ is very closely equal to the value $\tau = \tau_b$ obtained for the Maclaurin spheroids. Moreover, their work strongly suggested that dynamical instability with respect to a barlike mode always sets in beyond the point $\tau = \tau_i$, which again does not greatly depend on the particular sequence. Recent developments have shown that some of these propositions may need refinement, however.

To be specific, Imamura et al. (1995) have shown that for angular momentum distributions similar to those of the Maclaurin spheroids, there is a qualitative correspondence between the onset of secular instability for compressible and incompressible fluids. However, for angular momentum distributions that are more peaked toward the equatorial radius, their work indicates that secular instability with respect to a barlike mode sets in at lower values of τ, shifting from $\tau = 0.14$ to $\tau = 0.09$ over the range of angular momentum distributions considered. More recently, Toman et al. (1998) have shown that the onset of dynamical instability with respect to a barlike mode is not very sensitive to the compressibility or angular momentum distribution when the polytropic models are parameterized by τ. The eigenfunctions for the fastest growing barlike modes are, however, qualitatively different from the Maclaurin eigenfunctions in one important respect: They develop strong spiral arms. These spiral arms are stronger for larger values of the polytropic index and for configurations whose angular momentum distributions deviate significantly from those of the Maclaurin spheroids.

2.9 Bibliographical notes

The literature on classical hydrodynamics is very extensive. Among the many textbooks on the subject, my own preference goes to:

1. Batchelor, G. K., *An Introduction to Fluid Dynamics*, Cambridge: Cambridge University Press, 1967.
2. Landau, L. D., and Lifshitz, E. M., *Fluid Mechanics*, 2nd Edition, Oxford: Pergamon Press, 1987.
3. Tritton, D. J., *Physical Fluid Dynamics*, 2nd Edition, Oxford: Clarendon Press, 1988.

The only book devoted exclusively to the problem of rotation is:

4. Greenspan, H. P., *The Theory of Rotating Fluids*, Cambridge: Cambridge University Press, 1968 (reprinted by Breukelen Press, Brookline, MA, 1990).

Excellent introductions to geophysical fluid dynamics are:

5. Houghton, J. T., *The Physics of Atmospheres*, 2nd Edition, Cambridge: Cambridge University Press, 1986.
6. Holton, J. R., *An Introduction to Dynamic Meteorology*, 3rd Edition, New York: Academic Press, 1992.

At a more advanced level, the following monograph is particularly worth noting:

7. Pedlosky, J., *Geophysical Fluid Dynamics*, 2nd Edition, New York: Springer-Verlag, 1987.

See also:

8. Gill, A. E., *Atmosphere–Ocean Dynamics*, Orlando: Academic Press, 1982.
9. Apel, J. R., *Principles of Ocean Physics*, Orlando: Academic Press, 1987.

A general survey from the viewpoint of astrophysics will be found in:

10. Shore, S. N., *An Introduction to Astrophysical Hydrodynamics*, San Diego: Academic Press, 1992.

Sections 2.5 and 2.6. Ekman layers, at a rigid plane boundary and at the ocean–atmosphere interface, were originally discussed in:

11. Ekman, V. W., *Arkiv Mat. Astron. Fysik* (Stockholm), **2**, No 11, 1, 1905.

However, the first quantitative discussions of the so-called Ekman pumping mechanism were given by:

12. Bondi, H., and Lyttleton, R. A., *Proc. Cambridge Phil. Soc.*, **44**, 345, 1948.
13. Charney, J. G., and Eliassen, A., *Tellus*, **1**, No 2, 38, 1949.

A comprehensive discussion of Ekman pumping is given by:

14. Greenspan, H. P., and Howard, L. N., *J. Fluid Mech.*, **17**, 385, 1963.

An illustrative example will be found in:

15. Harada, A., *J. Meteorol. Soc. Japan*, **60**, 876, 1982.

Oceanic currents are discussed in:

16. Sverdrup, H. U., *Proc. Natl. Acad. Sci. U.S.A.*, **33**, 318, 1947.
17. Munk, W. H., *J. Meteorol.*, **7**, 79, 1950.
18. Bryan, K., *J. Atmos. Sci.*, **20**, 594, 1963.

See also Reference 7 for a comprehensive review of these matters.

Section 2.7. Reference is made to the following papers:

19. Charney, J. G., *J. Meteorol.*, **4**, 135, 1947.
20. Eady, E. T., *Tellus*, **1**, No 3, 33, 1949.
21. Green, J. S. A., *Quart. J. Roy. Meteorol. Soc.*, **86**, 237, 1960.

The analysis in this section is largely derived from:

22. Stone, P. H., *J. Atmos. Sci.*, **23**, 390, 1966; *ibid.*, **27**, 721, 1970; *ibid.*, **29**, 419, 1972.

Deviations from hydrostatic equilibrium in Eq. (2.141) are discussed in:

23. Tokioka, T., *J. Meteorol. Soc. Japan*, **48**, 503, 1970.
24. Stone, P. H., *J. Fluid Mech.*, **45**, 659, 1971.

The mass-divergence effect in the continuity equation was also considered by:

25. Hyun, J. M., and Peskin, R. L., *J. Atmos. Sci.*, **33**, 2054, 1976; *ibid.*, **35**, 169, 1978.

Sections 2.8.1 and 2.8.2. The standard reference on the virial equations and homogeneous ellipsoids is:

26. Chandrasekhar, S., *Ellipsoidal Figures of Equilibrium*, New Haven: Yale University Press, 1969 (reprinted, with Section 63 revised, by Dover Publications, New York, 1987).

The reference to Lebovitz is to his paper:

27. Lebovitz, N. R., *Astrophys. J.*, **134**, 500, 1961.

With the advent of fast computers, several authors have computed new equilibrium sequences that branch off the Maclaurin–Jacobi ellipsoids and have finite deformations from these configurations. The following papers are particularly worth noting:

28. Eriguchi, Y., and Hachisu, I., *Prog. Theor. Phys.*, **67**, 844, 1982.
29. Eriguchi, Y., Hachisu, I., and Sugimoto, D., *Prog. Theor. Phys.*, **67**, 1068, 1982.
30. Eriguchi, Y., and Hachisu, I., *Astron. Astrophys.*, **148**, 289, 1985.

See also Reference 33 (pp. 544–546).

Section 2.8.3. For a general account of rotating polytropes, see:

31. Tassoul, J. L., *Theory of Rotating Stars*, pp. 233–272, Princeton: Princeton University Press, 1978.

Other useful reviews of rotating fluid masses can be found in:

32. Lebovitz, N. R., *Annu. Rev. Fluid Mech.*, **11**, 229, 1979.
33. Durisen, R. H., and Tohline, J. E. in *Protostars & Planets II* (Black, D. C., and Matthews, M. S., eds.), p. 534, Tucson: University of Arizona Press, 1985.

The following references are quoted in the text:

34. Jeans, J. H., *Problems of Cosmogony and Stellar Dynamics*, pp. 165–186, Cambridge: Cambridge University Press, 1919.
35. James, R. A., *Astrophys. J.*, **140**, 552, 1964.
36. Tassoul, J. L., and Ostriker, J. P., *Astron. Astrophys.*, **4**, 423, 1970.
37. Bodenheimer, P., and Ostriker, J. P., *Astrophys. J.*, **180**, 159, 1973.
38. Ostriker, J. P., and Bodenheimer, P., *Astrophys. J.*, **180**, 171, 1973.

Sequences of uniformly rotating, nonaxisymmetric polytropes with $n < 0.8$ have been constructed in:

39. Ipser, J. R., and Managan, R. A., *Astrophys. J.*, **250**, 362, 1981.
40. Vandervoort, P. O., and Welty, D. E., *Astrophys. J.*, **248**, 504, 1981.
41. Hachisu, I., and Eriguchi, Y., *Prog. Theor. Phys.*, **68**, 206, 1982.

There is also a growing literature on the nonaxisymmetric instabilities in differentially rotating polytropes. The following papers may be noted:

42. Imamura, J. N., Toman, J., Durisen, R. H., Pickett, B. K., and Yang, S., *Astrophys. J.*, **444**, 363, 1995.
43. Pickett, B. K., Durisen, R. H., and Davis, G. A., *Astrophys. J.*, **458**, 714, 1996.
44. Smith, S. C., Houser, J. L., and Centrella, J. M., *Astrophys. J.*, **458**, 236, 1996.
45. Toman, J., Imamura, J. N., Pickett, B. K., and Durisen, R. H., *Astrophys. J.*, **497**, 370, 1998.

Other contributions may be traced to References 42–45.

3

Rotating stars

3.1 Introduction

Consider a single star that rotates about a fixed direction in space, with some assigned angular velocity. As we know, the star then assumes the shape of an *oblate* figure. However, we are at once faced with the following questions. What is the geometrical shape of the free boundary? What is the form of the surfaces upon which the physical variables (such as pressure, density, ...) remain a constant? To sum up, what is the actual stratification of a rotating star, and how does it depend on the angular velocity distribution? For rotating stars, we have no a priori knowledge of this stratification, which is itself an unknown that must be derived from the basic equations of the problem. This is in sharp contrast to the case of a nonrotating star, for which a spherical stratification can be assumed ab initio.*

In principle, by making use of the equations derived in Section 2.2, one should be able to calculate at every instant the angular momentum distribution and the stratification in a rotating star. Obviously, this is an impossible task at the present level of knowledge of the subject, even were the initial conditions known. Until very recently, the standard procedure was to calculate in an approximate manner an equilibrium structure that corresponds to some prescribed rotation law, ruling out those configurations that are dynamically or thermally unstable with respect to axisymmetric disturbances (see Sections 3.4.2 and 3.5). Unfortunately, the results presented in Section 3.3 indicate that, no matter whether radiation or convection is providing the energy transport, the large-scale motion in a star is always the combination of a pure rotation and a circulation in meridian planes passing through the rotation axis. Moreover, as we shall see in Section 3.4.3, no dynamically stable model can possibly exist when nonaxisymmetric disturbances are taken into account. These barotropic and baroclinic instabilities, which have their roots in the geophysical literature, are mild ones in the sense that they continuously generate small-scale, eddy motions that interact with the large-scale flow. Lacking any better description of these transient motions, we shall further assume that the eddy flux

* Following the publication of Newton's (1687) *Principia*, the effects of rotation upon the internal structure of a self-gravitating body were investigated mainly with a view to their possible applications to geodesy and planetary physics. Many a classical result derived during the period 1740–1940 still retains its usefulness today when applied to centrally condensed stars. For a brief historical account of these and related matters, see J. L. Tassoul, *Theory of Rotating Stars*, Section 1.3, Princeton: Princeton University Press, 1978.

momentum can be represented parametrically by means of suitable coefficients of eddy viscosity. This is known as the *eddy–mean flow interaction,* which is presented in Section 3.6. This approach has been familiar to geophysicists since the late 1940s. It is particularly convenient because it resolves in a very simple manner the many contradictions and inconsistencies that have beset the theory of rotating stars.

3.2 Basic concepts

Because molecular viscosity is negligible for large-scale motions in a star, the momentum equation (2.7) reduces to

$$\frac{D\mathbf{v}}{Dt} = -\operatorname{grad} V - \frac{1}{\rho}\operatorname{grad} p, \tag{3.1}$$

where $\mathbf{v}$ is the velocity in an inertial frame of reference, V is the gravitational potential, ρ is the density, and p is the pressure. The gravitational potential and the density are related by the Poisson equation

$$\nabla^2 V = 4\pi G\rho, \tag{3.2}$$

where G is the constant of gravitation. Mass conservation implies that

$$\frac{D\rho}{Dt} + \rho \operatorname{div}\mathbf{v} = 0. \tag{3.3}$$

Finally, neglecting the dissipation function Φ_v and thermal conductivity, we may recast Eq. (2.12) in the form

$$\rho T\frac{DS}{Dt} = \rho\epsilon_{\mathrm{Nuc}} - \operatorname{div}\mathcal{F}, \tag{3.4}$$

where ϵ_{Nuc} is the rate of energy released by the thermonuclear reactions per unit mass and unit time, and $\mathcal{F}$ is the radiative flux vector. If we except the outermost surface layers of a star, this vector is given by

$$\mathcal{F} = -\chi \operatorname{grad} T, \tag{3.5}$$

where

$$\chi = \frac{4ac}{3}\frac{T^3}{\kappa\rho} \tag{3.6}$$

is the coefficient of radiative conductivity. As usual, a is the radiation pressure constant, c is the speed of light, and κ is the coefficient of opacity per unit mass.

In the following we shall consider a mixture of blackbody radiation and a simple ideal gas. Neglecting the ionization and excitation energies, we thus replace Eqs. (2.13) and (2.14) by

$$U = c_V T + \frac{1}{\rho} a T^4 \tag{3.7}$$

and

$$p = \frac{\mathcal{R}}{\bar{\mu}}\rho T + \frac{1}{3} a T^4. \tag{3.8}$$

For isentropic motions, Eq. (2.11) becomes

$$T\frac{DS}{Dt}=\frac{DU}{Dt}+p\frac{D}{Dt}\left(\frac{1}{\rho}\right)=\frac{DU}{Dt}+\frac{p}{\rho}\,\mathrm{div}\,\mathbf{v}=0, \tag{3.9}$$

thus expressing that the specific entropy of each fluid particle remains a constant along its path (although this entropy may differ from one path line to another). Equation (3.9) can be written also in the form

$$\frac{Dp}{Dt}=\Gamma_1\frac{p}{\rho}\frac{D\rho}{Dt}. \tag{3.10}$$

Here we have let

$$\Gamma_1=\beta+\frac{(4-3\beta)^2(\gamma-1)}{\beta+12(\gamma-1)(1-\beta)}, \tag{3.11}$$

where $\beta=p_g/(p_g+p_r)$ is the ratio of gaseous pressure to total pressure. One can also write

$$\frac{1}{T}\frac{DT}{Dt}=(\Gamma_3-1)\frac{1}{\rho}\frac{D\rho}{Dt}=\frac{\Gamma_2-1}{\Gamma_2}\frac{1}{p}\frac{Dp}{Dt}, \tag{3.12}$$

where Γ_2 and Γ_3 are related to Γ_1 by the following relations:

$$\Gamma_3-1=\frac{\Gamma_1-\beta}{4-3\beta}=\Gamma_1\frac{\Gamma_2-1}{\Gamma_2}. \tag{3.13}$$

The variation of the ratio β is given by

$$\frac{1}{\beta}\frac{D\beta}{Dt}=(\Gamma_3-\Gamma_1)\frac{1}{\rho}\frac{D\rho}{Dt}. \tag{3.14}$$

The Γs reduce to the usual adiabatic exponent $\gamma=c_p/c_V$ in the limit $p_r \ll p_g$; they reduce to 4/3 for blackbody radiation alone ($p_g \ll p_r$). For a mixture of an ideal gas and blackbody radiation, the generalized adiabatic exponents are intermediate in value between 4/3 and γ.

3.2.1 The Poincaré–Wavre theorem

The simplest model of a rotating star we can make is to assume that the configuration rotates about a fixed direction in space with some assigned angular velocity. Assume further that the star is axially symmetric and that the motion is steady in time. Let the star rotate about the z axis, and take the center of our inertial frame of reference at the center of mass. Then, in cylindrical polar coordinates (ϖ, φ, z), the velocity $\mathbf{v}$ has the form

$$\mathbf{v}=\Omega\varpi\mathbf{1}_\varphi, \tag{3.15}$$

where $\mathbf{1}_\varphi$ is the unit vector in the azimuthal direction. By virtue of our assumptions, Eq. (3.3) is identically satisfied. Since we have neglected any large-scale motion in meridian planes passing through the rotation axis, the φ component of Eq. (3.1) is identically satisfied also. The remaining components of this equation imply that

$$\frac{1}{\rho}\frac{\partial p}{\partial\varpi}=-\frac{\partial V}{\partial\varpi}+\Omega^2\varpi \tag{3.16}$$

and

$$\frac{1}{\rho}\frac{\partial p}{\partial z} = -\frac{\partial V}{\partial z}. \tag{3.17}$$

Many useful properties can be deduced from these equations. For this purpose, let us define the effective gravity

$$\mathbf{g} = -\left(\frac{\partial V}{\partial \varpi} - \Omega^2 \varpi\right) \mathbf{1}_\varpi - \frac{\partial V}{\partial z} \mathbf{1}_z, \tag{3.18}$$

where $\mathbf{1}_\varpi$ and $\mathbf{1}_z$ are the unit vectors in the ϖ and z direction. Equations (3.16) and (3.17) become

$$\frac{1}{\rho} \operatorname{grad} p = \mathbf{g}. \tag{3.19}$$

It follows at once that *the effective gravity is everywhere orthogonal to the surfaces of constant pressure* (*i.e., the isobaric surfaces*). This is a general property, which is valid no matter whether one has $\Omega = \Omega(\varpi)$ or $\Omega = \Omega(\varpi, z)$.

Let us now assume the star rotates as a solid body. Equation (3.18) then reduces to

$$\mathbf{g} = -\operatorname{grad} \Phi, \tag{3.20}$$

where, except for an additive constant, one has

$$\Phi = V(\varpi, z) - \frac{1}{2}\Omega^2 \varpi^2. \tag{3.21}$$

Under what circumstances can we also derive the effective gravity from a potential in a differentially rotating star? By virtue of Eq. (3.18), this is possible if and only if Ω does not depend on z, that is, when the angular velocity is a constant over cylinders centered about the axis of rotation. Then, instead of Eq. (3.21), one has

$$\Phi = V(\varpi, z) - \int^{\varpi} \Omega^2(\varpi')\varpi'\, d\varpi'. \tag{3.22}$$

Various interesting conclusions can be inferred from the existence of such a potential.

First, by virtue of Eq. (3.19), one can always write

$$\frac{1}{\rho}\, dp = g_\varpi\, d\varpi + g_z\, dz. \tag{3.23}$$

Making use of Eqs. (3.20) and (3.22), we obtain

$$\frac{1}{\rho}\, dp = -\, d\Phi. \tag{3.24}$$

By definition, for any displacement on a level surface $\Phi =$ *constant* one has $d\Phi = 0$. Since Eq. (3.24) shows that $dp = 0$ on the same surface, it follows at once that the isobaric surfaces coincide with the level surfaces. If so, then, we can write

$$p = p(\Phi) \qquad \text{or} \qquad \Phi = \Phi(p). \tag{3.25}$$

By virtue of Eq. (3.24), one readily sees that

$$\frac{1}{\rho} = -\frac{d\Phi(p)}{dp} \qquad \text{or} \qquad \rho = \rho(p). \tag{3.26}$$

Accordingly, the density is also a constant over an isobaric surface. Thus, the surfaces upon which p, ρ, and Φ remain a constant all coincide. As a consequence, when a potential Φ does exist, the vector $\mathbf{g}$ is also normal to the surfaces of constant density (i.e., the isopycnic surfaces).

Reciprocally, let us consider a system for which the surfaces of constant pressure and constant density coincide. If we let

$$\Phi(p) = -\int \frac{dp}{\rho(p)}, \tag{3.27}$$

Eq. (3.23) then becomes

$$d\Phi = g_\varpi \, d\varpi + g_z \, dz. \tag{3.28}$$

As function of the coordinates, the differential $d\Phi$ is an exact total differential. Accordingly, Eq. (3.20) must hold true, and the vector $\mathbf{g}$ may be derived from a potential.

Finally, let us suppose that the effective gravity is everywhere normal to the isopycnic surfaces. By virtue of Eq. (3.23), any displacement over one of these surfaces gives $dp = 0$, so that the pressure is a constant over an isopycnic surface. The coincidence of the surfaces of constant pressure and constant density is thus established.

If we now collect all the pieces together, it is a simple matter to see that we have proved the equivalence of the following statements:

(a) *The angular velocity depends on ϖ only.*
(b) *The effective gravity can be derived from a potential.*
(c) *The effective gravity is normal to the isopycnic surfaces.*
(d) *The isobaric surfaces and the isopycnic surfaces coincide.*

Thus, any of these statements implies the three others. By definition, a system for which these statements hold true is called a *barotrope*.

Following current practice, we shall call a system for which these statements do not hold true a *barocline*. The major distinction between a barotrope and a barocline lies in their respective stratification. Of particular importance is the fact that the isopycnic surfaces are in general inclined to and cut the isobaric surfaces in a barocline.

Note that slow but inexorable meridional currents do exist in a rotating star. As we shall see in Chapters 4 and 5, however, these currents are so slow that they do not upset the mechanical balance defined by Eqs. (3.16) and (3.17). Hence, they do not modify the basic conclusions reported in this section.

3.3 Some tentative solutions

In Section 3.2.1 we have demonstrated some simple mechanical properties of an axially symmetric star that rotates with some assigned angular velocity. Yet, because we have hitherto circumvented the use of the condition of energy conservation, we do not know whether we can apply these results, without modification, to a radiating star. For example, is there any constraint imposed by the condition of radiative equilibrium on the angular velocity distribution in a barotropic star? Similarly, to what extent is it necessary to modify the conclusions of the Poincaré–Wavre theorem when turbulent friction is properly taken into account in a star in strict convective equilibrium? We shall devote this section to the study of these two questions.

3.3.1 *The case of radiative equilibrium*

Consider a barotropic star in *strict* radiative equilibrium. By making use of Eq. (3.4), we can write

$$\operatorname{div} \mathcal{F} = \rho \epsilon_{\rm Nuc}. \tag{3.29}$$

Equations (3.19) and (3.20) further imply that

$$\frac{1}{\rho} \operatorname{grad} p = - \operatorname{grad} \Phi, \tag{3.30}$$

where Φ is defined in Eq. (3.22). By virtue of the Poincaré–Wavre theorem, we immediately deduce that $p = p(\Phi)$ and $\rho = \rho(\Phi)$. If the chemical composition is constant (or a function of p and ρ only), one also has $T = T(\Phi)$. Hence, if we assume that $\epsilon_{\rm Nuc} = \epsilon_{\rm Nuc}(\rho, T)$ and $\kappa = \kappa(\rho, T)$, both the energy generation rate and the opacity coefficient depend on Φ only. It follows that

$$\mathcal{F} = -\frac{4ac}{3} \frac{T^3}{\kappa\rho} \frac{dT}{d\Phi} \operatorname{grad} \Phi \tag{3.31}$$

or

$$\mathcal{F} = f(\Phi) \operatorname{grad} \Phi, \tag{3.32}$$

where

$$f(\Phi) = -\frac{4ac}{3} \frac{T^3}{\kappa\rho} \frac{dT}{d\Phi}. \tag{3.33}$$

Let us consider next the divergence of Eq. (3.32). One obtains

$$\operatorname{div} \mathcal{F} = f'(\Phi) \left(\frac{d\Phi}{dn} \right)^2 + f(\Phi) \nabla^2 \Phi, \tag{3.34}$$

where dn is along the outward normal to a level surface, and a prime denotes a derivative with respect to Φ. We also have

$$\left(\frac{d\Phi}{dn} \right)^2 = \left(\frac{\partial \Phi}{\partial \varpi} \right)^2 + \left(\frac{\partial \Phi}{\partial z} \right)^2 = g^2. \tag{3.35}$$

Clearly, $d\Phi/dn$ is the magnitude of the effective gravity $\mathbf{g}$. Combining Eqs. (3.2) and (3.22), one can write

$$\nabla^2 \Phi = 4\pi G\rho - \frac{1}{\varpi} \frac{d}{d\varpi} (\Omega^2 \varpi^2). \tag{3.36}$$

By making use of Eqs. (3.34)–(3.36), we can thus recast Eq. (3.29) in the form

$$f'(\Phi) g^2 + f(\Phi) \left[4\pi G\rho - \frac{1}{\varpi} \frac{d}{d\varpi} (\Omega^2 \varpi^2) \right] = \rho \epsilon_{\rm Nuc}. \tag{3.37}$$

This is the condition of strict radiative equilibrium in a barotropic star.

The case of a uniformly rotating barotrope is particularly straightforward, and it was originally discussed by von Zeipel (1924). Equation (3.37) then reduces to

$$f'(\Phi) g^2 + f(\Phi)(4\pi G\rho - 2\Omega^2) = \rho \epsilon_{\rm Nuc}. \tag{3.38}$$

As we know, g is not constant over a level surface in a rotating body, because the distance from one level to the next one is not the same for every point on it. Accordingly, since ρ, $\epsilon_{\rm Nuc}$, and Ω are all constant on level surfaces, the coefficient of g in Eq. (3.38) must vanish separately. We have, therefore,

$$f'(\Phi) = 0 \qquad \text{or} \qquad f(\Phi) = constant. \tag{3.39}$$

Hence, Eq. (3.38) assumes the form

$$\epsilon_{\rm Nuc} \propto \left(1 - \frac{\Omega^2}{2\pi G\rho}\right), \tag{3.40}$$

and Eq. (3.32) reduces to

$$|\mathcal{F}| \propto g. \tag{3.41}$$

This is known as *von Zeipel's law of gravity darkening*. Obviously, condition (3.40) is never fulfilled in an actual star. It follows at once that *rigid rotation is impossible for a barotrope in static radiative equilibrium.*

Let us consider next the general conservative law $\Omega = \Omega(\varpi)$. In this case, as was pointed out by Rosseland (1926) and Vogt (1935), it is intuitively evident that the law $\Omega = \Omega(\varpi)$ is incompatible with condition (3.37). Indeed, while Ω will be constant over cylinders centered about the rotation axis, g will be constant over certain oblate surfaces. Therefore, by virtue of Eq. (3.37), conditions (3.39) and (3.41) still pertain, but we must impose the additional condition

$$\frac{1}{\varpi}\frac{d}{d\varpi}\left(\Omega^2\varpi^2\right) = constant. \tag{3.42}$$

After integrating, one obtains

$$\Omega^2 = c_1 + \frac{c_2}{\varpi^2}, \tag{3.43}$$

where c_1 and c_2 denote two arbitrary constants. If $c_2 = 0$, we simply recover the case of a uniformly rotating barotrope. Similarly, if $c_2 \neq 0$, the rotational law (3.43) becomes singular on the rotation axis; it must be disregarded because it also leads to an impossible constraint on $\epsilon_{\rm Nuc}$ (i.e., condition [3.40] with Ω^2 being replaced by c_1). This argument shows that *a differentially rotating barotrope cannot remain in static radiative equilibrium.*

It is not the usual energy generation rates that prevent the rotation laws $\Omega = constant$ or $\Omega = \Omega(\varpi)$ from being realized, but rather the condition of strict radiative equilibrium. Indeed, in the limit $\Omega = 0$, g is a constant over each spherical surface $\Phi \equiv V = constant$, and there is no requirement that some terms in Eq. (3.37) should vanish independently of the remaining terms. Therefore, this equation must be regarded as an indication that for nonspherical stars at least one of the assumptions leading to conditions (3.40) and (3.42) must be relaxed.

This problem can be solved in two different ways: Either we assume strict radiative equilibrium while allowing Ω to depend on both ϖ and z or we assume that strict radiative equilibrium breaks down in a rotationally distorted star. The latter solution leads to the formation of a large-scale meridional flow (and concomitant differential rotation) in the radiative zone of a rotating star. The former solution is mainly of academic interest, however, because there is no obvious reason why the angular velocity would adjust itself so as to prevent meridional currents. These matters will be discussed further in Chapter 4.

3.3.2 *The case of convective equilibrium*

In a region of efficient convection, the energy transport keeps the actual temperature gradient closely equal to the adiabatic lapse rate. Let us assume that such a region has reached a state of mechanical equilibrium, with no large-scale motions in meridian planes passing through the rotation axis. Since the fluid is essentially barotropic, it follows at once that

$$\Omega = \Omega(\varpi) \tag{3.44}$$

over the whole region where *strict* convective equilibrium prevails.

The actual form of the rotation law depends on the azimuthal forces. By assumption, there are no meridional motions; and there is no large-scale magnetic field. Accordingly, the only remaining force is the φ component of turbulent friction acting on the rotational motion. An explicit expression for this force will be presented in Section 3.6. Anticipating these results and introducing spherical polar coordinates (r, θ, φ), one finds that

$$\frac{1}{r^2}\frac{\partial}{\partial r}\left(\mu_V r^4 \frac{\partial \Omega}{\partial r} + \lambda_V r^3 \Omega\right) + \frac{1}{\sin^3\theta}\frac{\partial}{\partial \theta}\left(\mu_H \sin^3\theta \frac{\partial \Omega}{\partial \theta}\right) = 0, \tag{3.45}$$

where μ_V and μ_H are the vertical and horizontal coefficients of eddy viscosity, and λ_V is a parameter representing the influence of global rotation on the anisotropic convective elements (see Eq. [3.133]).

Equation (3.45) must be solved with appropriate boundary conditions. Because eddy viscosity is always much larger in a convective zone than in the surrounding regions, we shall merely prescribe that the tangential viscous stresses vanish at the boundaries of the convective zone (see Eq. [2.21]). For a slowly rotating solar-type star, these conditions become

$$\mu_V r \frac{\partial \Omega}{\partial r} + \lambda_V \Omega = 0 \qquad \text{at} \qquad r = R_i \quad \text{and} \quad r = R_o, \tag{3.46}$$

where R_i and R_o are the inner and outer radii of its (almost) spherical convective layer (see Eq. [3.131]).

For the sake of simplicity, let us assume that the parameter λ_V and the eddy viscosities are constant. Following Kippenhahn (1963), one can show that Eqs. (3.45) and (3.46) can be satisfied only if the angular velocity is constant on spheres, with the rotation law

$$\Omega = \Omega_0 r^{-\alpha}, \tag{3.47}$$

where Ω_0 is a constant and $\alpha = \lambda_V/\mu_V$. Because the parameter λ_V does not in general vanish in a convective zone, it prevents rigid rotation from being a solution of Eq. (3.45). Accordingly, conditions (3.44) and (3.47) cannot be satisfied simultaneously in a convective region, so that a pure rotation cannot be a solution of the problem. This result confirms Biermann's (1958) original finding that large-scale meridional currents are always present in a region of efficient convection.

This conclusion is very similar to the result obtained in Section 3.3.1. However, *in contrast to the case of a radiative zone, it is the necessity to conserve linear momentum, rather than energy, that drives the meridional flow in a convective zone.* A detailed discussion of these large-scale currents will be made in Section 5.2.

3.4 The dynamical instabilities

Consider an axially symmetric star and assume that it rotates with some prescribed angular velocity $\Omega = \Omega(\varpi, z)$. By making use of Eqs. (3.16) and (3.17), we can rewrite these conditions of mechanical equilibrium in the compact form

$$\text{grad}\, V + \frac{1}{\rho}\,\text{grad}\, p - \frac{j^2}{\varpi^3}\,\mathbf{1}_\varpi = \mathbf{0}, \tag{3.48}$$

where $j = \Omega\varpi^2$ is the angular momentum per unit mass. Then, under what conditions is this configuration stable with respect to small isentropic disturbances? Although no definitive answer can be given at the present time, some interesting results can be obtained for axially symmetric motions (i.e., motions for which the specific angular momentum of each fluid particle is preserved along its path). Departures from axial symmetry will be discussed briefly in Section 3.4.3.

Two types of description can be used to analyze the oscillations of a star about a known state of equilibrium: Either we specify the *Eulerian change* noted by an external observer who, at every instant t, views a given volume of fluid at a fixed location in space, or we describe the *Lagrangian change* within a given mass element, which is followed along its path in the course of time. Let $Q(\mathbf{r}, t)$ and $Q_0(\mathbf{r}, t)$ be the values of any physical quantity in the perturbed and unperturbed flows, respectively. Consider also the Lagrangian displacement $\boldsymbol{\xi}(\mathbf{r}, t)$, which describes any departure from the state of equilibrium. Given these definitions, one finds that

$$\delta Q = Q(\mathbf{r}, t) - Q_0(\mathbf{r}, t) \tag{3.49}$$

is the Eulerian variation of the quantity Q, whereas

$$\Delta Q = Q[\mathbf{r} + \boldsymbol{\xi}(\mathbf{r}, t),\, t] - Q_0(\mathbf{r}, t) \tag{3.50}$$

is its Lagrangian variation. In the linear approximation, we can thus write

$$\Delta Q = \delta Q + \boldsymbol{\xi} \cdot \text{grad}\, Q. \tag{3.51}$$

One also has

$$\delta \int_{\mathcal{V}} Q\, dm = \int_{\mathcal{V}} \Delta Q\, dm, \tag{3.52}$$

where $dm = \rho dv$ is the mass element and $\mathcal{V}$ is the total volume. Very much for the same reason, we can also write

$$\frac{d}{dt}\int_{\mathcal{V}} Q\, dm = \int_{\mathcal{V}} \frac{DQ}{Dt}\, dm. \tag{3.53}$$

3.4.1 *An energy principle*

As was originally shown by Fjørtoft (1946), one can derive the appropriate stability criterion on the basis of the energy equation. His analysis relies upon the fact that the total energy – which is extremal for configurations satisfying Eq. (3.48) – is a minimum for dynamically stable equilibria and fails to be a minimum for dynamically unstable ones. This can be seen as follows. Take the scalar product of Eq. (3.1) with the velocity $\mathbf{v}$, and integrate over the volume $\mathcal{V}$. After performing an integration by parts and using the fact that the pressure vanishes at the free surface, we obtain

$$\frac{d}{dt}\int_{\mathcal{V}}\frac{1}{2}|\mathbf{v}^2|\,dm = \int_{\mathcal{V}}\frac{p}{\rho}\,\mathrm{div}\,\mathbf{v}\,dm - \int_{\mathcal{V}}\mathbf{v}\cdot\mathrm{grad}\,V\,dm. \tag{3.54}$$

By making use of Eqs. (3.9) and (3.53), one finds that

$$\int_{\mathcal{V}}\frac{p}{\rho}\,\mathrm{div}\,\mathbf{v}\,dm = -\int_{\mathcal{V}}\frac{DU}{Dt}\,dm = -\frac{d}{dt}\int_{\mathcal{V}}U\,dm = -\frac{dU_T}{dt}, \tag{3.55}$$

where U_T is the total thermal energy. Similarly, we can write

$$\int_{\mathcal{V}}\mathbf{v}\cdot\mathrm{grad}\,V\,dm = -\frac{dW}{dt}, \tag{3.56}$$

where W is the gravitational potential energy. Using Eqs. (3.55) and (3.56), we can integrate Eq. (3.54) to obtain

$$\int_{\mathcal{V}}\frac{1}{2}|\mathbf{v}|^2\,dm + U_T + W = \mathit{constant}. \tag{3.57}$$

Suppose now that an axially symmetric motion is superimposed upon the state of equilibrium. Equation (3.57) then becomes

$$\int_{\mathcal{V}}\frac{1}{2}|\mathbf{v}_p|^2\,dm + E = \mathit{constant}, \tag{3.58}$$

where $\mathbf{v}_p$ is the velocity field of the axially symmetric pulsation, and E is the total energy

$$E = \frac{1}{2}\int_{\mathcal{V}}\frac{j^2}{\varpi^2}\,dm + \int_{\mathcal{V}}U\,dm + \frac{1}{2}\int_{\mathcal{V}}V\,dm. \tag{3.59}$$

Obviously, any increase of the kinetic energy of the axially symmetric motion must be supplied from the total energy E. Accordingly, this energy must be a minimum for stable, isentropic motions.

Let us now compute the first and second variations of the total energy E by keeping constant the total mass M and the total angular momentum J. Dynamically stable equilibria correspond to the conditions

$$\delta E = 0 \quad \text{and} \quad \delta^2 E > 0. \tag{3.60}$$

In the case of axially symmetric motions, the specific angular momentum is preserved for each fluid particle. We thus have $Dj/Dt = 0$, so that we can write

$$\delta\frac{1}{2}\int_{\mathcal{V}}\frac{j^2}{\varpi^2}\,dm = -\int_{\mathcal{V}}\frac{j^2}{\varpi^3}\,\xi_\varpi\,dm. \tag{3.61}$$

Similarly, for isentropic motions one has $DS/Dt = 0$. Equation (3.9) thus implies that

$$\Delta U = -\frac{p}{\rho}\,\mathrm{div}\,\boldsymbol{\xi}. \tag{3.62}$$

By making use of Eq. (3.52), we obtain

$$\delta \int_{\mathcal{V}} U\, dm = \int_{\mathcal{V}} \Delta U\, dm = -\int_{\mathcal{V}} \frac{p}{\rho} \operatorname{div} \boldsymbol{\xi}\, dm = \int_{\mathcal{V}} \frac{1}{\rho} \boldsymbol{\xi} \cdot \operatorname{grad} p\, dm. \tag{3.63}$$

Finally, the first variation of the potential energy is

$$\delta W = \int_{\mathcal{V}} \boldsymbol{\xi} \cdot \operatorname{grad} V\, dm. \tag{3.64}$$

From Eqs. (3.61), (3.63), and (3.64), it thus follows that

$$\delta E = \int_{\mathcal{V}} \boldsymbol{\xi} \cdot \left(\operatorname{grad} V + \frac{1}{\rho} \operatorname{grad} p - \frac{j^2}{\varpi^3} \mathbf{1}_{\varpi} \right) dm. \tag{3.65}$$

By making use of Eq. (3.48), one readily sees that the condition $\delta E = 0$ defines a state of mechanical equilibrium. Similarly, it is a simple matter to prove that

$$\delta^2 E = \int_{\mathcal{V}} \boldsymbol{\xi} \cdot \left[\Delta \left(\operatorname{grad} V + \frac{1}{\rho} \operatorname{grad} p \right) + \frac{3 j^2}{\varpi^4} \xi_{\varpi} \mathbf{1}_{\varpi} \right] dm. \tag{3.66}$$

By virtue of Eqs. (3.48) and (3.51), one obtains

$$\Delta \left(\operatorname{grad} V + \frac{1}{\rho} \operatorname{grad} p \right)$$
$$= \operatorname{grad} \delta V + \frac{1}{\rho} \operatorname{grad} \delta p - \frac{\delta \rho}{\rho^2} \operatorname{grad} p + \boldsymbol{\xi} \cdot \operatorname{grad} (\Omega^2 \varpi) \mathbf{1}_{\varpi}. \tag{3.67}$$

Hence, we can rewrite Eq. (3.66) in the compact form

$$\delta^2 E = -\int_{\mathcal{V}} \boldsymbol{\xi} \cdot \mathbf{L} \boldsymbol{\xi}\, dm, \tag{3.68}$$

where

$$\mathbf{L} \boldsymbol{\xi} = -\operatorname{grad} \delta V - \frac{1}{\rho} \operatorname{grad} \delta p + \frac{\delta \rho}{\rho^2} \operatorname{grad} p - \frac{1}{\varpi^3} \boldsymbol{\xi} \cdot \operatorname{grad} (\Omega^2 \varpi^4) \mathbf{1}_{\varpi}. \tag{3.69}$$

Equations (3.3) and (3.10) imply that

$$\delta \rho = -\rho \operatorname{div} \boldsymbol{\xi} - \boldsymbol{\xi} \cdot \operatorname{grad} \rho \tag{3.70}$$

and

$$\delta p = -\Gamma_1 p \operatorname{div} \boldsymbol{\xi} - \boldsymbol{\xi} \cdot \operatorname{grad} p. \tag{3.71}$$

Similarly, by assuming that the density vanishes at the free surface, we have

$$\delta V = -G \int_{\mathcal{V}} \frac{\delta \rho(\mathbf{r}', t)}{|\mathbf{r} - \mathbf{r}'|} dv', \tag{3.72}$$

where $\delta \rho$ is given in terms of $\boldsymbol{\xi}$ by Eq. (3.70). Thus, if there exists a displacement $\boldsymbol{\xi}$ such that

$$\int_{\mathcal{V}} \boldsymbol{\xi} \cdot \mathbf{L} \boldsymbol{\xi}\, dm > 0, \tag{3.73}$$

we have $\delta^2 E < 0$, and the system is dynamically unstable because its total energy fails to be an absolute minimum.

For the sake of completeness, let us briefly consider the small oscillations about the state of mechanical equilibrium defined in Eq. (3.48). A rigorous discussion was made by Lebovitz (1970), who derived the following equation:

$$\frac{\partial^2 \boldsymbol{\xi}}{\partial t^2} = \mathbf{L}\boldsymbol{\xi}, \tag{3.74}$$

where $\mathbf{L}$ is the time-dependent operator defined in Eq. (3.69) and $\boldsymbol{\xi}$ is a two-dimensional vector with components ξ_ϖ and ξ_z. This operator is symmetric in the sense that

$$\int_{\mathcal{V}} \boldsymbol{\xi} \cdot \mathbf{L}\boldsymbol{\eta}\, dm = \int_{\mathcal{V}} \boldsymbol{\eta} \cdot \mathbf{L}\boldsymbol{\xi}\, dm, \tag{3.75}$$

where $\boldsymbol{\xi}$ and $\boldsymbol{\eta}$ are two arbitrary vectors. As was shown by Lebovitz, the symmetry of $\mathbf{L}$ implies that the configuration is unstable if, for any trial function $\boldsymbol{\xi}$, condition (3.73) is satisfied. The strength of this result is that it avoids any assumptions about the existence of normal modes or about their properties when they do exist. It can be put into a more familiar form if we consider the normal-mode solution

$$\boldsymbol{\xi} = \boldsymbol{\xi}_0(\mathbf{r})e^{i\sigma t}, \tag{3.76}$$

for which

$$-\sigma^2 \boldsymbol{\xi}_0 = \mathbf{L}\boldsymbol{\xi}_0 \tag{3.77}$$

and

$$\sigma^2 = -\frac{\int_{\mathcal{V}} \boldsymbol{\xi}_0 \cdot \mathbf{L}\boldsymbol{\xi}_0\, dm}{\int_{\mathcal{V}} \boldsymbol{\xi}_0 \cdot \boldsymbol{\xi}_0\, dm}. \tag{3.78}$$

By virtue of Eq. (3.75), this equation provides a variational basis for the determination of the allowed values of σ^2, with the smallest eigenvalue being the minimum of the expression on the right-hand side of Eq. (3.78). Accordingly, if condition (3.73) is satisfied for some vector $\boldsymbol{\xi} = \boldsymbol{\xi}_0$, the right-hand side of Eq. (3.78) is negative for such a choice. This implies a negative value for the least eigenvalue σ^2, so that the mechanical equilibrium is dynamically unstable. The equivalence of the two methods is therefore demonstrated.

To proceed any further, we must now insert Eqs. (3.69)–(3.72) into Eq. (3.73). Integrating by parts and rearranging the various terms in this equation, we eventually obtain

$$\int_{\mathcal{V}} \boldsymbol{\xi} \cdot \mathbf{L}\boldsymbol{\xi}\, dm = -\int_{\mathcal{V}} \boldsymbol{\xi} \cdot \mathbf{M}\boldsymbol{\xi}\, dm - \int_{\mathcal{V}} \left[\frac{(\delta p)^2}{\Gamma_1 p\rho} + \boldsymbol{\xi} \cdot \operatorname{grad} \delta V\right] dm, \tag{3.79}$$

where the tensor $\mathbf{M}$ has the form

$$\mathbf{M} = \frac{1}{\rho} \operatorname{grad} p \left(\frac{1}{\rho} \operatorname{grad} \rho - \frac{1}{\Gamma_1 p} \operatorname{grad} p\right) + \frac{1}{\varpi^3} \operatorname{grad} \varpi \operatorname{grad} (\Omega^2 \varpi^4). \tag{3.80}$$

3.4.2 *The Solberg–Høiland conditions*

In order to discuss the implications of this stability criterion, we shall assume that the Eulerian changes δp and δV can be neglected in the configuration. The hypothesis $\delta p \equiv 0$ is valid whenever the characteristic time of the disturbances exceeds the travel time of a sound wave in the perturbed domain. (This amounts to filtering out the p-modes of oscillation.) By virtue of Eq. (3.72), the hypothesis $\delta V \equiv 0$ implies that we restrict our analysis to disturbances having many nodes, that is, perturbations with wavelengths much shorter than the star's radius.

Given these assumptions, it is immediately apparent from Eq. (3.79) that the stability of an equilibrium with respect to axially symmetric motions depends on the character of the quadratic form $\boldsymbol{\xi} \cdot \mathbf{M}\boldsymbol{\xi}$. By virtue of Eq. (3.73), a self-gravitating system is dynamically stable with respect to short-wavelength perturbations if and only if $\boldsymbol{\xi} \cdot \mathbf{M}\boldsymbol{\xi}$ is positive definite. Indeed, if this condition is not satisfied, it is always possible to find a Lagrangian displacement $\boldsymbol{\xi}$ for which the second variation,

$$\delta^2 E = \int_{\mathcal{V}} \boldsymbol{\xi} \cdot \mathbf{M}\boldsymbol{\xi}\, dm, \tag{3.81}$$

is negative. If so, then, the total energy E fails to be an absolute minimum, thus indicating an unstable state of equilibrium.

For further use, let us define the following vectors:

$$\boldsymbol{\Phi} = \frac{1}{\varpi^3} \operatorname{grad}(\Omega^2 \varpi^4) = 2\frac{\Omega}{\varpi} \operatorname{grad} j, \tag{3.82}$$

$$\boldsymbol{\Phi}_0 = \operatorname{grad} \varpi = \mathbf{1}_\varpi, \tag{3.83}$$

$$\boldsymbol{\Psi} = \frac{1}{\Gamma_1 p} \operatorname{grad} p - \frac{1}{\rho} \operatorname{grad} \rho = \frac{1}{c_p} \frac{\gamma - 1}{\Gamma_3 - 1} \operatorname{grad} S, \tag{3.84}$$

$$\boldsymbol{\Psi}_0 = -\frac{1}{\rho} \operatorname{grad} p = -\mathbf{g}, \tag{3.85}$$

where j is the angular momentum per unit mass and S is the entropy per unit mass. We can thus rewrite Eq. (3.80) in the compact form

$$\mathbf{M} = \boldsymbol{\Phi}_0 \boldsymbol{\Phi} + \boldsymbol{\Psi}_0 \boldsymbol{\Psi}. \tag{3.86}$$

Note that the vectors $\boldsymbol{\Phi}_0$ and $\boldsymbol{\Psi}_0$ are always directed along the outer normal to the surfaces $\varpi =$ *constant* and $p =$ *constant*, respectively. Similarly, the vectors $\boldsymbol{\Phi}$ and $\boldsymbol{\Psi}$ are orthogonal to the surfaces $j =$ *constant* and $S =$ *constant*, respectively, although we do not know a priori whether they are directed along the inner or outer normal.

Since the vectors (3.82)–(3.85) and the tensor (3.86) play an essential role in the subsequent discussion, we shall briefly summarize their main properties. First, taking the curl of Eq. (3.48), we obtain

$$\operatorname{grad} \frac{1}{\rho} \times \operatorname{grad} p = \frac{1}{\varpi^3} \operatorname{grad}(\Omega^2 \varpi^4) \times \mathbf{1}_\varpi. \tag{3.87}$$

This is the thermal wind relation, which relates the z dependence of the angular velocity Ω to the baroclinicity of the system (see Eq. [2.83]). By making use of Eqs. (3.82)–(3.85), we thus have

$$\boldsymbol{\Phi} \times \boldsymbol{\Phi}_0 + \boldsymbol{\Psi} \times \boldsymbol{\Psi}_0 = \mathbf{0}, \tag{3.88}$$

so that the rotation $\boldsymbol{\Phi} \to \boldsymbol{\Phi}_0$ is always opposite to the rotation $\boldsymbol{\Psi} \to \boldsymbol{\Psi}_0$. Equation (3.87) is also the condition that makes the tensor $\mathbf{M}$ symmetric. Indeed, since the sets $(\boldsymbol{\Phi}_0, \boldsymbol{\Psi}_0)$ and $(\boldsymbol{\Phi}, \boldsymbol{\Psi})$ can be interchanged, it follows at once that

$$\boldsymbol{\Phi}\, \boldsymbol{\Phi}_0 + \boldsymbol{\Psi}\, \boldsymbol{\Psi}_0 = \boldsymbol{\Phi}_0\, \boldsymbol{\Phi} + \boldsymbol{\Psi}_0\, \boldsymbol{\Psi}, \tag{3.89}$$

and the curves $\boldsymbol{\xi} \cdot \mathbf{M}\boldsymbol{\xi} = constant$ represent a family of concentric conics. At each point of the configuration, one can thus find their two (orthogonal) principal axes (x, y) so that

$$\boldsymbol{\xi} \cdot \mathbf{M}\boldsymbol{\xi} = \alpha_x \xi_x^2 + \alpha_y \xi_y^2, \tag{3.90}$$

where ξ_x and ξ_y are the components of $\boldsymbol{\xi}$ along the principal axes. Because the trace and the determinant of $\mathbf{M}$ are invariant with respect to a rotation of the axes, we also have

$$\alpha_x + \alpha_y = \operatorname{trace} \mathbf{M} = \boldsymbol{\Phi} \cdot \boldsymbol{\Phi}_0 + \boldsymbol{\Psi} \cdot \boldsymbol{\Psi}_0 \tag{3.91}$$

and

$$\alpha_x \alpha_y = \det \mathbf{M} = (\boldsymbol{\Phi}_0 \times \boldsymbol{\Psi}_0) \cdot (\boldsymbol{\Phi} \times \boldsymbol{\Psi}). \tag{3.92}$$

From Eqs. (3.90)–(3.92) we observe that $\boldsymbol{\xi} \cdot \mathbf{M}\boldsymbol{\xi}$ is positive definite if and only if α_x and α_y are both positive. Hence, the conditions of stability are

$$\operatorname{trace} \mathbf{M} > 0 \qquad \text{and} \qquad \det \mathbf{M} > 0, \tag{3.93}$$

or, returning to the original variables,

$$\frac{1}{\varpi^3} \frac{\partial j^2}{\partial \varpi} + \frac{1}{c_p} \frac{\gamma - 1}{\Gamma_3 - 1} (-\mathbf{g}) \cdot \operatorname{grad} S > 0 \tag{3.94}$$

and simultaneously

$$-g_z \left(\frac{\partial j^2}{\partial \varpi} \frac{\partial S}{\partial z} - \frac{\partial j^2}{\partial z} \frac{\partial S}{\partial \varpi} \right) > 0. \tag{3.95}$$

Equations (3.94) and (3.95) are often known as the *Solberg–Høiland conditions* for dynamical stability.

Now, as was shown by Holmboe (1948), the equation governing small axisymmetric oscillations can be brought to the form

$$\frac{\partial^2 \boldsymbol{\xi}}{\partial t^2} = -\boldsymbol{\Phi}_0 (\boldsymbol{\Phi} \cdot \boldsymbol{\xi}) - \boldsymbol{\Psi}_0 (\boldsymbol{\Psi} \cdot \boldsymbol{\xi}). \tag{3.96}$$

This equation gives the meridional acceleration in the perturbed motion as a result of *two* forces. The first term on the right-hand side of Eq. (3.96) represents the *centrifugal buoyancy*. It is directed opposite to the unit vector $\mathbf{1}_\varpi$, and it has the magnitude $(\boldsymbol{\Phi} \cdot \boldsymbol{\xi})$. Since the vector $\boldsymbol{\Phi}$ is perpendicular to the surfaces $j = constant$, it follows at once that only the component of $\boldsymbol{\xi}$ perpendicular to these surfaces is active in the generation of centrifugal buoyancy. The second term represents the *gravitational buoyancy*. It is in the same direction as the effective gravity $\mathbf{g}$, and it has the magnitude $|\mathbf{g}|(\boldsymbol{\Psi} \cdot \boldsymbol{\Phi})$. Thus, only the component of $\boldsymbol{\xi}$ perpendicular to the surfaces $S = constant$ contributes to the gravitational buoyancy. The stability of the system depends on the direction of the resultant buoyancy with reference to all permissible displacements $\boldsymbol{\xi}$.

In the limit $j \equiv 0$, stability conditions (3.94) and (3.95) reduce to the single inequality

$$(-\mathbf{g}) \cdot \operatorname{grad} S \equiv N^2 > 0, \tag{3.97}$$

which is the condition for the temperature lapse rate to be subadiabatic throughout the configuration (see Eq. [2.136]). Not unexpectedly, the solution of Eq. (3.96) then reduces to stable buoyancy oscillations.

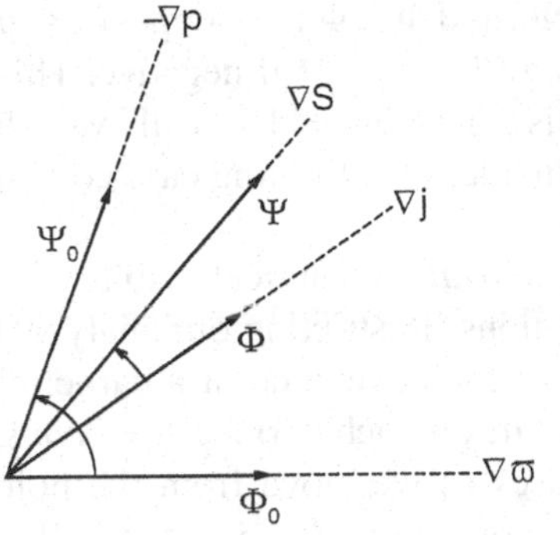

Fig. 3.1. A dynamically stable situation.

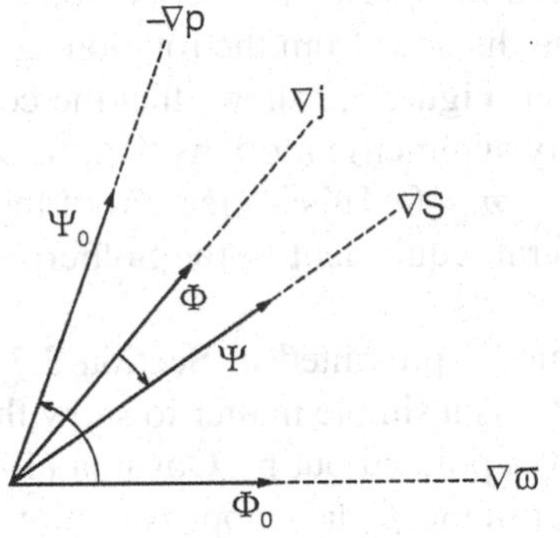

Fig. 3.2. A dynamically unstable situation.

In the limit $S \equiv$ *constant*, the configuration degenerates into a barotrope. In this case, the stability condition (3.94) becomes

$$\frac{dj}{d\varpi} > 0, \tag{3.98}$$

with the solution of Eq. (3.96) being stable inertial oscillations. Note that criterion (3.98) generalizes to homentropic fluids the well-known Rayleigh criterion for an incompressible fluid.*

Given these results, one would be tempted to conclude that, in the general case of a baroclinic star, stability conditions (3.94) and (3.95) are equivalent to conditions (3.97) and (3.98) simultaneously. This is not quite true, as will become apparent from the following discussion.

Since we are mainly interested in the radiative regions of a rotating barocline, let us restrict our discussion to the case for which trace $\mathbf{M} > 0$ (see Eq. [3.91]). Figures 3.1 and 3.2 depict, at any given point, two plausible orientations of the basic vectors. In Figure 3.1 the vector products $\mathbf{\Phi}_0 \times \mathbf{\Psi}_0$ and $\mathbf{\Phi} \times \mathbf{\Psi}$ both point along the same direction, so that the determinant of $\mathbf{M}$ is positive (see Eq. [3.92]). This implies stability. On the

* See, e.g., Chandrasekhar, S., *Hydrodynamic and Hydromagnetic Stability*, Section 66, Oxford: Clarendon Press, 1961 (New York: Dover Publications, 1981).

contrary, in Figure 3.2 the vector products $\mathbf{\Phi}_0 \times \mathbf{\Psi}_0$ and $\mathbf{\Phi} \times \mathbf{\Psi}$ have opposite signs. Their scalar product is therefore negative, and the determinant of **M** if negative. This implies instability. By virtue of Eqs. (3.82) and (3.84), this determinant identically vanishes when the surfaces $j = constant$ and $S = constant$ coincide. This limiting case corresponds to a neutral state of equilibrium.

In summary, in this section we have considered the dynamical stability of a baroclinic star with respect to axially symmetric motions. Restricting our analysis to short-wavelength disturbances, we have shown that the radiative zone of a baroclinic star is stable with respect to these motions if and only if, on each surface $S = constant$, the angular momentum per unit mass $\Omega\varpi^2$ increases as we move from the poles to the equator. In other words, if the specific angular momentum decreases radially outward on the surfaces $S = constant$, there exist unstable motions. In geophysics, this form of instability is called *symmetric instability*.

Not unexpectedly, in the radiative regions of a *barotropic* stellar model, this instability occurs whenever $\Omega\varpi^2$ decreases with increasing distance from the rotation axis. In the case of a stably stratified *baroclinic* star, however, Figure 3.2 shows that the configuration may become unstable with respect to axially symmetric motions (i.e., trace $\mathbf{M} > 0$ and det $\mathbf{M} < 0$) even when $N^2 > 0$ and $\partial(\Omega\varpi^2)/\partial\varpi > 0$. This is clear proof that stability conditions (3.97) and (3.98) are not, in general, equivalent to the Solberg–Høiland conditions (Eqs. [3.94] and [3.95]).

What is the exact link between the simple model presented in Section 2.7 and the more elaborate discussion made in this section? It is a simple matter to show that these two models are strictly equivalent. Indeed, as was pointed out by Ooyama (1966), the tensor **M** that corresponds to a rotating fluid layer in the f plane approximation is given by

$$\mathbf{M} = \begin{pmatrix} f^2 & f\,\dfrac{dU}{dz} \\ -\,\alpha g\,\dfrac{\partial\Theta}{\partial y} & \alpha g\,\dfrac{\partial\Theta}{\partial z} \end{pmatrix} \equiv \begin{pmatrix} f^2 & f\,\dfrac{dU}{dz} \\ f\,\dfrac{dU}{dz} & N^2 \end{pmatrix} \tag{3.99}$$

(see Eqs. [2.135] and [2.136]). Accordingly, we can write

$$\det\mathbf{M} = f^2\left(\frac{dU}{dz}\right)^2 (Ri - 1), \tag{3.100}$$

so that the condition $\det\mathbf{M} < 0$ implies $Ri < 1$, and conversely (see Eq. [2.137]). Condition (2.151) is therefore equivalent to the Solberg–Høiland conditions for dynamical instability.

To conclude, let us mention the work of Lorimer and Monaghan (1980), who have made a preliminary numerical investigation of the symmetric instability in differentially rotating polytropes. Following these authors, this instability is a violent one in the sense that, given an initially unstable j-distribution, a slowly rotating barotrope will at once generate meridional currents and nonaxisymmetric motions in the nonlinear regime, where the resulting flow becomes chaotic with a very slow trend to equilibrium. Further studies along these lines would be most welcome.

3.4.3 Nonaxisymmetric motions

In Section 2.7 we have considered a basic flow that has a shear in the vertical direction, that is, along the direction of the effective gravity (see Eq. [2.143]). Besides the symmetric instability, this very simple model exhibits two forms of dynamical instability with respect to nonaxisymmetric motions. One of them – the shear-flow instability – occurs when the Richardson number satisfies the condition

$$Ri < \frac{1}{4} \tag{3.101}$$

(see Eq. [2.137]). In this case, then, instability sets in when the vertical shear is so steep that the destabilizing effect of inertia overwhelms the stabilizing effect of buoyancy. Its maximum growth rates are associated with short-wavelength zonal disturbances. The other one – the baroclinic instability – occurs for almost all positive values of the Richardson number, and it is associated with zonal disturbances of *all* wavelengths (see Section 2.7.2). These nonaxisymmetric motions may become unstable because the isothermal surfaces and the isobaric surfaces do not coincide in a barocline. Hence, the potential energy of the basic flow can be converted into kinetic energy of baroclinic waves. This is quite different from the shear-flow instability, which is a form of barotropic instability, because it draws its energy mainly from the kinetic energy of the zonal motion.

Not unexpectedly, the case of a rotating star satisfying condition (3.48) is much more complex than the simple problem discussed in Section 2.7. For example, letting $\Omega = \Omega(\varpi)$ one can easily see that both vertical and latitudinal shears become possible. In general, for a star rotating with some assigned angular velocity $\Omega = \Omega(\varpi, z)$, the stability problem is complicated by the presence of a vertical shear as well as latitudinal variations of both the angular velocity and the temperature over an isobaric surface. In the simple barotropic case, the component of the rotational motion with latitudinal shear will become unstable to disturbances that transfer momentum down the meridional gradient in angular velocity, thus weakening the basic zonal flow. This is the reason why this instability is called a barotropic instability. It disappears only if the surfaces $\Omega = constant$ and $p = constant$ coincide. (Recall that the shear-flow instability is also a form of barotropic instability, drawing its energy from the component of the rotational motion with vertical shear.) In the general baroclinic case, thus, the basic zonal flow may develop all these instabilities with respect to nonaxisymmetric disturbances: (a) the *barotropic instability*, because there is a latitudinal gradient in angular velocity, (b) the *shear-flow instability*, because there is a vertical shear in the rotational motion, and (c) the *baroclinic instability*, because the isothermal surfaces are always inclined to the isobaric surfaces in a barocline.

Although the importance of shear-flow instability has long been recognized, the other two instabilities have received scant attention in the astronomical literature. Important progress has been made by Fujimoto (1987, 1988) and Hanawa (1987), who investigated the stability of a baroclinic star with respect to nonaxisymmetric, isentropic disturbances. Their calculations strongly suggest the prevalence of the barotropic and baroclinic instabilities in differentially rotating stars, *for all positive values of the Richardson number*, at least for short azimuthal wavelengths; the instabilities disappear only if the rotation is strictly uniform at every point. As we shall see in Section 3.6, this is an important result because shear-flow instability generates small-scale eddies wherever condition (3.101) is satisfied. Since there is no reason to expect this inequality to be satisfied at every point in a stellar radiative zone, it is evident that one can hardly justify the presence of turbulence in a rotating star on the basis of shear-flow instability alone.

3.5 The thermal instabilities

The stability criterion derived in Section 3.4.2 is based on the assumption that the displaced fluid particles move isentropically ($DS/Dt = 0$). Whereas viscous effects are negligible in a star, the effects of radiative conductivity may become important, at least for sufficiently small mass elements, because of the smoothing of temperature differences by radiative transfer. To be specific, consider a system that has a subadiabatic temperature gradient so that it is dynamically stable with respect to isentropic motions. If the fluid is unstable without the isentropic constraint due to a slightly adverse angular momentum distribution, the thermal conductivity will thus reduce the restoring force of thermal buoyancy. Hence, it will permit axially symmetric disturbances to grow, with their amplitudes being limited by the thermal conductivity that relaxes the isentropic constraint. As we shall see in this section, two types of thermally unstable motions can occur simultaneously in a baroclinic star.

Consider an axially symmetric star that rotates with the assigned angular velocity $\Omega = \Omega(\varpi, z)$. Assume that it satisfies the Solberg–Høiland conditions for dynamical stability with respect to axially symmetric disturbances. We shall consider a simple ideal gas with negligible radiation pressure. However, we shall make allowance for a gradient of chemical composition. By virtue of Eq. (3.8), we thus have $p \propto \rho T/\bar{\mu}$, where $\bar{\mu}$ is a function of position and time. Since we are chiefly interested in a dissipation mechanism (i.e., radiative conductivity), we may expect that the most unstable perturbations will be found to have wavelengths that are much smaller than the star's radius. It is therefore expedient to work with a simplified set of equations that approximate the exact equations in a small region of the star. The analysis is restricted to small axisymmetric motions, and we assume that their size is much smaller than any scale height of the equilibrium model. Then, the coefficients in the perturbation equations will be independent of ϖ, z, and t, so that the Eulerian changes may be expanded in plane waves of the form

$$\exp[nt + i(k_\varpi \varpi + k_z z)]. \tag{3.102}$$

Consistent with the above approximations, we may now take $\delta V \equiv 0$. We shall also make use of the Boussinesq approximation for compressibility effects; the pressure variations thus contribute little to the density variations.

By virtue of Eq. (3.102), the momentum equation reduces to

$$n^2 \boldsymbol{\xi} = -\frac{\delta\rho}{\rho}\boldsymbol{\Psi}_0 - (\boldsymbol{\xi}\cdot\boldsymbol{\Phi})\boldsymbol{\Phi}_0 - \frac{i}{\rho}\mathbf{k}\delta p, \tag{3.103}$$

where $\boldsymbol{\xi}$ is a two-dimensional vector with components ξ_ϖ and ξ_z (see Eq. [3.74]). Note that Eq. (3.103) already incorporates the conservation of angular momentum of each mass element along its path. (This property still holds because we can rightfully neglect viscous friction.) Similarly, by virtue of Eq. (3.3), our approximations lead to the condition $\mathbf{k}\cdot\boldsymbol{\xi} = 0$, thus implying that the wave vector $\mathbf{k}$ is transverse to the displacement $\boldsymbol{\xi}$. Letting next

$\boldsymbol{\xi} = \xi \mathbf{a}$ (where $\mathbf{a}$ is the unit vector along $\boldsymbol{\xi}$) and multiplying Eq. (3.103) by $\mathbf{a}$, we obtain

$$n^2 \xi = -\frac{\delta\rho}{\rho}(\mathbf{a} \cdot \boldsymbol{\Psi}_0) - \xi(\mathbf{a} \cdot \boldsymbol{\Phi})(\mathbf{a} \cdot \boldsymbol{\Phi}_0) \tag{3.104}$$

(see Eqs. [3.82], [3.83], and [3.85]).

Outside the central regions where thermonuclear reactions take place, Eq. (3.4) reduces to

$$\rho T \frac{DS}{Dt} = \mathrm{div}(\chi \, \mathrm{grad}\, T). \tag{3.105}$$

The small-perturbation counterpart of this equation can be brought to the form

$$n\rho(\delta S + \boldsymbol{\xi} \cdot \mathrm{grad}\, S) = -\chi k^2 \left[\frac{\delta(T/\bar{\mu})}{T/\bar{\mu}} + \frac{\delta\bar{\mu}}{\bar{\mu}} \right], \tag{3.106}$$

where $k^2 = k_\varpi^2 + k_z^2$. For a simple ideal gas, we have

$$S = c_V \log \frac{T}{\bar{\mu}\rho^{\gamma-1}} + \textit{constant}. \tag{3.107}$$

Accordingly, we can write

$$\mathrm{grad}\, S = c_V \left[\frac{1}{T}\, \mathrm{grad}\, T - (\gamma - 1)\frac{1}{\rho}\, \mathrm{grad}\, \rho - \frac{1}{\bar{\mu}}\, \mathrm{grad}\, \bar{\mu} \right] \tag{3.108}$$

and

$$\delta S = c_V \left[\frac{\delta(T/\bar{\mu})}{T/\bar{\mu}} - (\gamma - 1)\frac{\delta\rho}{\rho} \right] = -\, c_p \frac{\delta\rho}{\rho}, \tag{3.109}$$

where we made use of the fact that

$$\frac{\delta\rho}{\rho} + \frac{\delta(T/\bar{\mu})}{T/\bar{\mu}} = 0. \tag{3.110}$$

Since the rate of diffusion of chemical species is comparable to the (negligible) viscous diffusion rate, we shall also assume that

$$\frac{D\bar{\mu}}{Dt} = \frac{\partial\bar{\mu}}{\partial t} + \mathbf{v} \cdot \mathrm{grad}\, \bar{\mu} = 0. \tag{3.111}$$

It follows that

$$\frac{\delta\bar{\mu}}{\bar{\mu}} + \boldsymbol{\xi} \cdot \frac{\mathrm{grad}\, \bar{\mu}}{\bar{\mu}} = 0. \tag{3.112}$$

By making use of Eqs. (3.108)–(3.112), we can thus rewrite Eq. (3.106) in the form

$$\left(1 + \frac{\epsilon}{n}\right)\frac{\delta\rho}{\rho} = \xi\,(\mathbf{a} \cdot \boldsymbol{\Psi}) - \frac{\epsilon}{n}\,\xi \left(\mathbf{a} \cdot \frac{\mathrm{grad}\, \bar{\mu}}{\bar{\mu}}\right), \tag{3.113}$$

where

$$\epsilon = \chi k^2 / \rho c_p \tag{3.114}$$

and $\boldsymbol{\Psi} = \mathrm{grad}\, S/c_p$, with $\mathrm{grad}\, S$ being defined in Eq. (3.108).

It is now a simple matter to eliminate $\delta\rho/\rho$ between Eqs. (3.104) and (3.113) to obtain

$$n^3 + \epsilon n^2 + An + \epsilon B = 0, \tag{3.115}$$

where

$$A = (\mathbf{a} \cdot \boldsymbol{\Phi})(\mathbf{a} \cdot \boldsymbol{\Phi}_0) + (\mathbf{a} \cdot \boldsymbol{\Psi})(\mathbf{a} \cdot \boldsymbol{\Psi}_0) \tag{3.116}$$

and

$$B = (\mathbf{a} \cdot \boldsymbol{\Phi})(\mathbf{a} \cdot \boldsymbol{\Phi}_0) - \left(\mathbf{a} \cdot \frac{\operatorname{grad} \bar{\mu}}{\bar{\mu}}\right)(\mathbf{a} \cdot \boldsymbol{\Psi}_0). \tag{3.117}$$

In the limiting case $\epsilon = 0$, Eq. (3.115) provides the requisite dispersion relation for discussing dynamical stability (see Section 3.4.2). When $\epsilon = 0$ and $A > 0$, its three roots are $n = \pm i\sqrt{A}$, which describe stable oscillations, and the trivial root $n = 0$. When $\epsilon > 0$, the roots can be written in the forms

$$n = \pm i\sigma + a \qquad \text{and} \qquad n = b, \tag{3.118}$$

where σ, a, and b are real numbers (see Eqs. [3.102]). According to the Routh–Hurwitz criterion,* a and b are negative if and only if

$$B > 0 \qquad \text{and} \qquad A - B > 0. \tag{3.119}$$

These two inequalities can be rewritten in the form

$$(\mathbf{a} \cdot \boldsymbol{\Phi})(\mathbf{a} \cdot \boldsymbol{\Phi}_0) - \left(\mathbf{a} \cdot \frac{\operatorname{grad} \bar{\mu}}{\bar{\mu}}\right)(\mathbf{a} \cdot \boldsymbol{\Psi}_0) > 0 \tag{3.120}$$

and

$$(\mathbf{a} \cdot \boldsymbol{\Psi})(\mathbf{a} \cdot \boldsymbol{\Psi}_0) + \left(\mathbf{a} \cdot \frac{\operatorname{grad} \bar{\mu}}{\bar{\mu}}\right)(\mathbf{a} \cdot \boldsymbol{\Psi}_0) > 0. \tag{3.121}$$

These are the conditions for thermal stability with respect to axisymmetric motions, when both radiative conductivity and a gradient of chemical composition are taken into account.

Consider first the chemically homogeneous part of a stellar radiative zone. By virtue of Eq. (3.120), *thermal instability* occurs whenever a vector **a** can be found that will make $(\mathbf{a} \cdot \boldsymbol{\Phi})(\mathbf{a} \cdot \boldsymbol{\Phi}_0)$ negative (i.e., $b > 0$ in Eq. [3.118]). Figure 3.3 illustrates the case of a dynamically stable barocline (as illustrated in Figure 3.1). It is a simple matter to see that all vectors **a** that lie in the cross-hatched region make the body thermally unstable at that point. Obviously, the only way to prevent this instability in a star is to remove the cross-hatched region at every point. This can be done only if the vector $\boldsymbol{\Phi}$ points in the ϖ direction, that is, if

$$\frac{\partial \Omega}{\partial z} = 0 \qquad \text{and} \qquad \frac{d}{d\varpi}\left(\Omega \varpi^2\right) > 0 \tag{3.122}$$

at every point of the radiative interior. This result was originally obtained by Goldreich and Schubert (1967) and, independently, by Fricke (1968).

* See, e.g., *Handbook of Applied Mathematics* (Pearson, C. E., ed.), p. 929, New-York: Van Nostrand, 1974.

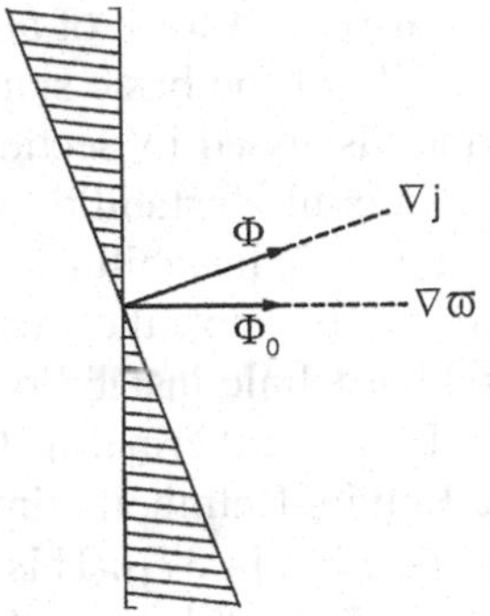

Fig. 3.3. A thermally unstable situation.

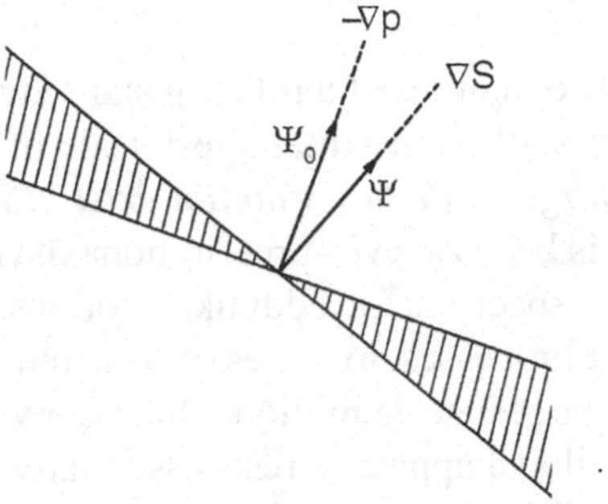

Fig. 3.4. A thermally overstable situation.

Following Shibahashi (1980), let us consider the implications of condition (3.121) in a chemically homogeneous region. In that case, if one can find a vector **a** that will make $(\mathbf{a} \cdot \mathbf{\Psi})(\mathbf{a} \cdot \mathbf{\Psi}_0)$ negative, *thermal overstability* occurs in the system (i.e., $a > 0$ in Eq. [3.118]). Figure 3.4 clearly shows that all vectors **a** lying in the cross-hatched region generate overstable motions at that point. This oscillatory instability can be removed only if the vectors $\mathbf{\Psi}$ and $\mathbf{\Psi}_0$ point to the same direction. By virtue of Eq. (3.88), this requirement also implies that the vector $\mathbf{\Phi}$ points in the ϖ direction. Again, this is true only if condition (3.122) is satisfied at every point of the radiative interior.

Now, is it possible to maintain the chemically *in*homogeneous part of a stellar radiative zone in static equilibrium with the steady rotation law $\Omega = \Omega(\varpi, z)$? Condition (3.120) shows that a stable gradient of chemical composition (i.e., grad $\bar{\mu} < 0$) often has a stabilizing influence on all unstable motions in the wedge between the surfaces $\varpi =$ *constant* and $j =$ *constant*. Accordingly, a suitable stratification of mean molecular weight might well prevent the *Goldreich–Schubert–Fricke instability* from occurring in a baroclinic star. However, by making use of Eqs. (3.108) and (3.121), one also sees that *Shibahashi's oscillatory instability* is probably little affected by a stable $\bar{\mu}$-gradient. This is a mere consequence of the fact that the overstable motions are located in the wedge between the surfaces $p =$ *constant* and $S =$ *constant*, which differ little from the surfaces $\bar{\mu} =$ *constant*. Obviously, further discussion of the effects of a $\bar{\mu}$-gradient in a baroclinic star necessarily requires the use of a particular model for the radiative interior.

To conclude, let us note that these thermal instabilities also are a form of *baroclinic instability*, since both of them are driven by the baroclinicity of the basic state. However, they differ from the baroclinic instability of the kind discussed in Sections 2.7.2 and 3.4.3 in two obvious ways. First, unlike the usual baroclinic instability, which is associated with nonaxisymmetric motions, they are axisymmetric instabilities. Second, because they depend upon the relaxation of the isentropic constraint, their time scale is certainly much longer than the time scale for the usual baroclinic instability. At this writing, the time scale for angular momentum transport by these thermal instabilities remains controversial, ranging in the literature from the Kelvin–Helmholtz time to the Eddington–Sweet time of large-scale meridional currents (see Eq. [4.37]). This is probably of no great consequence, however, because the dynamical instabilities with respect to nonaxisymmetric disturbances will generally dominate in a rotating star.

3.6 The eddy–mean flow interaction

The first step toward understanding the dynamics of a rotating star requires that we simplify the basic equations so that they describe only the largest scale of motion. As was pointed out in Section 2.4, however, *large-scale flows do not exist in isolation in a huge natural system, such as a star*. This is because ever-present nonaxisymmetric instabilities in a rotating star generate a wide spectrum of eddylike motions.* These small-scale disturbances give rise, by nonlinear processes, to fluxes of heat and momentum and, hence, influence the dynamics of the largest scale motions. In geophysics, this is called the *eddy–mean flow interaction*. This global approach rests essentially on a dynamical linkage between the ever-present eddylike motions (which we call "anisotropic turbulence" because effective gravity and rotation define two preferential directions in a star) and the mean flow (that is, the overall rotation and concomitant motions in meridian planes passing through the rotation axis). As usual, the role of these eddylike motions is simply parameterized in frictional form through the use of eddy viscosities and related coefficients. Not unexpectedly, these coefficients attain much larger values in a convectively unstable region than in a stellar radiative interior.

Neglecting molecular viscosity and omitting the overbars, we can thus rewrite Eq. (2.58) in the form

$$\frac{D\mathbf{v}}{Dt} = -\operatorname{grad} V - \frac{1}{\rho}\operatorname{grad} p + \frac{1}{\rho}\mathbf{F}(\mathbf{v}), \tag{3.123}$$

where $\mathbf{F}$ is the turbulent viscous force per unit volume, which can be written as the vectorial divergence of Reynolds stresses (see Eq. [2.59]). In spherical polar coordinates (r, θ, φ), the mean velocity $\mathbf{v}$ is

$$\mathbf{v} = u_r\,\mathbf{1}_r + u_\theta\,\mathbf{1}_\theta + \Omega r \sin\theta\,\mathbf{1}_\varphi, \tag{3.124}$$

where u_r and u_θ are the components of the two-dimensional meridional velocity $\mathbf{u}$.

* As was noted by Balbus and Hawley (1998) and others, small-scale magneto-rotational instabilities play an important role in generating turbulence in accretion disks. Under very specific circumstances, similar instabilities might be relevant to the study of turbulent motions in stellar radiative zones.

For axisymmetric motions, the poloidal part of Eq. (3.123) has the components

$$\rho\left(\frac{\partial u_r}{\partial t}+u_r\,\frac{\partial u_r}{\partial r}+\frac{u_\theta}{r}\,\frac{\partial u_r}{\partial \theta}-\frac{u_\theta^2}{r}\right)=-\,\rho\,\frac{\partial V}{\partial r}-\frac{\partial p}{\partial r}+\rho\Omega^2 r\sin^2\theta$$
$$+\,\frac{1}{r\sin\theta}\left[\frac{\sin\theta}{r}\,\frac{\partial}{\partial r}(r^2\sigma_{rr})+\frac{\partial}{\partial\theta}\,(\sigma_{r\theta}\sin\theta)\right]-\frac{1}{r}\,(\sigma_{\theta\theta}+\sigma_{\varphi\varphi}) \qquad (3.125)$$

and

$$\rho\left(\frac{\partial u_\theta}{\partial t}+u_r\,\frac{\partial u_\theta}{\partial r}+\frac{u_\theta}{r}\,\frac{\partial u_\theta}{\partial \theta}+\frac{u_r u_\theta}{r}\right)=-\,\frac{\rho}{r}\,\frac{\partial V}{\partial \theta}-\frac{1}{r}\,\frac{\partial p}{\partial \theta}+\rho\Omega^2 r\sin\theta\cos\theta$$
$$+\,\frac{1}{r\sin\theta}\left[\frac{\sin\theta}{r}\,\frac{\partial}{\partial r}(r^2\sigma_{r\theta})+\frac{\partial}{\partial\theta}(\sigma_{\theta\theta}\sin\theta)\right]+\frac{1}{r}\,(\sigma_{r\theta}-\sigma_{\varphi\varphi}\cot\theta). \qquad (3.126)$$

They depend on the Reynolds stresses σ_{rr}, $\sigma_{\theta\theta}$, $\sigma_{\varphi\varphi}$, and $\sigma_{r\theta}$. These quantities can be simply expressed as

$$\sigma_{rr}=2\mu_V\,\frac{\partial u_r}{\partial r}, \qquad (3.127)$$

$$\sigma_{\theta\theta}=2\mu_H\left(\frac{u_r}{r}+\frac{1}{r}\,\frac{\partial u_\theta}{\partial\theta}\right), \qquad (3.128)$$

$$\sigma_{\varphi\varphi}=2\mu_H\left(\frac{u_r}{r}+\frac{u_\theta\cot\theta}{r}\right), \qquad (3.129)$$

$$\sigma_{r\theta}=\sigma_{\theta r}=\mu_H\,\frac{1}{r}\,\frac{\partial u_r}{\partial\theta}+\mu_V\left(\frac{\partial u_\theta}{\partial r}-\frac{u_\theta}{r}\right), \qquad (3.130)$$

where μ_V and μ_H are the vertical and horizontal coefficients of eddy viscosity. Equations (3.125) and (3.126) thus depend on two parameters. Of course, this can only be a very crude model, but it does make allowance for a difference in momentum transfer between the vertical (i.e., along the effective gravity) and horizontal directions.

For axisymmetric motions, the toroidal part of Eq. (3.123) depends on the Reynolds stresses $\sigma_{r\varphi}$ and $\sigma_{\theta\varphi}$. Following Rüdiger (1980) and others, we shall let

$$\sigma_{r\varphi}=\sigma_{\varphi r}=\mu_V r\,\frac{\partial\Omega}{\partial r}\,\sin\theta+\lambda_V\Omega\sin\theta \qquad (3.131)$$

and

$$\sigma_{\theta\varphi}=\sigma_{\varphi\theta}=\mu_H\,\frac{\partial\Omega}{\partial\theta}\,\sin\theta+\lambda_H\Omega\cos\theta. \qquad (3.132)$$

These relations depend on the eddy viscosities and two additional parameters, λ_V and λ_H, which represent the influence of global rotation on anisotropic turbulence. The free parameter λ_H identically vanishes whenever the eddylike motions have horizontal symmetry, being then isotropic in planes perpendicular to the effective gravity. To a good degree of approximation, the λ_H term can be neglected in a slowly rotating star. In that case, by making use of Eqs. (3.131) and (3.132), one can show that the φ component of Eq. (3.123) has the form

$$\rho \frac{\partial \Omega}{\partial t} + \rho u_r \left(\frac{\partial \Omega}{\partial r} + 2\frac{\Omega}{r} \right) + \rho \frac{u_\theta}{r} \left(\frac{\partial \Omega}{\partial \theta} + 2\Omega \cot\theta \right)$$
$$= \frac{1}{r^4} \frac{\partial}{\partial r} \left(\mu_V r^4 \frac{\partial \Omega}{\partial r} + \lambda_V r^3 \Omega \right) + \frac{1}{r^2} \frac{1}{\sin^3\theta} \left(\mu_H \sin^3\theta \frac{\partial \Omega}{\partial \theta} \right), \quad (3.133)$$

which depends on three independent parameters only. As we shall see in Section 5.2.1, however, the λ_H term makes a nonvanishing contribution to the toroidal viscous force acting in the solar convective envelope. In that sense, thus, the Sun is not a slowly rotating star.

Now, as was originally pointed out by Schatzman (1969), anisotropic turbulence generated by the nonaxisymmetric instabilities may contribute to the diffusion of chemical elements within a stellar radiative zone. More recently, Press (1981) suggested that internal waves generated by chaotic motions at the boundary of a convective zone might also lead to species mixing in stably stratified regions. As usual, lacking any better description of all these eddy and/or wave events, we shall lay emphasis on the mean properties, using gross parameterizations of the smallest scale motions. For axisymmetric motions, the turbulent transport of a chemical element with concentration c can be described by the following equation:

$$\frac{D}{Dt}(\rho c) = \frac{1}{r^2} \frac{\partial}{\partial r} \left(\rho D_V r^2 \frac{\partial c}{\partial r} \right) + \frac{1}{r^2 \sin\theta} \frac{\partial}{\partial \theta} \left(\rho D_H \sin\theta \frac{\partial c}{\partial \theta} \right), \quad (3.134)$$

where ρ is the density and D_V and D_H are the vertical and horizontal coefficients of eddy diffusivity. (D/Dt is the *total* derivative.) As was noted by Fujimoto (1988) and others, however, vertical mixing is probably much less efficient than horizontal mixing, especially in a strongly stratified system. Indeed, for element mixing, work has to be done against gravity, so that the vertical displacements may be easily inhibited by the buoyancy force. In contrast, the instabilities responsible for horizontal turbulence are the barotropic and baroclinic instabilities, which are caused by *latitudinal* variations of angular velocity and temperature along the isobaric surfaces (see Section 3.4.3). Recall that these instabilities are operative for all positive values of the Richardson number Ri whereas the usual shear-flow instability, which is associated with a *vertical* shear in the rotational motion, is operative only when condition (3.101) is satisfied.

Various measurements in the laboratory and in the Earth's atmosphere indicate that, under stable conditions, the eddy diffusivities of matter and momentum decrease with increasing stability. These studies also show that *the turbulent diffusion of matter is a much less effective process than the turbulent diffusion of momentum in a stably stratified system*. Specifically, it is found that the ratio of eddy diffusivity to eddy viscosity, $\rho D_V/\mu_V$, is of the order of a few tenths for $Ri < 1$, whereas for $Ri > 1$ this ratio steadily decreases to zero as $Ri \to \infty$ (e.g., Turner 1973). These results are quite interesting because they strongly suggest that the ratio $\rho D_V/\mu_V$ can also be assumed to be much smaller than one in a stellar radiative interior. This matter will be discussed further in Section 5.4.1.

In this section we have developed a theoretical framework that describes the largest scale of motion in a rotating star. In particular, whereas the poloidal part of the momentum equation depends on two independent parameters (i.e., μ_V and μ_H), it is found that its toroidal part depends on at least three independent parameters (i.e., μ_V, μ_H, and λ_V). As was seen in Section 3.3, the parameter λ_V is of paramount importance because it prevents solid-body rotation in a convective envelope. Note also that the equation governing turbulent diffusion of matter in an axially symmetric star depends on two additional parameters (i.e., D_V and D_H). Equations (3.127)–(3.132) specify the Reynolds stresses in such a way that Eqs. (3.125), (3.126), and (3.133) represent a closed set of equations for the large-scale flow. (Compare with Eqs. [2.60]–[2.65].) Unfortunately, because there is no a priori justification for this particular model, it must be borne in mind that *the eddy-viscosity coefficients cannot be calculated from first principles alone.* A similar remark can be made about the eddy-diffusivity coefficients, D_V and D_H, since the ad hoc nature of the underlying model precludes a deterministic calculation of their values in a rotating star.

As was noted in Section 2.4, measurements in the Earth's atmosphere and in the oceans show that the eddy viscosities greatly exceed their molecular counterparts (see Eqs. [2.66] and [2.67]). In the astrophysical literature, it is usually accepted that one can write, for example, $D_V = L_c V_c$, where L_c is some typical length and V_c is some typical speed of the turbulent motions. Unfortunately, although this expression is dimensionally correct, it is not possible at this writing to calculate unequivocally the quantities L_c and V_c from results obtained on the basis of a linear stability analysis. A linear theory by its nature can say nothing about the process by which unstable eddylike or wavelike motions achieve some finite amplitude in the full nonlinear regime. Accordingly, no matter what kind of instability is assumed to be responsible for the small-scale motions, *the magnitude of the eddy coefficients cannot quantitatively be given by a measure of the instability of the mean flow.* That is to say, regardless of the spatial form that is assigned to the eddy coefficients, their overall magnitude can be determined only by fitting the chosen empirical formulae to the observational data.

It is not known at this writing whether one can find a better way of closing the equations for the large-scale flow in a rotating star. In any case, perhaps the greatest value of these parameterized models is that they give at least a reasonable global picture of the large-scale dynamics of the flow. They also provide a new perspective from which more elaborate models can be viewed.

3.7 Bibliographical notes

Section 3.2.1. The restriction imposed upon the angular velocity in a barotrope was originally derived by Poincaré:

1. Poincaré, H., *Théorie des tourbillons*, pp. 176–178, Paris: Georges Carré, 1893.

An exhaustive discussion of barotropes and baroclines will be found in:

2. Wavre, R., *Figures planétaires et géodésie*, pp. 25–33, Paris: Gauthier-Villars, 1932.

Section 3.3.1. The reference to von Zeipel is to his paper:

3. von Zeipel, H., *Mon. Not. R. Astron. Soc.*, **84**, 665, 1924.

The generalization to differentially rotating barotropes was made in:

4. Rosseland, S., *Astrophys. J.*, **63**, 342, 1926.
5. Vogt, H., *Astron. Nachr.*, **255**, 109, 1935.

An independent derivation of Vogt's result will be found in:

6. Roxburgh, I. W., *Mon. Not. R. Astron. Soc.*, **132**, 201, 1966.

Section 3.3.2. The case of convective equilibrium was considered by:

7. Biermann, L., in *Electromagnetic Phenomena in Cosmic Physics* (Lehnert, B., ed.), I.A.U. Symposium No 6, p. 248, Cambridge: Cambridge University Press, 1958.
8. Kippenhahn, R., *Astrophys. J.*, **137**, 664, 1963.

Kippenhahn's discussion is based on the Lebedinski–Wasiutyński equation, which is similar to our equation (3.45) but in which λ_V is replaced by $2(\mu_V - \mu_H)$. It was therefore concluded that anisotropic eddy viscosity should prevent uniform rotation. As was shown in Reference 34, however, the correct equation should depend on three independent parameters: μ_V, μ_H, and λ_V, which is not identically equal to $2(\mu_V - \mu_H)$. Since λ_V does not in general vanish in a convective zone, his method of solution thus remains essentially unchanged.

Sections 3.4.1 and 3.4.2. The classical references on the subject are those of:

9. Solberg, H., *Procès-Verbaux Ass. Météor., U.G.G.I.*, 6ème Assemblée Générale (Edinburgh), *Mém. et Disc.*, **2**, 66, 1936.
10. Høiland, E., *Archiv Mat. Naturv.* (Oslo), **42**, No. 5, 1, 1939.
11. Høiland, E., *Avhandl. Norske Videnskaps-Akademi i Oslo*, I, *Mat.-Naturv. Klasse*, No. 11, 1, 1941.

In their general form, conditions (3.94) and (3.95) were originally obtained in Reference 11. The analysis in these sections is taken from:

12. Fjørtoft, R., *Geofysiske Publikasjoner (Oslo)*, **16**, No. 5, 1, 1946.
13. Holmboe, J., *J. Marine Research*, **7**, 163, 1948.
14. Ooyama, K., *J. Atmos. Sci.*, **23**, 43, 1966.
15. Lebovitz, N. R., *Astrophys. J.*, **160**, 701, 1970.

See especially Holmboe's paper. A preliminary discussion of finite-amplitude motions is given by:

16. Lorimer, G. S., and Monaghan, J. J., *Proc. Astron. Soc. Australia*, **4**, 45, 1980.

Section 3.4.3. The barotropic and baroclinic instabilities are discussed at length in Reference 7 of Chapter 2. Interesting studies of nonaxisymmetric motions in baroclinic stars will be found in:

17. Fujimoto, M. Y., *Astron. Astrophys.*, **176**, 53, 1987.
18. Hanawa, T., *Astron. Astrophys.*, **179**, 383, 1987.

See also:

19. Fujimoto, M. Y., *Astron. Astrophys.*, **198**, 163, 1988.

Other pertinent comments on the literature will be found in Reference 26 (p. 392n) of Chapter 5.

Section 3.5. Reference is made to the following papers:

20. Goldreich, P., and Schubert, G., *Astrophys. J.*, **150**, 571, 1967.
21. Fricke, K., *Zeit. Astrophys.*, **68**, 317, 1968.
22. Shibahashi, H., *Publ. Astron. Soc. Japan*, **68**, 341, 1980.

Shibahashi's oscillatory instability is sometimes called "*a*xisymmetric *b*aroc-linic *d*iffusive (ABCD) instability." (This is somewhat confusing, however, because the Goldreich–Schubert–Fricke instability is also an *a*xisymmetric *b*aroclinic *d*iffusive instability.) The role of a $\bar{\mu}$-gradient is further discussed in:

23. Knobloch, E., and Spruit, H. C., *Astron. Astrophys.*, **125**, 59, 1983.

Various evaluations of the time scale of the Goldreich–Schubert–Fricke instability will be found in:

24. Colgate, S. A., *Astrophys. J. Letters*, **153**, L81, 1968.
25. Kippenhahn, R., *Astron. Astrophys.*, **2**, 309, 1969.
26. James, R. A., and Kahn, F. D., *Astron. Astrophys.*, **5**, 232, 1970; *ibid.*, **12**, 332, 1971.
27. Kippenhahn, R., Ruschenplatt, G., and Thomas, H. C., *Astron. Astrophys.*, **91**, 181, 1980.
28. Knobloch, E., *Geophys. Astrophys. Fluid Dyn.*, **22**, 133, 1982.
29. Korycansky, D. G., *Astrophys. J.*, **381**, 515, 1991.

See also Reference 16 (pp. 341–343) of Chapter 4. A rigorous derivation of Eq. (3.115) will be found in:

30. Lifshitz, A., and Lebovitz, N. R., *Astrophys. J.*, **408**, 603, 1993.

Section 3.6. Application of anisotropic eddy viscosity to astronomical problems was originally made by:

31. Lebedinski, A. I., *Astron. Zh.*, **18**, No. 1, 10, 1941.
32. Wasiutyński, J., *Astrophys. Norvegica*, **4**, 1, 1946.

The λ_V-effect has its roots in the work of Biermann:

33. Biermann, L., *Zeit. Astrophys.*, **28**, 304, 1951.

A modern reference on these and related matters is:

34. Rüdiger, G., *Geophys. Astrophys. Fluid Dyn.*, **16**, 239, 1980; *ibid.*, **21**, 1, 1982; *ibid.*, **25**, 213, 1983.

See also Reference 7 of Chapter 5. The references to Schatzman and Press are to their papers:

35. Schatzman, E., *Astron. Astrophys.*, **3**, 331, 1969.
36. Press, W. H., *Astrophys. J.*, **245**, 286, 1981.

The following book is particularly worth noting:

37. Turner, J. S., *Buoyancy Effects in Fluids*, Cambridge: Cambridge University Press, 1973.

See also Reference 19. The inhibition of vertical mixing is further discussed in:

38. Vincent, A., Michaud, G., and Meneguzzi, M., *Phys. Fluids*, **8**, 1312, 1996.

For a lucid discussion of the eddy coefficients, see:

39. Canuto, V. M., and Battaglia, A., *Astron. Astrophys.*, **193**, 313, 1988.

Quite different *empirical* formulae for the eddy coefficients will be found in the current literature; see, for example:

40. Pinsonneault, M. H., Kawaler, S. D., and Demarque, P., *Astrophys. J. Suppl.*, **74**, 501, 1990.
41. Zahn, J. P., *Space Sci. Review*, **66**, 285, 1994.

Compare Figure 16 in Reference 40 (p. 548) with the ad hoc formulae suggested in Reference 41. Such a comparison is useful because it clearly indicates that the practical evaluation of an eddy coefficient is at least partly an art, not just a science.

Reference is also made to:

42. Balbus, S. A., and Hawley, J. F., *Rev. Modern Phys.*, **70**, 1, 1998.

Other papers dealing with magnetohydrodynamical effects in accretion disks and rotating stars may be traced to Reference 42. For the interested reader, a penetrating discussion of the weak-field shearing instability presented in Reference 42 (pp. 30–32) will be found in:

43. Acheson, D. J., and Hide, R., *Rep. Prog. Phys.*, **36**, 159, 1973.

See especially their Section 4.3 (pp. 182–185).

4

Meridional circulation

4.1 Introduction

In Section 3.3.1 we noted that the conditions of mechanical and radiative equilibrium are, in general, incompatible in a rotating barotrope. This paradox can be solved in two different ways: Either one makes allowance for a slight departure from barotropy and chooses the angular velocity $\Omega = \Omega(\varpi, z)$ so that strict radiative equilibrium prevails at every point or one makes allowance for large-scale motions in meridian planes passing through the rotation axis. The first alternative is mainly of academic interest because there is no reason to expect rotating stars to select zero-circulation configurations. Moreover, these baroclinic models are thermally unstable with respect to axisymmetric motions, as well as dynamically unstable with respect to nonaxisymmetric motions (see Sections 3.4 and 3.5). Hence, the slightest disturbance will generate three-dimensional motions and, as a result, a large-scale meridional circulation will commence. The second alternative was independently suggested by Vogt (1925) and Eddington (1925), who pointed out that the breakdown of strict radiative equilibrium in a barotrope tends to set up slight rises in temperature and pressure over some areas of any given level surface and slight falls over other areas. The ensuing pressure gradient between the poles and the equator thereby causes a flow of matter. In fact, it is the small departures from spherical symmetry in a rotating star that lead to unequal heating along the polar and equatorial radii, which in turn causes large-scale currents in meridian planes. Slow but inexorable, thermally driven currents also exist in a tidally distorted star, as well as in a magnetic star, since the tidal interaction with a companion and the Lorentz force both generate small departures from spherical symmetry in a star. Obviously, it is the causal relation between nonsphericity and meridional circulation that makes the stellar problem entirely different from those expounded in Sections 2.5 and 2.6. This fact strongly suggests that well-known results obtained in geophysics (such as geostrophy and Ekman layers) should not be applied indiscriminately to a stellar radiative zone. I shall comment further on these important matters in Section 4.8.

In Section 4.2.1 we will obtain the steady circulation pattern in the radiative envelope of a uniformly rotating, *frictionless* star. Following Sweet (1950), we shall thus calculate the meridional flow generated by the nonsphericity of a chemically homogeneous region in slow uniform rotation. Section 4.2.2 presents a critical reassessment of his solution, which becomes infinite both at the free surface and at the core–envelope interface, and which also fails to take into account the transport of specific angular momentum by the meridional flow. In Sections 4.3 and 4.4, by making use of the *eddy–mean flow interaction*, which takes place continuously in a stellar radiative envelope, we obtain a simple but adequate description of the mean state of motion in a rotating star. This solution, which is free of the objections that can be made about Sweet's frictionless solution, satisfies all the boundary conditions and all the basic equations. Thermally driven currents in cooling white dwarfs are considered next in Section 4.5. Section 4.6 is devoted to the circulatory currents in the radiative envelope of an early-type star, which is a detached component of a close binary, and whose surface is nonuniformly heated by the radiation of its companion. Meridional flows in magnetic stars are considered further in Section 4.7. We conclude the chapter with a general overview of the problem, pointing out the differences and similarities between the large-scale currents that are encountered in geophysics and astrophysics.

4.2 A frictionless solution

Consider a single, nonmagnetic star that has a fully convective core, in which hydrogen burning is taking place, and a chemically homogeneous radiative envelope. Assume also that the axially symmetric star is slowly rotating with the constant angular velocity Ω_0. We shall also neglect viscosity and the inertia of the circulation itself. Then, in an inertial frame of reference, the equations governing steady motions in the radiative envelope are

$$\frac{\partial p}{\partial r} = -\rho \frac{\partial V}{\partial r} + \rho\, \Omega_0^2 r \sin^2\theta, \tag{4.1}$$

$$\frac{\partial p}{\partial \theta} = -\rho \frac{\partial V}{\partial \theta} + \rho\, \Omega_0^2 r^2 \sin\theta \cos\theta, \tag{4.2}$$

$$\nabla^2 V = 4\pi G \rho, \tag{4.3}$$

$$\mathrm{div}(\rho \mathbf{u}) = 0, \tag{4.4}$$

$$\rho T \mathbf{u} \cdot \mathrm{grad}\, S = \mathrm{div}(\chi\, \mathrm{grad}\, T), \tag{4.5}$$

$$p = \frac{\mathcal{R}}{\bar{\mu}} \rho T, \tag{4.6}$$

where p is the pressure, ρ is the density, V is the inner gravitational potential, $\mathbf{u}$ is the two-dimensional circulation velocity, T is the temperature, and $\mathcal{R}$ is the perfect gas constant. For electron-scattering opacity, the coefficient of radiative conductivity has the form

$$\chi = \frac{4ac}{3} \frac{T^3}{\kappa \rho}, \tag{4.7}$$

where κ is a constant. For a simple ideal gas, we also have

$$S = c_V \log \frac{p}{\rho^{5/3}} + constant, \tag{4.8}$$

where c_V is the specific heat at constant volume.

4.2.1 *Sweet's meridional circulation*

Since it is the lack of spherical symmetry that causes the meridional flow, we shall first derive from Eqs. (4.1)–(4.3) an expression for the distortion of the level surfaces due to the slow but uniform rotation. Following Milne (1923) and Chandrasekhar (1933), we shall expand about hydrostatic equilibrium in powers of the nondimensional parameter

$$\epsilon = \frac{\Omega_0^2 R^3}{GM}, \tag{4.9}$$

where M is the total mass and R is the equatorial radius. (In a realistic main-sequence model, ϵ does not exceed the critical value $\epsilon_c \approx 0.4$, at which point equatorial breakup is likely to occur.) Letting $P_2(\mu)$ denote the Legendre polynomial of degree two, we thus write, in spherical polar coordinates $(r, \mu = \cos\theta, \varphi)$,

$$p(r, \mu) = p_0(r) + \epsilon[p_{1,0}(r) + p_{1,2}(r)P_2(\mu)] \tag{4.10}$$

and a similar truncated expansion for the density. The inner gravitational potential is

$$V(r, \mu) = V_0(r) - \frac{GM}{R} + \epsilon[V_{1,0}(r) + c_{1,0} + V_{1,2}(r)P_2(\mu)], \tag{4.11}$$

whereas the potential that is appropriate to the surrounding vacuum has the form

$$V_{\text{ext}}(r, \mu) = -\frac{GM}{r} + \epsilon\left[\frac{B_0}{r} + \frac{B_2}{r^3}P_2(\mu)\right], \tag{4.12}$$

where B_0 and B_2 are constants.

Now, by making use of Eqs. (4.1)–(4.3), one can easily show that the nonradial functions (i.e., $p_{1,2}$, $\rho_{1,2}$, and $V_{1,2}$) satisfy the following set of equations:

$$\rho' \, p_{1,2} - p' \, \rho_{1,2} = 0, \tag{4.13}$$

$$p_{1,2} = -\rho V_{1,2} - \frac{1}{3}\rho\,\omega_0^2 r^2, \tag{4.14}$$

$$\frac{d^2V_{1,2}}{dr^2} + \frac{2}{r}\frac{dV_{1,2}}{dr} - \frac{6}{r^2}V_{1,2} = 4\pi G\rho_{1,2}, \tag{4.15}$$

where $\omega_0^2 = GM/R^3$. A prime denotes a derivative with respect to r. Without confusion, we have omitted the subscript "0" from the functions p_0 and ρ_0 that define the (known) model corresponding to $\epsilon = 0$. The continuity of the gravitational field across the free surface, which is a slightly oblate surface, further implies that

$$V_{1,2}(R) = \frac{B_2}{R^3} \quad \text{and} \quad \left(\frac{dV_{1,2}}{dr}\right)_R = -3\frac{B_2}{R^4}. \tag{4.16}$$

We shall not write down the relations between the radial functions (i.e., $p_{1,0}$, $\rho_{1,0}$, and $V_{1,0}$) since, to first order in ϵ, they are not relevant to the circulation problem.

To solve Eqs. (4.13)–(4.15), we shall let

$$V_{1,2} = A_2\Phi_2(r) - \frac{1}{3}\omega_0^2 r^2, \tag{4.17}$$

where A_2 is a constant. Thence, it is a simple matter to show that the function Φ_2 satisfies the following equation:

$$\frac{d^2\Phi_2}{dr^2} + \frac{2}{r}\frac{d\Phi_2}{dr} - \frac{6}{r^2}\Phi_2 + 4\pi G\frac{\rho\rho'}{p'}\Phi_2 = 0, \tag{4.18}$$

with $\Phi_2(0) = \Phi_2'(0) = 0$. Boundary conditions (4.16) now become

$$A_2\Phi_2(R) - \frac{B_2}{R^3} = \frac{1}{3}\omega_0^2 R^2 \tag{4.19}$$

and

$$A_2\Phi_2'(R) + 3\,\frac{B_2}{R^4} = \frac{2}{3}\,\omega_0^2 R. \tag{4.20}$$

Solving for A_2 and B_2, one obtains

$$A_2 = \frac{1}{3}\,\omega_0^2\,\frac{5R^2}{3\Phi_2(R) + R\Phi_2'(R)} \tag{4.21}$$

and

$$B_2 = \frac{1}{3}\,\omega_0^2 R^5\,\frac{2\Phi_2(R) - R\Phi_2'(R)}{3\Phi_2(R) + R\Phi_2'(R)}, \tag{4.22}$$

thus ensuring that the inner potential (4.11) smoothly joins the outer potential (4.12).

Thus, by letting

$$h = \frac{5}{3}\,\frac{GM}{R}\,\frac{\Phi_2(r)}{3\Phi_2(R) + R\Phi_2'(R)}, \tag{4.23}$$

we have shown that

$$p_{1,2} = -\rho h \tag{4.24}$$

and

$$\rho_{1,2} = -\frac{\rho\rho'}{p'}\,h, \tag{4.25}$$

where the function Φ_2 can be obtained from Eq. (4.18).

Following Sweet (1950), we now turn to Eqs. (4.4)–(4.6). By making use of the equation of state, one readily sees that the temperature can be expanded as was done for the pressure and density. Hence, we can write

$$\frac{T_{1,2}}{T} = \frac{p_{1,2}}{p} - \frac{\rho_{1,2}}{\rho}, \tag{4.26}$$

where we have also omitted the subscript "0" from the temperature in the spherical model. Combining Eqs. (4.24)–(4.26), one finds that

$$T_{1,2} = -T\left(\frac{\rho}{p} - \frac{\rho'}{p'}\right) h. \tag{4.27}$$

If we now make use of Eqs. (4.4) and (4.5), it is a simple matter to show that, correct to first order in ϵ, the circulation velocity has the form

$$\mathbf{u} = \epsilon\, u(r)\, P_2(\mu)\, \mathbf{1}_r + \epsilon\, v(r)\,(1-\mu^2)\,\frac{dP_2(\mu)}{d\mu}\,\mathbf{1}_\mu, \tag{4.28}$$

where

$$v = \frac{1}{6}\,\frac{1}{\rho r^2}\,\frac{d}{dr}(\rho r^2 u). \tag{4.29}$$

(As usual, we also have $u_\theta = -r u_\mu/\sin\theta$.) By virtue of Eq. (4.29), the meridional circulation depends on the single function u only.

Inserting our truncated expansions into Eq. (4.5), one finds that

$$\frac{d}{dr}\left[\frac{1}{T'}\,\frac{dT_{1,2}}{dr} + 3\,\frac{T_{1,2}}{T} - \frac{\rho_{1,2}}{\rho}\right] - \frac{6}{r^2}\,\frac{T_{1,2}}{T'} + \frac{4\pi c_V}{L}\, r^2 \rho T\left(\frac{p'}{p} - \frac{5}{3}\,\frac{\rho'}{\rho}\right) u = 0, \tag{4.30}$$

where L is the total luminosity of the model corresponding to $\epsilon = 0$. In establishing this equation, we have made use of the fact that

$$4\pi r^2 \chi T' = -L, \tag{4.31}$$

where χ is the coefficient of radiative conductivity in the spherical model. Inserting next solutions (4.25) and (4.27) into Eq. (4.30), we obtain an algebraic equation for the function u. Its solution, $u = u_S$ (say), has the form

$$u_S = \frac{2Lr^4}{G^2 m^3} \frac{n+1}{n-3/2} \left[h' + \left(\frac{2}{r} - \frac{m'}{m} \right) h \right], \tag{4.32}$$

where m is the mass contained within the sphere of radius r, $m' = 4\pi \rho r^2$, and

$$n = \frac{\rho' T}{\rho T'} \tag{4.33}$$

is the effective polytropic index. This is Sweet's (1950) solution for the meridional flow in the radiative envelope of a star in slow uniform rotation. Equation (4.32) can also be written in the form

$$u_S = \frac{2Lr^4}{G^2 m^3} \frac{\nabla_{\rm ad}}{\nabla_{\rm ad} - \nabla} \left[h' + \left(\frac{2}{r} - \frac{m'}{m} \right) h \right], \tag{4.34}$$

where

$$\nabla = \frac{\partial \ln T}{\partial \ln p} \quad \text{and} \quad \nabla_{\rm ad} = \left(\frac{\partial \ln T}{\partial \ln p} \right)_S . \tag{4.35}$$

Combining Eqs. (4.23) and (4.34), one readily sees that $|u_S| \approx LR^2/GM^2$ in the bulk of a stellar radiative zone. Hence, we have

$$|u_r| \approx \epsilon \frac{LR^2}{GM^2} = \frac{\epsilon R}{t_{\rm KH}}, \tag{4.36}$$

where $t_{\rm KH}$ is the Kelvin–Helmholtz time and ϵ is the ratio of centrifugal force to gravity at the equator (see Eq. [4.9]). This result implies at once that the time scale of the meridional flow in the bulk of a radiative envelope ($t_{\rm ES}$, say) is

$$t_{\rm ES} = \frac{t_{\rm KH}}{\epsilon}, \tag{4.37}$$

which is known as the Eddington–Sweet time.

4.2.2 The classical objections

Table 4.1 gives a detailed solution for a Cowling point-source model with electron-scattering opacity.* Here we have $T_c = 1.7606 \times 10^7 \, \bar{\mu} \, \bar{M}/\bar{R}$ and $p_c = 4.0779 \times 10^{17} \, \bar{M}^2/\bar{R}^4$, where $\bar{\mu}$ is the mean molecular weight and the remaining barred quantities

* This simple numerical model, with power-law opacity and point-source energy generation, was originally discussed by Thomas George Cowling (1906–1990) in 1935. It consists of a convective core that contains all the energy sources and a radiative envelope. See, e.g., Cox, J. P., and Giuli, R. T., *Principles of Stellar Structure*, Sections 19.2a and 23.4, New York: Gordon and Breach, 1968; Tayler, R. J., *Quart. J. R. Astron. Soc.*, **32**, 201, 1991.

Table 4.1. *Physical properties of a Cowling point-source model.*

r (R)	m (M)	T (T_c)	p (p_c)	n	h ($\bar{M}/\bar{R}$)	u_S ($\bar{L}\bar{R}^2/\bar{M}^2$)	rv_S ($\bar{L}\bar{R}^2/\bar{M}^2$)
0.00000	0.	1.0000E+0	1.0000E+00	1.5000	0.	...	...
0.05000	2.4440E−3	9.8719E−1	9.6829E−01	1.5000	5.8690E+12	...	...
0.10000	1.8886E−2	9.4965E−1	8.7883E−01	1.5000	2.2905E+13	...	...
0.15000	6.0178E−2	8.8988E−1	7.4702E−01	1.5000	4.9487E+13	...	...
0.20000	1.3169E−1	8.1176E−1	5.9370E−01	1.5000	8.3212E+13	...	...
0.25000	2.3234E−1	7.2002E−1	4.3991E−01	1.5000	1.2126E+14	...	...
0.28318	3.1197E−1	6.5413E−1	3.4607E−01	1.5000	1.4751E+14	infinite	infinite
0.28319	3.1199E−1	6.5412E−1	3.4605E−01	1.5001	1.4752E+14	3.7196E−1	−2.2226E+3
0.28320	3.1201E−1	6.5409E−1	3.4602E−01	1.5002	1.4752E+14	1.6414E−1	−4.3284E+2
0.28400	3.1403E−1	6.5247E−1	3.4388E−01	1.5070	1.4816E+14	3.5870E−3	−2.0748E−1
0.29000	3.2927E−1	6.4042E−1	3.2802E−01	1.5577	1.5291E+14	4.2638E−4	−3.0082E−3
0.30000	3.5507E−1	6.2072E−1	3.0244E−01	1.6400	1.6081E+14	1.7107E−4	−5.0179E−4
0.35000	4.8749E−1	5.2904E−1	1.9237E−01	2.0069	1.9897E+14	4.4700E−5	−3.5569E−5
0.40000	6.1480E−1	4.4806E−1	1.1381E−01	2.2980	2.3356E+14	3.0266E−5	−1.3809E−5

0.45000	7.2584E−1	3.7703E−1	6.3121E−02	2.5195	2.6420E+14	2.7805E−5	−8.6574E−6
0.50000	8.1507E−1	3.1511E−1	3.3055E−02	2.6823	2.9180E+14	3.0232E−5	−7.4173E−6
0.55000	8.8192E−1	2.6136E−1	1.6412E−02	2.7982	3.1793E+14	3.6467E−5	−8.5870E−6
0.60000	9.2897E−1	2.1478E−1	7.7220E−03	2.8779	3.4421E+14	4.6910E−5	−1.3113E−5
0.65000	9.6012E−1	1.7436E−1	3.4200E−03	2.9306	3.7207E+14	6.2653E−5	−2.4106E−5
0.70000	9.7944E−1	1.3919E−1	1.4046E−03	2.9636	4.0262E+14	8.5282E−5	−4.8436E−5
0.75000	9.9054E−1	1.0842E−1	5.2052E−04	2.9829	4.3665E+14	1.1677E−4	−1.0071E−4
0.80000	9.9631E−1	8.1376E−2	1.6572E−04	2.9932	4.7471E+14	1.5940E−4	−2.1316E−4
0.85000	9.9888E−1	5.7460E−2	4.1255E−05	2.9979	5.1714E+14	2.1564E−4	−4.6505E−4
0.90000	9.9979E−1	3.6182E−2	6.4894E−06	2.9996	5.6410E+14	2.8805E−4	−1.0998E−3
0.95000	9.9999E−1	1.7139E−2	3.2677E−07	3.0000	6.1562E+14	3.7930E−4	−3.3448E−3
0.99000	1.0000E+0	3.2894E−3	4.4332E−10	3.0000	6.6006E+14	4.6775E−4	−2.2836E−2
0.99900	1.0000E+0	3.2597E−4	4.2756E−14	3.0000	6.7044E+14	4.8973E−4	−2.4429E−1
0.99990	1.0000E+0	3.2568E−5	4.2602E−18	3.0000	6.7149E+14	4.9198E−4	−2.4593E+0
0.99999	1.0000E+0	3.2565E−6	4.2587E−22	3.0000	6.7160E+14	4.9220E−4	−2.4609E+1
1.00000	1.0000E+0	0.	0.	3.0000	6.7161E+14	4.9222E−4	infinite

Source: Tassoul, J. L., and Tassoul, M., *Astrophys. J. Suppl.*, **49**, 317, 1982.

are expressed in solar units instead of in cgs units. The sixth column must be multiplied by $\bar{M}/\bar{R}$ to obtain the values of the function h in cgs units. Similarly, once the last two columns have been multiplied by $\bar{L}\,\bar{R}^2/\bar{M}^2$, they provide Sweet's solution – u_S and rv_S – in cgs units. His solution for the meridional flow consists of a single cell, with interior upwelling at the poles and interior downwelling at the equator (see Figure 4.3). Unfortunately, as was expected from Eqs. (4.29) and (4.32), one finds that $u_S \propto 1/(n-3/2)$ and $v_S \propto 1/(n-3/2)^2$ at the core boundary, whereas $u_S \neq 0$ and $v_S \propto \rho'/\rho$ at the free surface. This implies at once that the frictionless solution does not stream along the boundaries. To be specific, without mass loss, a consistent solution of the problem must be such that

$$\mathbf{n}\cdot\mathbf{u} = 0, \qquad \text{with} \qquad |\mathbf{u}| \text{ finite}, \tag{4.38}$$

at the boundary $r = R$ (see Eq. [2.20]). A similar condition applies at the core boundary $r = R_c$ if we assume that the circulatory currents do not penetrate into the convective region. Yet, one finds that

$$u_r \propto 1 \qquad \text{and} \qquad u_\theta \propto (R-r)^{-1}, \tag{4.39}$$

near the free surface, and

$$u_r \propto (r-R_c)^{-1} \qquad \text{and} \qquad u_\theta \propto (r-R_c)^{-2}, \tag{4.40}$$

near the core–envelope interface.

As was shown by Baker and Kippenhahn (1959), the situation is even worse when the prescribed rotation law is nonuniform. In that case, neglecting viscous friction and the inertial terms $\mathbf{u}\cdot\operatorname{grad}\mathbf{u}$, they found that Eq. (4.36) must be replaced by

$$|u_r| = \epsilon\,\frac{LR^2}{GM^2}\left(\alpha_0 + \beta_0\,\frac{\bar{\rho}}{\rho}\,\frac{\Delta\Omega}{\Omega_0}\right), \tag{4.41}$$

where α_0 and β_0 are constants of order unity, $\bar{\rho}$ is the mean density, and $\Delta\Omega$ is a measure of the prescribed nonuniform rotation rate. Hence, for electron-scattering opacity, Eq. (4.39) must be replaced by

$$u_r \propto (R-r)^{-3} \qquad \text{and} \qquad u_\theta \propto (R-r)^{-3}, \tag{4.42}$$

near the free surface. As they noted, in radiative regions near the surface of a differentially rotating star one can thus expect much higher meridional velocities than are calculated on the assumption of strict uniform rotation. This matter will be considered further in Section 4.4.1.

From the viewpoint of astronomy, Eqs. (4.36) and (4.41) are quite satisfactory, since they provide an order of magnitude of the circulation velocities in the bulk of a radiative envelope. They also point to an apparent difference between solid-body rotation and differential rotation, the latter causing a definite intensification of the meridional currents in the surface layers of an early-type star. Unfortunately, these two formulae are not directly applicable in the surface regions, because none of them satisfies the kinematic boundary condition (4.38) at the outer boundary. Moreover, one readily sees that the $1/\rho$ singularity in Eq. (4.41) implies that one has $|\rho\mathbf{u}\cdot\operatorname{grad}\mathbf{u}| \propto 1/\rho$, thus invalidating the method of solution in the surface layers. Note also that in both solutions one has neglected the inexorable transport of angular momentum by the meridional currents.

Another serious objection was raised by Öpik (1951), who noted that Sweet's solution for the radial component of the circulation velocity,

$$u_r = \epsilon\, u_S(r)\, P_2(\mu), \tag{4.43}$$

should be replaced by

$$u_r = \epsilon\, u_S(r) \left(1 - \frac{\Omega_0^2}{2\pi G\rho}\right) P_2(\mu). \tag{4.44}$$

If so, then, the meridional flow consists of two distinct cells (or gyres, as they say in geophysics) separated by the level surface with density $\rho = \rho^*$ (say) given by $\Omega_0^2 = 2\pi G\rho^*$. The following analytical proof of this property was broached by Gratton (1945) and Mestel (1966). Consider a chemically homogeneous radiative envelope in uniform rotation. Neglect friction and the inertia of the meridional currents. Then, by making use of Eqs. (3.31)–(3.36), one can rewrite Eq. (4.5) in the form

$$\rho A(\Phi)\, \mathbf{u} \cdot \text{grad}\, \Phi = -f(\Phi)\,(4\pi G\rho - 2\Omega_0^2) - f'(\Phi)\, g^2, \tag{4.45}$$

where

$$A(\Phi) = c_V \left(\frac{dT}{d\Phi} - \frac{2}{3}\frac{T}{\rho}\frac{d\rho}{d\Phi}\right), \tag{4.46}$$

and $g = d\Phi/dn$ is the magnitude of the effective gravity. (Remember that g varies over a level surface!) Dividing Eq. (4.45) by g and integrating over a level surface, we obtain

$$f(\Phi)\,(4\pi G\rho - 2\Omega_0^2)\,\langle g^{-1}\rangle + f'(\Phi)\langle g\rangle = 0, \tag{4.47}$$

since in a steady state there can be no flux of matter across a level surface. (Angular brackets designate a mean value over a level surface.) From Eqs. (4.45) and (4.47), it is clear that one has

$$\rho A(\Phi)\, \mathbf{u} \cdot \text{grad}\, \Phi = f' \left(\frac{\langle g\rangle}{\langle g^{-1}\rangle} - g^2\right). \tag{4.48}$$

If the function $f'(\Phi)$ vanishes for a value $\Phi = \Phi^*$ (say), this equation implies that the meridional currents do not cross the corresponding level surface. By virtue of Eq. (4.47), one has $f'(\Phi) = 0$ on the level surface with density $\rho^*(\Phi^*)$ given by $\Omega_0^2 = 2\pi G\rho^*$. This concludes the analytical proof that there apparently exists a double-cell pattern in a uniformly rotating radiative envelope.

As we shall see in Section 4.4.1, the Gratton–Mestel proof of the double-cell pattern is incorrect; Öpik's equation (4.44) is also quite inadequate for describing the meridional flow in a radiative envelope.

4.3 A consistent first-order solution

In Sections 2.5.1 and 2.6.2 we have presented frictionless solutions that describe large-scale flows in the Earth's atmosphere and in the oceans (see Eqs. [2.79] and [2.113]). In both cases, however, these solutions fail to satisfy the appropriate boundary conditions. This is the reason why turbulent friction had to be retained in narrow layers near the natural boundaries (see Eqs. [2.87]–[2.88] and [2.119]). The importance of eddy viscosity near the boundaries is directly related to the fact that the viscous force contains second-order derivatives in the velocities (see Eq. [2.65]). Hence, if eddy viscosity is neglected altogether in Eq. (2.64), the order of this equation is reduced so that its solutions can no longer satisfy all the boundary conditions that are required by the nature of the problem. As we know, the only way to satisfy all these conditions is to retain turbulent friction in thin boundary layers, where the velocities may vary rapidly in space. Then, the frictional force will be of the same order as the nonfrictional terms, notwithstanding the smallness of the coefficients of eddy viscosity. This is the key idea involved in *boundary-layer theory*. Not unexpectedly, a boundary-layer analysis of the thermally driven currents in the radiative envelope of a nonspherical star is a much more complex problem because it involves both the momentum equation and the energy equation. This will become apparent in the following pages.

In Section 4.2 we calculated the thermally driven currents in a stellar radiative envelope that we compel to rotate as a solid body. To obtain a fully consistent solution in a nonmagnetic star, we shall retain turbulent friction in Eqs. (3.125), (3.126), and (3.133). Hence, it is no longer necessary to prescribe the rotation rate, since the transport of angular momentum by the meridional flow can now be adjusted steadily so as to balance the effects of friction on the angular velocity. By virtue of Eq. (3.133), neglecting the λ_V effect, we thus have

$$\frac{\sin^2\theta}{r^2}\frac{\partial}{\partial r}\left(\mu_V r^4\frac{\partial\Omega}{\partial r}\right)+\frac{1}{\sin\theta}\frac{\partial}{\partial\theta}\left(\mu_H\sin^3\theta\,\frac{\partial\Omega}{\partial\theta}\right)$$
$$=\rho\mathbf{u}\cdot\mathrm{grad}(r^2\sin^2\theta\,\Omega). \tag{4.49}$$

Similarly, we shall replace Eqs. (4.1) and (4.2) by the following equations:

$$\frac{\partial p}{\partial r}=-\rho\,\frac{\partial V}{\partial r}+\rho\,\Omega^2 r\sin^2\theta+\rho F_r, \tag{4.50}$$

$$\frac{\partial p}{\partial\theta}=-\rho\,\frac{\partial V}{\partial\theta}+\rho\,\Omega^2 r^2\sin\theta\cos\theta+\rho F_\theta, \tag{4.51}$$

where F_r and F_θ are the poloidal components of the turbulent viscous force per unit volume (see Eqs. [3.125] and [3.126]). Equations (4.3)–(4.6) remain unaffected by eddy viscosity. Equations (4.49)–(4.51) and (4.3)–(4.6) thus provide seven relations among the seven unknown functions Ω, $\mathbf{u}$, p, ρ, T, and V.

Because the angular velocity is in general a function of both r and θ, let us write

$$\Omega(r,\theta)=\langle\Omega\rangle+[\bar\Omega(r)-\langle\Omega\rangle]+\hat\Omega(r,\theta), \tag{4.52}$$

where $\bar\Omega$ is a suitable mean value of Ω on a meridian (i.e., a mean with respect to θ) and $\langle\Omega\rangle$ is a suitable mean value of $\bar\Omega$ along the radius (i.e., a mean with respect to both θ and r), with $\hat\Omega$ describing the θ variations of Ω. Given this decomposition of the angular velocity, Eq. (4.49) implies that

$$\frac{\mathcal{O}(\bar\Omega-\langle\Omega\rangle)}{t_V}+\frac{\mathcal{O}(\hat\Omega)}{t_H}\approx\frac{\mathcal{O}(\bar\Omega+\hat\Omega)}{t_{\mathrm{ES}}}, \tag{4.53}$$

where t_{ES} is the circulation time of the meridional flow (see Eq. [4.37]). We have also let

$$t_V = \frac{\mathcal{O}(\rho)}{\mathcal{O}(\mu_V)}\, R^2 \qquad \text{and} \qquad t_H = \frac{\mathcal{O}(\rho)}{\mathcal{O}(\mu_H)}\, R^2, \tag{4.54}$$

where $\mathcal{O}(\,)$ is the order of magnitude symbol.

Let us assume next that in Eq. (4.53) the three terms are of the same order of magnitude. Then, comparing the advection term on the right-hand side to the horizontal dissipation term on the left-hand side, one readily sees that the θ dependence can be neglected (i.e., $\hat{\Omega} \ll \bar{\Omega}$) if and only if one has $t_H \ll t_{\text{ES}}$. Similarly, comparing the advection term on the right-hand side to the vertical dissipation term on the left-hand side, one notices that the r dependence can be neglected (i.e., $\bar{\Omega} \approx \langle\Omega\rangle$) if and only if one has $t_V \ll t_{\text{ES}}$. Thus, if one has simultaneously $t_H \ll t_{\text{ES}}$ and $t_V \ll t_{\text{ES}}$ in a slowly rotating star, the angular velocity is nearly constant throughout the radiative envelope. The latter case is particularly simple because, as we shall see in Section 4.3.1, one can then expand the unknown function Ω in powers of the small parameter ϵ (see Eq. [4.9]). The former case, which is much more involved, will be considered in Section 4.3.2.

4.3.1 *The linear case ($t_V < t_{\text{ES}}$)*

As was originally pointed out by Krogdahl (1944), the condition that $\mathbf{u}$ vanishes with Ω plus the obvious properties that $\mathbf{u}$ must be an even function of $\epsilon^{1/2}$, whereas Ω is to be odd in $\epsilon^{1/2}$, suggest the following choice for the velocities:

$$\Omega = \Omega_0(1 + \epsilon w_1 + \epsilon^2 w_2 + \cdots), \tag{4.55}$$

$$\mathbf{u} = \epsilon \mathbf{u}_1 + \epsilon^2 \mathbf{u}_2 + \cdots. \tag{4.56}$$

Correct to $\mathcal{O}(\epsilon)$, it follows at once from Eq. (4.55) that $\Omega^2 = \Omega_0^2 + \mathcal{O}(\epsilon^2)$. Thus, to that order of approximation, Eqs. (4.50) and (4.51) do not depend on w_1, so that it is possible to calculate $\mathbf{u}_1$ from Eqs. (4.3)–(4.6) and (4.50)–(4.51), replacing of course the *function* Ω by the *constant* Ω_0. Thence, one calculates the function w_1 from Eq. (4.49). Correct to $\mathcal{O}(\epsilon^{3/2})$, this equation becomes

$$\frac{1}{r^4}\frac{\partial}{\partial r}\left(\mu_V r^4 \frac{\partial w_1}{\partial r}\right) + \frac{\mu_H}{r^2}\frac{1}{1-\mu^2}\frac{\partial}{\partial \mu}\left[(1-\mu^2)^2 \frac{\partial w_1}{\partial \mu}\right]$$
$$= 2\rho\left(\frac{1}{r} u_{1r} - \frac{\mu}{1-\mu^2} u_{1\mu}\right), \tag{4.57}$$

where $\mu = \cos\theta$. (The quantities μ_V and μ_H refer to the spherical model corresponding to $\epsilon = 0$.) The problem of finding the meridional flow (i.e., $\mathbf{u}_1$) is thus separated from that of evaluating the reaction of these currents on the overall rotation rate (i.e., w_1). In other words, the overall rotation of $\mathcal{O}(\epsilon^{1/2})$ forces a small departure from spherical symmetry, which generates large-scale meridional motions of $\mathcal{O}(\epsilon)$; these, in turn, react back on the driving mechanism, giving rise to differential rotation of $\mathcal{O}(\epsilon^{3/2})$.

Correct to $\mathcal{O}(\epsilon)$, the circulation velocity $\mathbf{u}$ can be represented by Eq. (4.28). However, because we have retained turbulent friction in Eqs. (4.50) and (4.51), the functions $p_{1,2}$,

$\rho_{1,2}$, and $T_{1,2}$ must be replaced by the following relations:

$$p_{1,2} = -\rho h + \mathcal{G}, \tag{4.58}$$

$$\rho_{1,2} = -\frac{\rho\rho'}{p'} h + \frac{\rho}{p'} \mathcal{G}', \tag{4.59}$$

$$T_{1,2} = -T \left(\frac{\rho}{p} - \frac{\rho'}{p'} \right) h + \frac{T}{p} \mathcal{G} - \frac{T}{p'} \mathcal{G}', \tag{4.60}$$

where $\mathcal{G}$ represents the contribution from turbulent friction. Since we must retain the dominant part of the viscous force near the boundaries only, we shall let $\mathcal{G} \equiv 0$ in the bulk of the radiative zone, and we shall write

$$\mathcal{G} = r^2 \frac{d}{dr} \left(\mu_V \frac{dv}{dr} \right) + \cdots \tag{4.61}$$

near the core–envelope interface and the free surface. Note that this function is nothing but the dominant term $\mu_V(\partial u_\theta / \partial r)$ in Eq. (3.130). By making use of Eq. (4.29), one readily sees that $\mathcal{G}$ contains the third-order derivative of the radial function u. Since it is not yet known how to model the variations of μ_V with any confidence, we shall closely follow the examples set in Eqs. (2.66) and (2.67). Thus, we shall let $\mu_V = 10^N \mu_{\rm rad}$, where N (≥ 0) is a constant and $\mu_{\rm rad}$ is the coefficient of radiative viscosity,

$$\mu_{\rm rad} = \frac{4a}{15c} \frac{T^4}{\kappa\rho}, \tag{4.62}$$

where κ is the coefficient of opacity per unit mass.

Inserting next Eqs. (4.59) and (4.60) into Eq. (4.30), we obtain, after collecting and rearranging terms,

$$\mathcal{L}^{VI} u - \frac{4\pi G^3 m^3 \rho^3}{L\, p r^4} \frac{n - 3/2}{n+1} (u - u_S) = 0, \tag{4.63}$$

where u_S is defined in Eq. (4.32), and where $\mathcal{L}^{VI} u$ is a sixth-order differential operator acting on the function u. Since $\mathcal{L}^{VI} u \equiv 0$ in the bulk of a radiative envelope, we thus recover Sweet's frictionless solution $u = u_S$. Near the two boundaries, however, one must explicitly solve Eq. (4.63) together with appropriate boundary conditions (see Section 2.2.2). In particular, we must ensure that matter is flowing *along* the free surface (see Eq. [4.38]). Moreover, the components of the stress vector acting on the outer boundary,

$$n_k(-p\delta_{ik} + \sigma_{ik}), \tag{4.64}$$

must identically vanish. At the core–envelope boundary, however, the components defined in Eq. (4.64) must be continuous *across* that boundary. For the sake of simplicity, we shall also prescribe that the core boundary acts as an effective $\bar{\mu}$-barrier (although another boundary condition could easily be conceived).* Hence, we shall also apply condition (4.38) at the lower boundary.

* Short of a better theory for the convective core, we have thus assumed that the core is a uniformly rotating, isentropic fluid (i.e., a polytrope of index $n = 3/2$, which is rotating at the constant angular velocity Ω_0). Strictly speaking, if convective core overshooting was properly taken into account, one should then solve for both the convective core and the radiative envelope. In practice, however, given some ad hoc description for the overshooting, one could either apply condition (4.38) at a (somewhat larger) effective core radius or prescribe some penetration velocity at the core radius $r = R_c$.

Near the core boundary, one finds that

$$\mathcal{L}^{VI}u = \mu_V(n+1)r^2\frac{d^5v}{dr^5} + \cdots = \frac{1}{6}\mu_V(n+1)r^2\frac{d^6u}{dr^6} + \cdots. \tag{4.65}$$

We can also expand the effective polytropic index in the form

$$n = \frac{3}{2} + n'(R_c)(r - R_c) + \cdots \tag{4.66}$$

(see Eq. [4.33]). Equation (4.63) then becomes

$$\delta_c^7\frac{d^6u}{dr^6} - (r - R_c)u = -\nu, \tag{4.67}$$

where

$$\delta_c = \left\{\frac{LR_c^6}{24\pi G^3}\left[\frac{\mu_V p(n+1)^2}{m^3\rho^3 n'}\right]_{R_c}\right\}^{1/7} \tag{4.68}$$

and

$$\nu = \left[\frac{(n - 3/2)u_S}{n'}\right]_{R_c}. \tag{4.69}$$

One can easily show that $\delta_c/R_c \ll 1$ so that δ_c may be taken as a measure of the boundary-layer thickness. Letting next

$$x = \frac{r - R_c}{\delta_c} \quad \text{and} \quad y = \frac{\delta_c u}{\nu}, \tag{4.70}$$

we can rewrite Eq. (4.17) in the form

$$\frac{d^6y}{dx^6} - xy = -1; \tag{4.71}$$

the origin $x = 0$ (i.e., $r = R_c$) becomes, therefore, a simple turning point for the equation. One can also show that our boundary conditions are

$$y = \frac{dy}{dx} = \frac{d^2y}{dx^2} = 0 \quad \text{at} \quad x = 0. \tag{4.72}$$

Finally, since the solution of Eq. (4.71) should match the frictionless solution at a distance from the core boundary, we must also have

$$y \to \frac{1}{x}, \quad \text{as} \quad x \to \infty. \tag{4.73}$$

Figure 4.1 illustrates the solution of Eq. (4.71) that satisfies conditions (4.72) and (4.73).

In order to discuss the motions in the surface boundary layer, we shall prescribe the usual radiative-zero boundary conditions on the spherical model. Hence, letting $z = R - r$, we have $p = p_b z^{n+1}$, $\rho = \rho_b z^n$, $T = T_b z$, and $\mu_V = 10^N \mu_b z$. As usual, one has $n = 3$ for electron-scattering opacity and $n = 3.25$ for Kramers' opacity law. To exhibit the differences between the core and surface boundary layers, we shall let, without confusion,

$$x = \frac{R - r}{\delta} \quad \text{and} \quad y = \frac{u}{u_S(R)}, \tag{4.74}$$

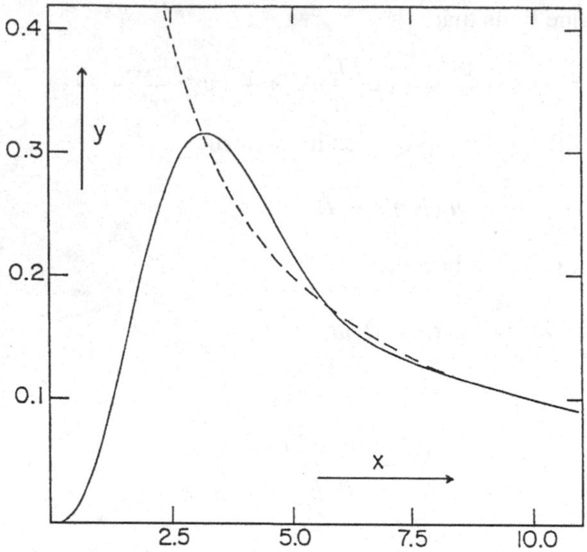

Fig. 4.1. Function $y(x)$ in the core boundary layer. The frictionless solution, $y = 1/x$, is indicated by a dashed curve. *Source:* Tassoul, J. L., and Tassoul, M., *Astrophys. J. Suppl.*, **49**, 317, 1982.

where

$$\delta = \left[10^N \frac{L\,R^6}{24\pi G^3 M^3} \frac{\mu_b\, p_b\, (n+1)^2}{\rho_b^3\,(n-3/2)} \right]^{1/(2n+4)}. \tag{4.75}$$

Again, one has $\delta/R \ll 1$ so that δ may be regarded as a measure of the boundary layer thickness. It then becomes a simple matter to show that Eq. (4.63) reduces to

$$\sum_{k=0}^{6} \frac{a_k}{x^k} \frac{d^{6-k}y}{dx^{6-k}} - x^{2n-2} y = -x^{2n-2}, \tag{4.76}$$

where the a_ks are numerical coefficients that depend on the effective polytropic index in the surface layers. One can also show that our boundary conditions reduce to $u = 0$, $\mu_V v' = 0$, and $(\mathcal{G}/\rho)' = 0$, at $r = R$. These three conditions become

$$y = 0, \tag{4.77}$$

$$x \frac{d^2y}{dx^2} + n \frac{dy}{dx} - \frac{n}{x} y = 0, \tag{4.78}$$

and

$$x^2 \frac{d^4y}{dx^4} + 2x \frac{d^3y}{dx^3} - n(n+2) \frac{d^2y}{dx^2} + \frac{n(n+2)}{x} \frac{dy}{dx} - \frac{n(n+2)}{x^2} y = 0 \tag{4.79}$$

at $x = 0$. Finally, since the solution of Eq. (4.76) must join smoothly the frictionless solution at some depth below the free surface, we must also prescribe that

$$y \to 1, \qquad \text{as} \qquad x \to \infty. \tag{4.80}$$

Figure 4.2 illustrates the solution of Eq. (4.76) that satisfies conditions (4.77)–(4.80).

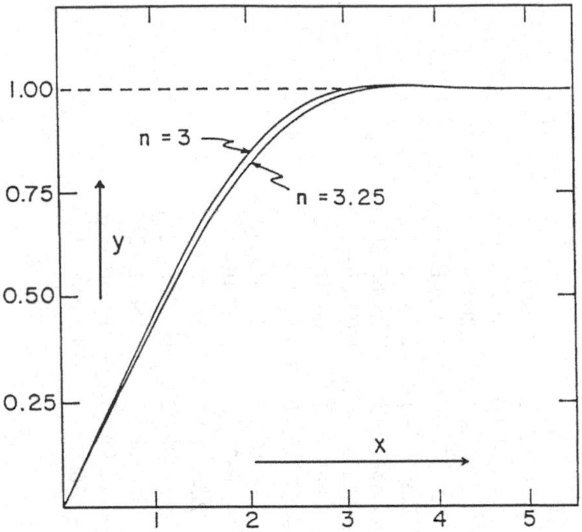

Fig. 4.2. Function $y(x)$ in the surface boundary layer. The frictionless solution, $y = 1$, is indicated by a dashed line. *Source:* Tassoul, J. L., and Tassoul, M., *Astrophys. J. Suppl.*, **49**, 317, 1982.

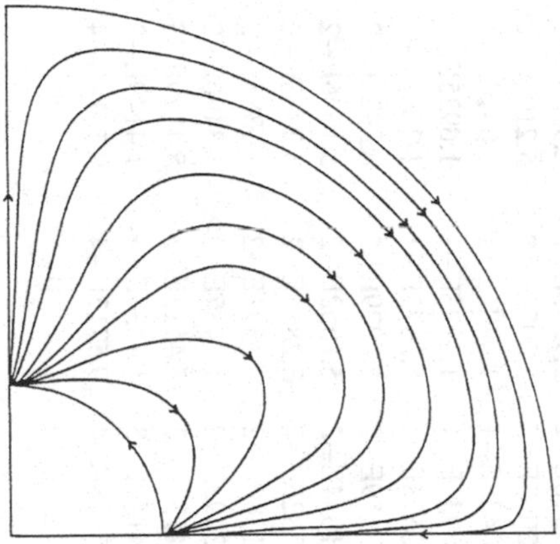

Fig. 4.3. First-order solution for the meridional flow in a Cowling point-source model, with electron-scattering opacity, $M = 3M_{\odot}$, and $N = 6$ in the boundary layers. The streamlines do not penetrate into the convective core, but there is an accumulation of streamlines in the core boundary layer. *Source:* Tassoul, J. L., and Tassoul, M., *Astrophys. J. Suppl.*, **49**, 317, 1982.

Figure 4.3 illustrates the streamlines of the meridional flow in a Cowling point-source model, with electron-scattering opacity, $M = 3M_{\odot}$, and $N = 6$ in the boundary layers (see also columns 2 and 5 in Table 4.2). To this order of approximation, the circulation pattern consists of a single cell (or gyre) extending from the core to the surface, with rising motions at the poles and sinking motions at the equator. Figure 4.3 gives the false

Table 4.2. *The first-order velocity field in a $3M_{\odot}$ stellar model ($N = 6$).*

	u			rv		
r/R	$\alpha = 0$	$\alpha = +10^{-3}$	$\alpha = -10^{-3}$	$\alpha = 0$	$\alpha = +10^{-3}$	$\alpha = -10^{-3}$
0.283182	0.	0.	0.	0.	0.	0.
0.283200	6.2797E−5	6.2843E−5	6.2884E−5	4.8677E−1	4.8657E−1	4.8689E−1
0.283250	2.8603E−3	2.8590E−3	2.8609E−3	5.4638E+0	5.4614E+0	5.4650E+0
0.283300	1.2381E−2	1.2375E−2	1.2384E−2	1.2605E+1	1.2599E+1	1.2607E+1
0.283350	2.9318E−2	2.9305E−2	2.9325E−2	1.9126E+1	1.9117E+1	1.9130E+1
0.283400	5.2093E−2	5.2070E−2	5.2105E−2	2.3471E+1	2.3461E+1	2.3476E+1
0.283500	1.0408E−1	1.0404E−1	1.0410E−1	2.3853E+1	2.3843E+1	2.3858E+1
0.283750	1.6931E−1	1.6923E−1	1.6935E−1	−9.0272E−1	−9.0233E−1	−9.0292E−1
0.284000	1.2678E−1	1.2673E−1	1.2681E−1	−1.0672E+1	−1.0667E+1	−1.0674E+1
0.284500	6.8959E−2	6.8929E−2	6.8975E−2	−1.8782E+0	−1.8774E+0	−1.8786E+0
0.285000	5.1074E−2	5.1052E−2	5.1086E−2	−1.4116E+0	−1.4110E+0	−1.4119E+0
0.286000	3.2826E−2	3.2811E−2	3.2833E−2	−5.5661E−1	−5.5637E−1	−5.5674E−1
0.287500	2.1380E−2	2.1371E−2	2.1385E−2	−2.3642E−1	−2.3631E−1	−2.3647E−1
0.290000	1.3493E−2	1.3487E−2	1.3496E−2	−9.5189E−2	−9.5147E−2	−9.5210E−2
0.300000	5.4137E−3	5.4113E−3	5.4149E−3	−1.5877E−2	−1.5870E−2	−1.5881E−2
0.350000	1.4145E−3	1.4138E−3	1.4149E−3	−1.1254E−3	−1.1250E−3	−1.1256E−3
0.400000	9.5778E−4	9.5712E−4	9.5814E−4	−4.3693E−4	−4.3686E−4	−4.3689E−4

0.450000	8.7991E−4	8.7913E−4	8.8045E−4	−2.7395E−4	−2.7406E−4	−2.7377E−4
0.500000	9.5670E−4	9.5554E−4	9.5760E−4	−2.3472E−4	−2.3506E−4	−2.3431E−4
0.550000	1.1540E−3	1.1521E−3	1.1557E−3	−2.7174E−4	−2.7246E−4	−2.7093E−4
0.600000	1.4845E−3	1.4810E−3	1.4876E−3	−4.1497E−4	−4.1625E−4	−4.1354E−4
0.650000	1.9827E−3	1.9763E−3	1.9886E−3	−7.6285E−4	−7.6458E−4	−7.6085E−4
0.700000	2.6988E−3	2.6874E−3	2.7097E−3	−1.5328E−3	−1.5329E−3	−1.5322E−3
0.750000	3.6954E−3	3.6767E−3	3.7134E−3	−3.1869E−3	−3.1738E−3	−3.1991E−3
0.800000	5.0444E−3	5.0257E−3	5.0622E−3	−6.7455E−3	−6.6620E−3	−6.8274E−3
0.850000	6.8240E−3	6.8950E−3	6.7522E−3	−1.4717E−2	−1.4254E−2	−1.5178E−2
0.900000	9.1156E−3	1.0170E−2	8.0606E−3	−3.4804E−2	−3.1660E−2	−3.7943E−2
0.925000	1.0479E−2	1.4505E−2	6.4533E−3	−5.7473E−2	−4.7300E−2	−6.7641E−2
0.950000	1.2003E−2	3.2423E−2	−8.4177E−3	−1.0585E−1	−5.9876E−2	−1.5182E−1
0.975000	1.3690E−2	2.4357E−1	−2.1619E−1	−2.5801E−1	−3.3609E−1	−1.7991E−1
0.980000	1.4105E−2	5.0824E−1	−4.8003E−1	−3.3131E−1	3.8968E+0	−4.5594E+0
0.985000	1.4427E−2	1.4525E+0	−1.4237E+0	−4.8428E−1	−3.6741E+0	2.7055E+0
0.990000	1.2628E−2	2.6253E+0	−2.6000E+0	−7.4853E−1	−1.1203E+2	1.1054E+2
0.995000	7.2169E−3	2.0576E+0	−2.0432E+0	−9.4255E−1	−2.5817E+2	2.5629E+2
0.997500	3.6952E−3	1.1078E+0	−1.1005E+0	−9.7945E−1	−2.9149E+2	2.8954E+2
0.999000	1.4913E−3	4.5329E−1	−4.5031E−1	−9.9211E−1	−3.0098E+2	2.9900E+2
1.000000	0.	0.	0.	−9.9954E−1	−3.0601E+2	3.0401E+2

Source: Tassoul, M., and Tassoul, J. L., *Astrophys. J.*, **440**, 789, 1995.

impression that the streamlines penetrate into the core. Actually, they are closed curves, but there is such an accumulation of streamlines in the core boundary layer ($R_c < r < R_c + \delta_c$) that a clear depiction is impossible without enlarging this narrow band. As was expected, because we have made allowance for turbulent friction in the radiative envelope, *matter is now flowing freely along its upper and lower boundaries*. Moreover, there are no mathematical singularities in the meridional flow; the circulation velocities remain uniformly small everywhere in the radiative envelope. This is a definite improvement over Sweet's frictionless solution.

By making use of Eq. (4.28), we can now solve Eq. (4.57) for the function w_1. One finds that

$$w_1 = \beta_1(r)\,\frac{dP_1}{d\mu} + \beta_3(r)\,\frac{dP_3}{d\mu}, \tag{4.81}$$

where $P_1(\mu) = \mu$ and $2P_3(\mu) = 5\mu^3 - 3\mu$. The nondimensional functions β_1 and β_3 are governed by the following inhomogeneous equations:

$$\frac{1}{r^4}\frac{d}{dr}\left(\mu_V r^4 \frac{d\beta_1}{dr}\right) = -\frac{2}{5}\,\rho\left(3v + \frac{u}{r}\right), \tag{4.82}$$

$$\frac{1}{r^4}\frac{d}{dr}\left(\mu_V r^4 \frac{d\beta_3}{dr}\right) - \frac{10}{r^2}\,\mu_H\,\beta_3 = -\frac{2}{5}\,\rho\left(2v - \frac{u}{r}\right). \tag{4.83}$$

Since the component $\sigma_{r\varphi}$ of the Reynolds stresses must vanish at the free surface, one has

$$\left(\mu_V \frac{d\beta_1}{dr}\right)_{r=R} = 0 \qquad \text{and} \qquad \left(\mu_V \frac{d\beta_3}{dr}\right)_{r=R} = 0. \tag{4.84}$$

Assuming that the convective core is uniformly rotating with angular velocity Ω_0, we shall also let

$$\beta_1(R_c) = 0 \qquad \text{and} \qquad \beta_3(R_c) = 0. \tag{4.85}$$

Thus, once we have obtained the functions u and v, Eqs. (4.82)–(4.85) can be solved to give a unique solution. The nonuniform rotation rate follows at once from Eq. (4.55).

It is immediately apparent from Eqs. (4.82) and (4.83) that $\epsilon|\beta_1|$ and $\epsilon|\beta_3|$ are of the order of $\langle \epsilon \rho u r / \mu_V \rangle$, where brackets indicate a suitable mean value. By virtue of Eqs. (4.37) and (4.54), one readily sees that $\epsilon|w_1| \approx t_V/t_{\text{ES}}$. To first order in ϵ, then, the convergence of expansion (4.55) implies that $\epsilon|w_1| < 1$ or $t_V < t_{\text{ES}}$. Letting $\mu_V = 10^N \mu_{\text{rad}}$, one can show that this requirement implies that $\epsilon 10^{6-N} < 1$ in a $3M_\odot$ star. In a typical rotating star having $\epsilon \approx 10^{-2}$, one must thus let $N \approx 5$–6. This value is quite similar to those encountered in geophysics (see Eqs. [2.66] and [2.67]). If the condition $\epsilon 10^{6-N} < 1$ is not met, however, one can no longer make use of expansion (4.55); that is to say, the full nonlinearity of Eq. (4.49) must be retained in the calculations.

4.3.2 *The nonlinear case* ($t_V \gtrsim t_{\text{ES}}$)

For the sake of simplicity, we shall consider a slowly rotating star for which one has $t_H \ll t_{\text{ES}}$ in its radiative envelope. Hence, we can essentially let $\Omega = \bar{\Omega}(r) + \hat{\Omega}(r, \theta)$, with $|\hat{\Omega}| \ll \bar{\Omega}$. In this case, by virtue of Eqs. (4.3)–(4.6) and (4.50)–(4.51), the circulation velocity $\mathbf{u}$ can still be represented by Eq. (4.28). Accordingly, if we let

$$\bar{\Omega} = \Omega_0 w(r), \tag{4.86}$$

Eq. (4.49) becomes

$$\frac{d^2 w}{dr^2} + \left(\frac{4}{r} + \frac{\mu'}{\mu}\right)\frac{dw}{dr} = -\frac{2}{5}\epsilon\frac{\rho}{\mu_V}\left[\left(3v + \frac{u}{r}\right)w + \frac{1}{2}u\frac{dw}{dr}\right], \tag{4.87}$$

where we have neglected the contributions to the function $\hat{\Omega}$. At the free surface, we have

$$\left(\mu_V \frac{dw}{dr}\right)_{r=R} = 0. \tag{4.88}$$

Parenthetically note that condition (4.88) is not automatically satisfied if μ_V vanishes at the surface, since this also implies that Eq. (4.87) has a singular point at $r = R$. In fact, Eq. (4.87) has a first integral that is quite convenient for our purposes. Setting the constant of integration equal to zero, one obtains the nonlinear equation

$$-5\mu_V \frac{dw}{dr} = \epsilon\rho u w, \tag{4.89}$$

therefore ensuring that boundary condition (4.88) is satisfied provided the product ρu vanishes at $r = R$.

A second relation between u and w can be obtained from Eqs. (4.3)–(4.6) and (4.50)–(4.51). In this case, Eq. (4.58) remains valid but Eqs. (4.59) and (4.60) must be replaced by

$$\rho_{1,2} = -\frac{\rho\rho'}{p'}h + \frac{\rho}{p'}\mathcal{G}' + \rho f \tag{4.90}$$

and

$$T_{1,2} = -T\left(\frac{\rho}{p} - \frac{\rho'}{p'}\right)h + \frac{T}{p}\mathcal{G} - \frac{T}{p'}\mathcal{G}' - Tf, \tag{4.91}$$

where

$$f = -\frac{1}{3}\frac{GM}{R^3}\frac{\rho}{p'}\frac{dw^2}{dr}r^2. \tag{4.92}$$

Again inserting these relations into Eq. (4.30), we obtain

$$\mathcal{L}^{VI}u - \frac{4\pi G^3 m^3 \rho^3}{L\,pr^4}\frac{n - 3/2}{n+1}[u - (u_S + u_f)] = 0. \tag{4.93}$$

Assuming electron-scattering opacity, one has

$$u_S = \frac{2Lr^4}{G^2 m^3}\frac{n+1}{n - 3/2}\left[h'(f) + \left(\frac{2}{r} - \frac{m'}{m}\right)h(f)\right] \tag{4.94}$$

and

$$u_f = \frac{L}{4\pi G m\rho}\frac{n+1}{n - 3/2}\frac{T}{T'}\mathcal{D}_2(f). \tag{4.95}$$

The function h is governed by the following inhomogeneous equation:

$$\frac{d^2h}{dr^2} + \frac{2}{r}\frac{dh}{dr} - \frac{6}{r^2}h + 4\pi G\frac{\rho\rho'}{p'}h = -\frac{p'}{\rho}\left[\frac{df}{dr} + 2\left(\frac{m'}{m} + \frac{1}{r}\right)f\right]. \tag{4.96}$$

(Note that $f \equiv 0$ in the convective core, where one assumes that $\bar{\Omega} \equiv \Omega_0$.) As explained in Section 4.2.1, when solving this equation one must always ensure the continuity of gravity across the core–envelope interface and across the free surface. One also has

$$\mathcal{D}_2(f) = \frac{d^2 f}{dr^2} + \left[\frac{2}{r} + (8-n)\frac{T'}{T}\right]\frac{df}{dr} + \left[-2\left(\frac{m'}{m} + \frac{1}{r}\right)\frac{T'}{T} - \frac{6}{r^2}\right]f. \tag{4.97}$$

Equations (4.89), (4.93), and (4.96) form a coupled system for the functions u, w, and h. Away from the boundaries, turbulent friction acting on the meridional flow is negligible so that one can replace Eq. (4.93) by

$$u = u_S + u_f \tag{4.98}$$

in the bulk of the radiative envelope. Near the core boundary, one can solve Eq. (4.93) along the lines presented in Section 4.3.1 (see Figure 4.1). Near the free surface, however, one readily sees from Eqs. (4.94)–(4.97) that the frictionless solution $u_S + u_f$ behaves as $1/\rho$. Following closely Eqs. (4.74)–(4.80), we shall thus let

$$x = \frac{R - r}{\delta} \quad \text{and} \quad y = \frac{\delta^n \rho_b u}{(\rho u_f)_R} \tag{4.99}$$

in the surface boundary layer. With this new definition for y, Eq. (4.93) becomes

$$\sum_{k=0}^{6} \frac{a_k}{x^k}\frac{d^{6-k}y}{dx^{6-k}} - x^{2n-2}y = -x^{n-2}. \tag{4.100}$$

Note that Eq. (4.100) is very similar in structure to Eq. (4.76), with x^{n-2} merely replacing x^{2n-2} on the right-hand side. Conditions (4.77)–(4.79) remain unchanged but Eq. (4.80) must be replaced by

$$y \to \frac{1}{x^n}, \quad \text{as} \quad x \to \infty, \tag{4.101}$$

since the solution of Eq. (4.100) should match the frictionless solution at the bottom of the surface boundary layer. Figure 4.4 illustrates the solution of Eq. (4.100) that satisfies conditions (4.77)–(4.79) and (4.101). *A uniformly valid solution of Eqs. (4.89), (4.93), and (4.96) can thus be obtained, all the way from the outer boundary to the core–envelope interface.*

The above formulation corresponds to the case for which one has $t_H \ll t_{\mathrm{ES}}$ so that we can let $\Omega = \bar{\Omega}(r)$. As explained in Section 4.3.1, if one also assumes that $t_V \ll t_{\mathrm{ES}}$, the function Ω remains nearly equal to a constant. In that case, correct to order ϵ, Eq. (4.86) can be rewritten in the form

$$\bar{\Omega} = \Omega_0[1 + \epsilon\beta_1(r)]. \tag{4.102}$$

After linearizing Eq. (4.89), we obtain

$$-5\mu_V \frac{d\beta_1}{dr} = \rho u. \tag{4.103}$$

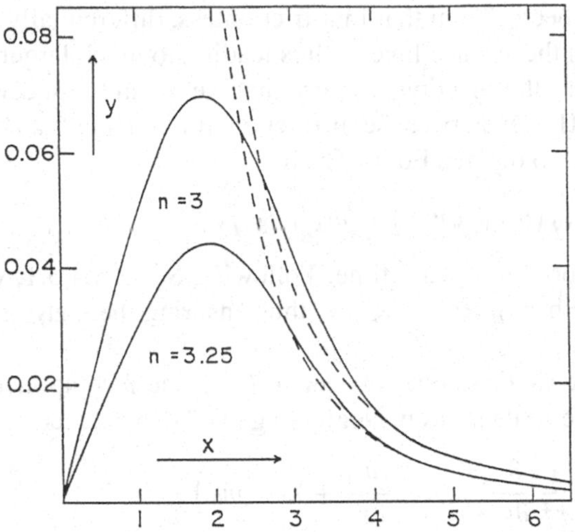

Fig. 4.4. Function $y(x)$ in the surface boundary layer. The frictionless solutions, $y = 1/x^n$, are indicated by dashed curves. *Source:* Tassoul, J. L., and Tassoul, M., *Astrophys. J. Suppl.*, **49**, 317, 1982.

This equation must be solved with the condition $\beta_1(R_c) = 0$ so that $\bar{\Omega} = \Omega_0$ at the core boundary $r = R_c$. (Condition [4.88] is automatically satisfied since one has $\rho u = 0$ at $r = R$.) This is a major simplification because it implies that $f \equiv u_f \equiv 0$; the right-hand side of Eq. (4.96) is thus identically equal to zero. This, in turn, implies that Eq. (4.90) no longer depends on rotation. Hence, the function u can be calculated along the lines presented in Section 4.3.1. Thence, one can solve Eq. (4.103) to obtain the function β_1. This is exactly the problem presented in Eq. (4.81), neglecting of course the θ dependence of the function w_1. Indeed, by making use of Eq. (4.29), one can easily show that the derivative of Eq. (4.103) is strictly equivalent to Eq. (4.82).

4.4 A consistent second-order solution

In Section 4.3.1 we have calculated the meridional velocity **u** and the angular velocity Ω in a slowly rotating star, when the departures from solid-body rotation are uniformly small throughout the whole radiative zone. The circulation pattern consists of a single cell (or gyre) extending from the convective core boundary to the free surface, with interior upwelling at the poles that is compensated by interior downwelling at the equator (see Figure 4.3). Although turbulent friction acting on the circulation is negligible in the bulk of the radiative envelope, there exist thin layers in which turbulent friction prevents the formation of unwanted singularities near the inner and outer boundaries. Such boundary-layer solutions satisfy all the basic equations and all the boundary conditions, with the circulation velocities remaining uniformly small throughout the radiative envelope. Of course, in the boundary layers these velocities depend on the coefficient μ_V. Fortunately, because they depend, respectively, on $(\mu_V)^{1/7}$ and $(\mu_V)^{1/10}$ in the core and surface boundary layers, their dependence on this poorly known parameter is considerably reduced (see Eqs. [4.68] and [4.75]).

Now, as was noted in Section 4.2.2, the claim has been made that there always exist two distinct cells separated by the level surface with density $\rho = \rho^*$ (say) given by

$\Omega_0^2 = 2\pi G\rho^*$. Moreover, it has been shown that in a frictionless, differentially rotating star one always has $u_r \propto 1/\rho$ in the surface layers, thus leading to much larger surface velocities than Sweet's (1950). Both objections require that we retain the second-order terms in Eqs. (4.55) and (4.56). However, because differential rotation plays an essential role in the discussion, we shall also replace Eq. (4.55) by

$$\Omega = \Omega_0 \left(w_0 + \epsilon w_1 + \epsilon^2 w_2 + \cdots\right), \tag{4.104}$$

where w_0 is a function of the coordinates and time. Following Section 4.3.1, we shall consider the case for which one has $t_H \ll t_V \ll t_{\rm ES}$, thus ensuring the convergence of expansion (4.104).

The general strategy is as follows. First, one solves to $\mathcal{O}(\epsilon^{1/2})$ the φ component of the momentum equation for the large-scale motion. Neglecting the θ dependence, we obtain

$$\rho \frac{\partial w_0}{\partial t} = \frac{1}{r^4}\frac{\partial}{\partial r}\left(\mu_V r^4 \frac{\partial w_0}{\partial r} + \lambda_V r^3 w_0\right), \tag{4.105}$$

where we have retained the λ_V effect (see Eq. [3.133]). Thus, unless the parameter λ_V identically vanishes, the solution of Eq. (4.105) does not correspond to a solid-body rotation. For steady motions, we have

$$w_0 = \exp\left(-\int^r \frac{\lambda_V}{\mu_V}\frac{dr}{r}\right). \tag{4.106}$$

Since μ_V and λ_V are poorly known quantities, we shall merely prescribe that

$$w_0 = 1 + \alpha(1 - r/R)^2, \tag{4.107}$$

where α is a constant. Second, one calculates the first-order velocity $\mathbf{u}_1$, which can be obtained from Eqs. (4.28) and (4.93), replacing w by w_0 in definition (4.92). Third, once the problem has been solved to that order, one calculates the back reaction w_1 (see Eq. [4.81]). Finally, collecting all the pieces together, one calculates the second-order velocity $\mathbf{u}_2$. To this order of approximation, however, one must retain the inertial terms $\mathbf{u}_1 \cdot \operatorname{grad} \mathbf{u}_1$ in the poloidal part of the momentum equation.

Correct to $\mathcal{O}(\epsilon^2)$, one has

$$\begin{aligned}\rho = \rho_0(r) &+ \epsilon[\rho_{1,0}(r) + \rho_{1,2}(r)P_2(\mu)] \\ &+ \epsilon^2[\rho_{2,0}(r) + \rho_{2,2}(r)P_2(\mu) + \rho_{2,4}(r)P_4(\mu)]\end{aligned} \tag{4.108}$$

and similar expressions for p, T, and V. (Henceforth we shall omit the subscripts "0" from the function ρ_0.) With the help of Eq. (4.4), we can also describe the meridional flow by means of a stream function. Thus, we let

$$\mathbf{u} = -\frac{1}{\rho r^2}\frac{\partial \Psi}{\partial \mu}\mathbf{1}_r + \frac{1}{\rho r^2}\frac{\partial \Psi}{\partial r}\mathbf{1}_\mu. \tag{4.109}$$

(One also has $u_\theta = -r u_\mu/\sin\theta$.) To the same order of approximation, one finds that

$$\begin{aligned}\Psi = {}& \epsilon\Psi_0(r)(1-\mu^2)\frac{dP_2(\mu)}{d\mu} \\ &+ \epsilon^2\left[\Psi_2(r)(1-\mu^2)\frac{dP_2(\mu)}{d\mu} + \Psi_4(r)(1-\mu^2)\frac{dP_4(\mu)}{d\mu}\right],\end{aligned} \tag{4.110}$$

where we have defined the following functions:

$$\Psi_0 = \frac{1}{6} r^2 \rho u(r), \tag{4.111}$$

$$\Psi_2 = \frac{1}{6} r^2 \left[\rho u_2(r) + \left(\rho_{1,0} + \frac{2}{7} \rho_{1,2} \right) u(r) \right], \tag{4.112}$$

$$\Psi_4 = \frac{1}{20} r^2 \left[\rho u_4(r) + \frac{18}{35} \rho_{1,2} u(r) \right]. \tag{4.113}$$

It is to be noted that, to $\mathcal{O}(\epsilon^2)$, the streamlines $\Psi = constant$ do depend on ϵ, whereas they are independent of this small parameter in the first-order approximation. Correct to $\mathcal{O}(\epsilon^2)$, the angular velocity can be brought to the form

$$\Omega = \Omega_0 \left[w_0 + \epsilon \sum_{i=1,3} \beta_i(r) \frac{dP_i(\mu)}{d\mu} + \epsilon^2 \sum_{i=1,3,5} \gamma_i(r) \frac{dP_i(\mu)}{d\mu} \right], \tag{4.114}$$

where the γ_is are governed by a set of inhomogeneous equations.

In Table 4.2 we list the first-order functions u and rv (in cm s^{-1}) for three values of α, in a Cowling point-source model with electron-scattering opacity ($M = 3M_\odot$, $R = 1.75R_\odot$, $L = 93L_\odot$, and $N = 6$ in the boundary layers). Evidently, the case $\alpha = 0$ corresponds to Sweet's problem, with $u_f \equiv 0$ and $u = u_S$ in the frictionless interior (see Eqs. [4.94] and [4.95]). In contrast, any model for which $\alpha \neq 0$ has $u = u_S + u_f$ in the frictionless interior, since $u_f \not\equiv 0$ when $dw_0/dr \not\equiv 0$. One can show that $u_f > 0$ when $\alpha > 0$ (i.e., when $dw_0/dr < 0$); similarly, one finds that $u_f < 0$ when $\alpha < 0$ (i.e., when $dw_0/dr > 0$). Since $u_S > 0$, it follows at once that the sum $u_S + u_f$ is always positive when $\alpha > 0$ but may change its sign along the radius when $\alpha < 0$. Therefore, to first order in ϵ, the meridional flow consists of a single cell when $\alpha > 0$, whereas it may consist of two cells when $\alpha < 0$. This property is immediately apparent from the fourth column in Table 4.2.

From the solutions presented in Table 4.2, one readily sees that there is a definite intensification of the function u near the surface of models for which $\alpha \neq 0$. Obviously, such an intensification does not occur in the limiting case $\alpha = 0$. Close scrutiny of the second-order corrections indicates that there always exists a surface intensification of the radial component $u_r = \epsilon u_{1r} + \epsilon^2 u_{2r}$, no matter whether one has $w_1 \not\equiv 0$ or $w_1 \equiv 0$. To be specific, in a frictionless model having $w_0 \not\equiv 1$, one has $u_{1r} \propto 1/\rho$ and $u_{2r} \propto 1/\rho$ in the surface layers. In contrast, letting $w_0 \equiv 1$ in a frictionless model, one finds that $u_{1r} \propto 1$ and $u_{2r} \propto 1/\rho$ in these layers. Strictly speaking, then, the case $w_0 \equiv 1$ is mathematically singular since $\epsilon^2 |u_{2r}|$ may become larger than $\epsilon |u_{1r}|$ in the surface layers.* Accordingly, a consistent expansion method requires a small amount of differential rotation to $\mathcal{O}(\epsilon^{1/2})$, so that one has $u_{1r} \propto 1/\rho$ in the frictionless solution near the surface. Of course, when turbulent friction is properly taken into account to all orders in the small parameter ϵ, there are no singularities in the components of the

* The case $w_0 \equiv 1$ is the only one for which the function u_{1r} has no $1/\rho$ singularity in the surface layers. This can happen only if there exists a centrifugal potential that is proportional to $r^2[1 - P_2(\cos\theta)]$, that is to say, in the case of *strict* uniform rotation to $\mathcal{O}(\epsilon^{1/2})$. Note that such a mathematical complication does not occur when the thermally driven currents are caused by disturbing forces other than the centrifugal force of rotation.

circulation velocity **u**. Figure 4.4 clearly illustrates how the frictional force acts to prevent the appearance of inordinately large radial velocities near the outer boundary. Because of the $1/\rho$ term in the frictionless solution that remains valid in the deep interior only, the function u_r at first increases toward the surface and then drops rapidly to zero at the free boundary. To be specific, there is an intensification of the radial velocities below the surface, typically by two or three orders of magnitude (see Table 4.2). However, given the extreme smallness of the meridional currents in the bulk of a stellar radiative envelope, *the maximum radial speed below the free surface does not exceed* $1\ cm\ s^{-1}$, which is a far remove from the various evaluations that can be found in the literature.

Figures 4.5 and 4.6 illustrate two second-order solutions for the meridional flow, respectively for $\alpha = +10^{-3}$ and $\alpha = -10^{-3}$ ($N = 6$, $\mu_H/\mu_V = 10^2$, and $\epsilon = 10^{-4}$). These curves are quite independent of the parameter ϵ in the deep interior, where the second-order terms make a negligible contribution to the first-order solution. In contrast, it is immediately apparent that, even for a rather low value of ϵ, the second-order terms make a sizeable contribution in the surface layers, where two or even three cells may occur. Note especially the cell in the equatorial belt, when the basic angular velocity decreases with depth ($\alpha < 0$). Obviously, there is a definite interplay between the meridional flow and the spatial variations of the angular velocity in the surface layers of a stellar radiative envelope. This is quite unfortunate because the actual run of the angular velocity depends on the eddy viscosities, which are poorly known parameters.

4.4.1 Answer to the classical objections

Consider again a uniformly rotating, nonmagnetic barotrope. Neglecting viscosity and the inertial terms $\mathbf{u} \cdot \operatorname{grad} \mathbf{u}$ in the momentum equation, one readily sees that the velocity **u** is present only in the equations expressing conservation of mass and energy (Eqs. [4.4] and [4.5]). In the case of a nonspherical star, then, Eqs. (4.3), (4.6), and (3.30) provide four scalar relations among the four functions p, ρ, T, and V. Indeed, letting $p = p_0 + \epsilon p_1$, etc., and linearizing these equations, one can calculate unequivocally the four nonspherical corrections (i.e., p_1, ρ_1, T_1, and V_1) to a given spherical model. These four corrections are independent of the velocity **u**. Hence, the potential Φ is also completely determined, and it does not depend on the velocity **u** either. Because Eq. (4.47) is derived from Eq. (4.5), which is independent of the remaining equations, there is thus no reason to believe that the constraint (4.47) will be satisfied by the functions $\langle g \rangle$ and $\langle g^{-1} \rangle$ that one has derived from the known potential Φ. Prima facie, this raises serious questions about the validity of Eq. (4.48).

As was pointed out in Section 4.2.2, the Gratton–Mestel proof of the double-cell pattern rests on the fact that their frictionless, nonmagnetic body remains strictly barotropic in spite of the inexorable meridional flow. We may therefore ask the following question: Is it actually possible to obtain such a flow in a uniformly rotating body that has all the properties of a barotrope? Specifically, given the approximations made, Eq. (4.49) implies that one has $\mathbf{u} \cdot \operatorname{grad}(r^2 \sin^2\theta) = 0$, that is, the streamlines of the meridional flow must coincide with the straight lines $r \sin\theta = constant$ (i.e., lines parallel to the rotation axis). This is an impossible requirement, since Eq. (4.38) implies that the streamlines must be closed curves. Moreover, because the meridional velocities in a frictionless system have unwanted singularities at the upper and lower boundaries, there is no reason

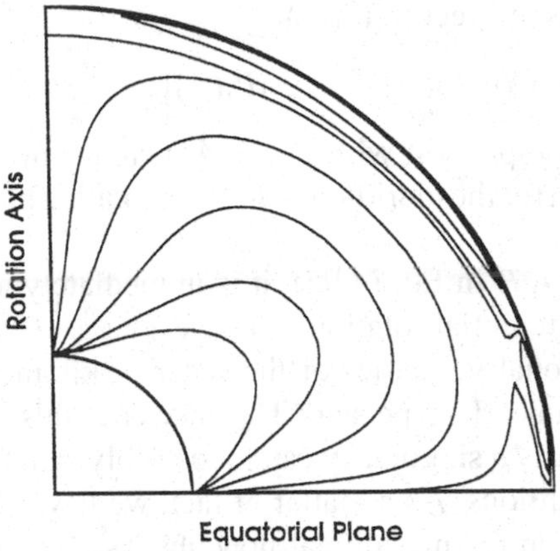

Fig. 4.5. Second-order solution for the meridional flow in a Cowling point-source model, with electron-scattering opacity, $M = 3M_\odot$, $N = 6$, $\mu_H/\mu_V = 10^2$, $\epsilon = 10^{-4}$, and $\alpha = +10^{-3}$. In the inner cell, interior upwelling along the rotation axis is compensated by interior downwelling in the equatorial belt. The sense of circulation is reversed in the outer cell that is adjacent to the rotation axis. Note that there are two cells in the outer layers: One of them is adjacent to the rotation axis, and the other is located in the equatorial belt. *Source:* Tassoul, M., and Tassoul, J. L., *Astrophys. J.*, **440**, 789, 1995.

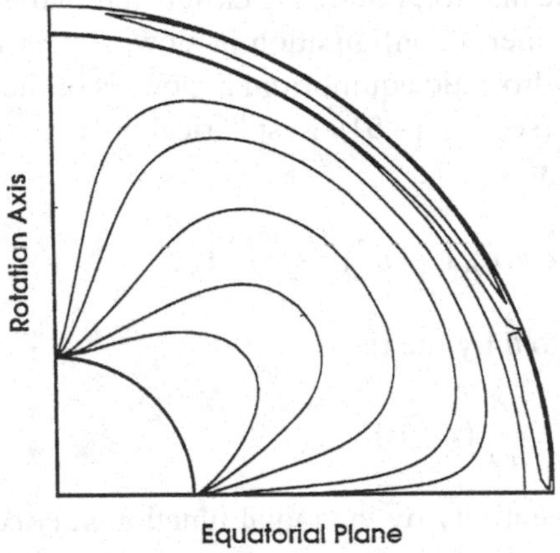

Fig. 4.6. Same as Figure 4.5, but for $\alpha = -10^{-3}$. *Source:* Tassoul, M., and Tassoul, J. L., *Astrophys. J.*, **440**, 789, 1995.

to believe that one can apply the condition

$$\oint_S \rho \mathbf{n} \cdot \mathbf{u}\, dS = 0 \tag{4.115}$$

on each level surface S, since this integral relation implicitly assumes that the velocity $\mathbf{u}$ is everywhere finite. We therefore conclude that the Gratton–Mestel result is the consequence of an excessively large number of conflicting assumptions that cannot be met in a realistic model.

Making use of Eq. (4.109), one also has, correct to $\mathcal{O}(\epsilon^2)$,

$$u_r = \epsilon u(r)P_2(\mu) + \epsilon^2 [u_0(r) + u_2(r)P_2(\mu) + u_4(r)P_4(\mu)], \tag{4.116}$$

where $u_0 = -(1/5)(\rho_{1,2}/\rho)u$. A mere comparison with Eq. (4.44) shows that Öpik's formula does not provide a reliable solution for the meridional flow in a rotating barotrope. It must therefore be disregarded.

Yet, because one has $u_f \propto 1/\rho$ as $r \to R$ in Eq. (4.95), it is immediately apparent from Eq. (4.98) that a small amount of differential rotation can have a large effect in the surface layers. Does it imply that meridional velocities of the order of kilometers per second are the rule in the outer layers of an early-type star? The answer to this question is flatly *no*, because any formula that has a $1/\rho$ singularity cannot possibly satisfy all the basic equations and all the boundary conditions. As a matter of fact, we have shown in this chapter that turbulent friction acting on the meridional flow always prevents huge surface velocities, having $|u_r| \lesssim 1$ cm s^{-1} and $|u_\theta| \lesssim 10^2$ cm s^{-1} in the surface layers of a $3M_\odot$ star in almost uniform rotation. Obviously, these speeds are much slower than those predicted on the basis of the formulae $u_r \propto \epsilon/\rho$ or $u_r \propto \epsilon^2/\rho$, which are completely inadequate in the outermost surface layers of a rotating star.

4.5 Meridional circulation in a cooling white dwarf

Consider a single, nonmagnetic white dwarf that produces its luminosity by cooling of its almost isothermal, degenerate interior. Following closely the analysis given in Sections 4.2.1 and 4.3.1, we shall consider a configuration in slow, almost uniform rotation. Hence, we shall expand about hydrostatic equilibrium in powers of the ratio of centrifugal force to gravity at the equator (see Eq. [4.9]). In spherical polar coordinates $(r, \mu = \cos\theta, \varphi)$, the meridional velocity $\mathbf{u}$ is

$$\mathbf{u} = \epsilon\, u(r)\, P_2(\mu)\mathbf{1}_r + \epsilon\, v(r)(1-\mu^2)\frac{dP_2(\mu)}{d\mu}\,\mathbf{1}_\mu, \tag{4.117}$$

where, by virtue of Eq. (4.4), v is related to u by the relation

$$v = \frac{1}{6}\frac{1}{\rho r^2}\frac{d}{dr}(\rho r^2 u). \tag{4.118}$$

The meridional flow is thus characterized entirely by the radial function u. (Recall that $u_\theta = -r u_\mu/\sin\theta$.) We also have

$$\Omega = \Omega_0\left\{1 + \epsilon\left[\beta_1(r)\frac{dP_1(\mu)}{d\mu} + \beta_3(r)\frac{dP_3(\mu)}{d\mu}\right]\right\}, \tag{4.119}$$

where β_1 and β_3 verify Eqs. (4.82)–(4.84) and the condition that both functions remain finite at $r = 0$. From Eqs. (4.117) and (4.119) one readily sees that the large-scale motion consists of a constant overall rotation of $\mathcal{O}(\epsilon^{1/2})$, a meridional flow of $\mathcal{O}(\epsilon)$, and a back reaction of the currents of $\mathcal{O}(\epsilon^{3/2})$.

The structure of this solution is very similar to that of a nondegenerate star. Of course, Eqs. (4.5) and (4.6) need to be modified. First, allowance must be made for a more general equation of state in the degenerate interior. Second, because energy is released throughout the star, Eq. (4.5) must be replaced by

$$\rho T\mathbf{u}\cdot \mathrm{grad}\, S = \mathrm{div}(\chi\, \mathrm{grad}\, T) + \rho\mathcal{E}, \tag{4.120}$$

where $\mathcal{E} = -T\,(\partial S/\partial t)$ is the energy released by cooling (per gram and per second) and χ is the coefficient of thermal conductivity in the degenerate interior (or radiative conductivity in the nondegenerate envelope).

Given these two modifications, it is a simple matter to calculate the function u in the bulk of a cooling white dwarf. Away from the surface layers, the frictionless solution, $u = u_S$ (say), has the form

$$u_S = \frac{2lr^4}{G^2m^3}\,\frac{1 - m\mathcal{E}/l}{\Delta}\,\frac{\nabla_{\rm ad}}{\nabla_{\rm ad} - \nabla}\left[h' + \left(\frac{2}{r} - \frac{m'}{m}\right)h\right], \tag{4.121}$$

where

$$l = -4\pi r^2 \chi T', \tag{4.122}$$

which is the net amount of energy crossing the spherical surface of radius r per second. (As usual, we have omitted the subscript "0" from the functions $\mathcal{E}_0$, χ_0, and T_0 in the spherical model.) We have also let

$$\nabla_{\rm ad} = \left(\frac{\partial \ln T}{\partial \ln p}\right)_S = \frac{p\Delta}{c_p \rho T}. \tag{4.123}$$

The second equality defines the parameter Δ. Remaining symbols have their standard meanings (see Eq. [4.34]).

As one moves toward the free surface, Eq. (4.121) merely reduces to Sweet's function (4.34), since we have $l \equiv L$, $\mathcal{E} \equiv 0$, and $\Delta \equiv 1$ in the nondegenerate envelope. As we know, this frictionless solution is not acceptable near the surface because it does not satisfy the kinematic boundary condition (4.38). We are thus forced to retain turbulent friction in the surface layers and, hence, to make explicit use of the sixth-order equation (4.63) for the function u. By making use of the radiative-zero boundary conditions, one can easily show that Eqs. (4.74)–(4.80) are the appropriate equations for the problem being considered. Once the function u has been calculated from $r = 0$ to $r = R$, one can solve Eqs. (4.82) and (4.83) for the functions β_1 and β_3.

Table 4.3 gives a detailed solution for a $0.8M_\odot$ white-dwarf model. The functions u and rv are given in cm s^{-1}. They were obtained using the formula $\mu_V = 10^N \mu_{\rm rad}$, with $N = 2$, in the nondegenerate envelope (see Eq. [4.62]). This choice of N is unimportant since u and rv depend on $(\mu_V)^{1/10}$. In Table 4.3 we also list the functions β_1 and β_3. They were obtained using the viscosity of a degenerate electron gas in the deep interior and the above formula in the outer layers. Figure 4.7 illustrates the meridional flow, which breaks down into three regions with motions in opposite senses. This situation arises because the factor $(1 - m\mathcal{E}/l)$ changes its sign twice along the radius. Accordingly, this triple-circulation pattern is a mere consequence of the stratification of the spherical models that were used to obtain the function u. For a typical white dwarf, with equatorial velocity $v_{\rm eq} \approx 50$ km s^{-1}, we have $\epsilon \approx 10^{-4}$. Hence, from Table 4.3 one readily sees that $|u_r|$ ($\approx \epsilon|u|$) $< 10^{-13}$ cm s^{-1} and $|u_\theta|$ ($\approx \epsilon r|v|$) $< 10^{-9}$ cm s^{-1}! Moreover, since $|\beta_1|$ and $|\beta_3|$ are both of order unity, $\epsilon|\beta_1|$ and $\epsilon|\beta_3|$ remain in general much smaller than one, so that Eq. (4.119) provides an acceptable solution for the azimuthal motion. As regards practical applications, such as large-scale mixing and microscopic diffusion in the surface layers, we therefore conclude that the meridional currents are utterly negligible in a cooling white dwarf in a state of slow, almost uniform rotation.

Table 4.3. *The velocity field in a* $0.8M_\odot$ *cooling white dwarf.*

r/R	$\log(1-m/M)$	u	rv	β_1	β_3
0.000000	0.	0.	0.	0.	0.
0.096517	−0.00413	4.0440E−12	1.9762E−12	−7.5979E−5	−2.2525E−4
0.172414	−0.02228	7.7119E−12	3.6337E−12	−2.5375E−4	−7.4405E−4
0.292340	−0.09691	1.5276E−11	6.4466E−12	−8.2887E−4	−2.3373E−3
0.374016	−0.18709	2.3923E−11	1.0428E−11	−1.5421E−3	−4.1660E−3
0.473461	−0.34679	4.4661E−11	1.9598E−11	−3.2013E−3	−7.7537E−3
0.583683	−0.60206	9.5232E−11	3.7716E−11	−7.6818E−3	−1.5030E−2
0.698923	−1.00000	2.0127E−10	1.0222E−11	−2.1271E−2	−3.0718E−2
0.847698	−2.00000	3.6225E−10	−1.1825E−09	−8.6396E−2	−8.2736E−2
0.884567	−2.50000	8.4421E−11	−2.3871E−09	−1.1174E−1	−9.8335E−2
0.908588	−3.00000	−2.9180E−10	−2.3595E−12	−9.7037E−2	−8.5441E−2
0.924756	−3.50000	−4.5561E−10	3.0146E−09	4.6999E−2	−4.8949E−2
0.936511	−4.00000	−5.8203E−10	9.5082E−09	5.1515E−2	2.0577E−2
0.950625	−4.50000	−5.7619E−10	7.3009E−09	2.4517E−1	1.5650E−1
0.962850	−5.00000	−4.3912E−10	8.3732E−09	3.5125E−1	2.2961E−1
0.978424	−6.00000	−1.4982E−10	6.9378E−09	3.9344E−1	2.5786E−1
0.987204	−7.00000	2.6621E−11	2.2088E−09	3.9532E−1	2.5892E−1
0.992409	−8.00000	1.3858E−10	−6.4318E−09	3.9504E−1	2.5866E−1
0.995597	−9.00000	2.3133E−10	−2.2086E−08	3.9492E−1	2.5855E−1
0.997503	−10.00000	3.0559E−10	−5.2426E−08	3.9489E−1	2.5853E−1
0.998598	−11.00000	3.6040E−10	−1.2305E−07	3.9488E−1	2.5852E−1
0.999212	−12.00000	3.7125E−10	−2.3394E−07	3.9488E−1	2.5852E−1
0.999712	−14.00000	3.7399E−10	−6.4598E−07	3.9488E−1	2.5852E−1
0.999838	−15.00000	3.6121E−10	−1.2407E−06	3.9488E−1	2.5852E−1
0.999909	−16.00000	2.4886E−10	−1.7597E−06	3.9488E−1	2.5852E−1
0.999949	−17.00000	1.4499E−10	−1.8792E−06	3.9488E−1	2.5852E−1
1.000000	infinite	0.	−1.8986E−06	3.9488E−1	2.5852E−1

Source: Tassoul, M., and Tassoul, J. L., *Astrophys. J.*, **267**, 334, 1983.

4.6 Meridional circulation in a close-binary component

Consider a system of two rotating stars revolving in circular orbits about their common center of gravity. We have a chemically homogeneous, early-type star of mass M (the primary) acted on by the tidal force originating from its companion of mass M' (the secondary). We shall assume that the radii of the components are much smaller than their mutual distance d, so that the secondary may be treated as a point mass when studying the tidal distortion of the primary. Assuming that the overall rotation of the primary is synchronized with revolution, we have

$$\Omega_0^2 = \frac{G(M+M')}{d^3}, \qquad (4.124)$$

where Ω_0 is the angular velocity of the primary.

4.6.1 *The tidally driven currents*

Since we want to fix our attention on the primary, it is convenient to choose a rotating frame of reference in which the origin is at the center of gravity of the mass M.

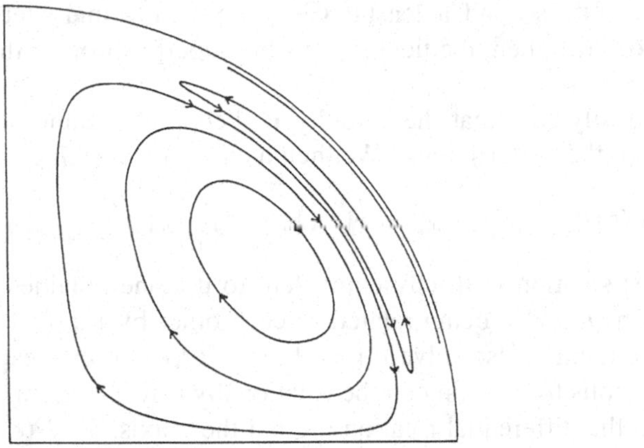

Fig. 4.7. Streamlines of meridional circulation in a $0.8M_\odot$ cooling white dwarf. The three circulation zones are separated by the spherical surfaces $r = 0.88975R$ and $r = 0.98590R$. The outer circulation pattern ($0.98590R < r < R$) is schematically depicted by a single curve. *Source:* Tassoul, M., and Tassoul, J. L., *Astrophys. J.*, **267**, 334, 1983.

The x axis points toward the center of the secondary, and the z axis is parallel to the overall angular velocity of the primary. Neglecting the inertial terms $\mathbf{u}_R \cdot \operatorname{grad} \mathbf{u}_R$, we thus have

$$2\boldsymbol{\Omega}_0 \times \mathbf{u}_R = -\operatorname{grad}(V - W) - \frac{1}{\rho} \operatorname{grad} p + \frac{1}{\rho} \mathbf{F}(\mathbf{u}_R), \tag{4.125}$$

where the three-dimensional velocity $\mathbf{u}_R$ is measured in our rotating frame of reference, and $\mathbf{F}$ is the turbulent viscous force per unit volume. In spherical polar coordinates (r, θ, φ), the potential W is given by

$$W = \frac{1}{3} \Omega_0^2 r^2 [1 - P_2(\mu)] + \frac{GM'}{d} \sum_{k=2}^{4} \frac{r^k}{d^k} P_k(\nu), \tag{4.126}$$

where $\mu = \cos\theta$ and $\nu = \sin\theta\cos\varphi$. The P_ks are the Legendre polynomials. Since our basic assumptions are identical to those made in Section 4.3.1, Eq. (4.125) must be combined with Eqs. (4.3)–(4.6).

Following standard practice, we shall expand about hydrostatic equilibrium in powers of the nondimensional parameter

$$\epsilon = \frac{\Omega_0^2 R^3}{GM} = \frac{M + M'}{M} \left(\frac{R}{d}\right)^3. \tag{4.127}$$

In particular, we shall let $p = p_0 + \epsilon p_1 + \cdots$, etc. In the frame rotating with the angular velocity Ω_0, the three-dimensional velocity $\mathbf{u}_R$ has the form

$$\mathbf{u}_R = \epsilon \mathbf{u}_1 + \epsilon^{3/2} \mathbf{u}_{3/2} + \cdots, \tag{4.128}$$

since the Coriolis force is of $\mathcal{O}(\epsilon^{3/2})$. By virtue of Eqs. (4.125) and (4.126), this three-dimensional velocity is the superposition of two different kinds of currents. One of them is caused by the small *oblateness* due to rotation around the z axis; the other one is caused by the small *prolateness* due to tidal action in the direction of the x axis. Hence, the general problem can be decomposed into two subproblems. In the first subproblem, the flow is caused by the rotational distortion only; for the second, the velocity $\mathbf{u}_R$ is due to the tidal distortion only. To evaluate the effects of the first we shall solve the equations with $M' = 0$; for the effects of the second, at least to $\mathcal{O}(\epsilon)$, we shall formally let $\Omega_0 = 0$ in the equations. Correct to $\mathcal{O}(\epsilon)$, then, the tidal flow is the superposition of these two circulation patterns.

Letting $M' = 0$, one readily sees that the velocity $\mathbf{u}_1$ becomes symmetrical with respect to both the z axis and the $(z = 0)$-plane. We then have, in the rotating frame,

$$u_{1r} = u(r)P_2(\mu), \qquad u_{1\mu} = v(r)(1-\mu^2)\frac{dP_2}{d\mu}, \tag{4.129}$$

and $u_{1\varphi} \equiv 0$. To $\mathcal{O}(\epsilon)$, this solution is strictly equivalent to the one obtained in Section 4.3.1, with the functions u and v being related to each other by Eq. (4.118). For the sake of completeness, one must also solve Eq. (4.125) to $\mathcal{O}(\epsilon^{3/2})$, thus expressing the balance between the Coriolis force acting on the rotationally driven currents and the turbulent friction acting on the differential rotation around the z axis. To $\mathcal{O}(\epsilon^{3/2})$, this equation is strictly equivalent to Eq. (4.57).

If we *formally* disregard the centrifugal and Coriolis forces in Eq. (4.125), the velocity $\mathbf{u}_1$ becomes symmetrical with respect to the x axis (but not with respect to the $(x = 0)$-plane!). In order to describe this part of the solution, it is convenient to use the radial variable r, the cosine of the colatitude from the x axis $\nu = \sin\theta\cos\varphi$, and the azimuthal angle ϕ around the x axis. Using these coordinates, one can show that the tidal contribution to the circulatory currents can be written in the form

$$u_{1r} = \sum_{k=2}^{4} u_k(r)P_k(\nu), \qquad u_{1\nu} = \sum_{k=2}^{4} v_k(r)(1-\nu^2)\frac{dP_k(\nu)}{d\nu}, \tag{4.130}$$

and $u_{1\phi} \equiv 0$. Equation (4.4) implies that

$$v_k = \frac{1}{k(k+1)}\frac{1}{\rho r^2}\frac{d}{dr}(\rho r^2 u_k) \tag{4.131}$$

$(k = 2, 3, 4)$. This motion depends, therefore, on the three functions u_2, u_3, and u_4.

When both the rotational and tidal terms are retained in Eq. (4.126), it is a simple matter to prove that the three-dimensional velocity $\mathbf{u}_1$ has the following components:

$$u_{1r} = \sum_{k=2}^{4} u_k(r)P_k(\nu) + u(r)P_2(\mu), \tag{4.132}$$

$$u_{1\mu} = \sum_{k=2}^{4} v_k(r)(1-\mu^2)\frac{\partial P_k(\nu)}{\partial\mu} + v(r)(1-\mu^2)\frac{dP_2(\mu)}{d\mu}, \tag{4.133}$$

$$u_{1\varphi} = \sum_{k=2}^{4} \frac{r v_k(r)}{(1-\mu^2)^{1/2}}\frac{\partial P_k(\nu)}{\partial\varphi}, \tag{4.134}$$

where, as we recall, $\nu = (1-\mu^2)^{1/2}\cos\varphi$. This solution is actually the vectorial sum of the velocity fields (4.129) and (4.130), with both solutions now being written in the rotating frame of reference (r, μ, φ). Note that the functions u_k and v_k are still related to each other by Eq. (4.131) because

$$\frac{\partial}{\partial\mu}\left[(1-\mu^2)\frac{\partial P_k(\nu)}{\partial\mu}\right] + \frac{1}{1-\mu^2}\frac{\partial^2 P_k(\nu)}{\partial\varphi^2} = \frac{d}{d\nu}\left[(1-\nu^2)\frac{dP_k(\nu)}{d\nu}\right] = -k(k+1)P_k(\nu). \tag{4.135}$$

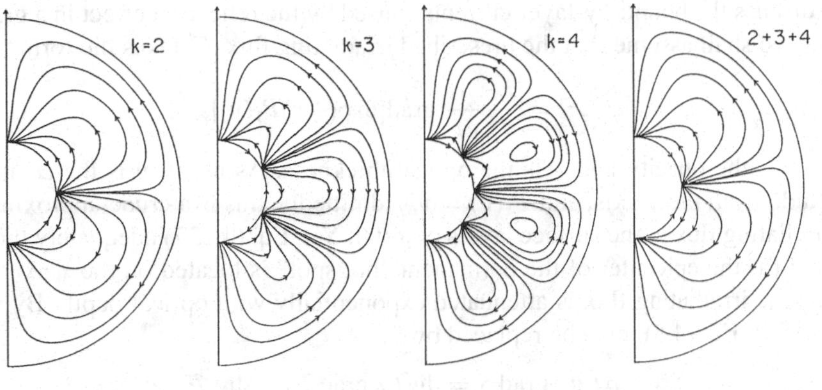

Fig. 4.8. Tidally driven currents in a synchronously rotating star. The vertical arrow points toward the companion; the tidal distortion of the model is not depicted. From left to right: Contributions from the P_2, P_3, and P_4 terms in Eq. (4.130), and the sum of these three contributions (when $M = M' = 3M_\odot$ and $\epsilon = 0.25$; that is, $d/R = 2$, $v_{eq} = 290$ km s^{-1}, and $P = 2\pi R/v_{eq} = 0.31$ day). Even for this relatively large value of ϵ, it is the P_2 term that dominates in the expansion. The streamlines do not penetrate into the convective core, but there is an accumulation of streamlines in the core boundary layer. *Source:* Tassoul, J. L., and Tassoul, M., *Astrophys. J.*, **261**, 265, 1982.

As usual, once the functions u and u_k have been obtained, Eq. (4.125) can be solved to $\mathcal{O}(\epsilon^{3/2})$ to give a unique solution for the velocity fields $\mathbf{u}_{3/2}$.

Figure 4.8 illustrates the pure tidally driven component of the circulation (see Eq. [4.130]). Following Section 4.3.1, we have considered a $3M_\odot$ Cowling point-source model, with electron-scattering opacity and $\mu_V = 10^6 \mu_{rad}$ in the boundary layers (see Eq. [4.62]). Although this large-scale motion is the combination of three terms, it is immediately apparent that the contribution from the $P_2(\nu)$ term dominates over the two others. Their time scale is of the order of the Kelvin–Helmholtz time, t_{KH}, divided by the ratio of the tidal force to gravity at the equator, $(M'/M)(R/d)^3$. These axially symmetric motions are the strict analog of the rotationally driven currents depicted in Figure 4.3.

4.6.2 The reflection effect in close binaries

To the best of my knowledge, Hosokawa (1959) was the first to point out that the mutual heating of the components in a close binary generates large-scale circulatory currents in their superficial layers. To illustrate the problem, we shall calculate the photospheric flow caused by the presence of a permanent "hot spot" on the surface of an early-type star that is a synchronously rotating component of a close binary.

Again consider a rotating frame of reference in which the origin is at the center of the primary (of mass M, radius R, and luminosity L when neglecting the "hot spot"). The x axis points toward the point-mass secondary (of mass M' and luminosity L'), and the z axis is parallel to the overall angular velocity of the primary. In this case, the appropriate expansion parameter, η (say), is the ratio of the fluxes at $r = R$,

$$\eta = \frac{L'}{L}\left(\frac{R}{d}\right)^2, \tag{4.136}$$

where d is the separation between the two centers of mass.

To discuss the boundary-layer currents caused by the reflection effect in a gray atmosphere, we shall assume that the prescribed irradiating flux $\mathcal{F}'$ takes the form

$$\mathcal{F}' = \frac{L'}{d^2} \frac{1}{\kappa\rho} \operatorname{grad}[\exp(-\tau) P_1(\nu)], \tag{4.137}$$

where κ is the opacity and τ is the optical thickness. As usual, ν is the cosine of the colatitude from the x axis, and $P_1(\nu) = \nu$. Admittedly, this is a crude approximation of the irradiating flux in the surface layers of a star. Yet, Eq. (4.137) adequately models the fact that (a) the epicenter of the permanent "hot spot" is located on the x axis ($\nu = 1$) and (b) the irradiating flux is attenuated exponentially with optical depth. By virtue of Eq. (4.137), Eq. (4.5) must be replaced by

$$\rho T \mathbf{u} \cdot \operatorname{grad} S = \operatorname{div}(\chi \operatorname{grad} T) + \operatorname{div} \mathcal{F}'. \tag{4.138}$$

It is a simple matter to prove that, correct to $\mathcal{O}(\eta)$, the velocity of the currents can be written in the form

$$u_r = \eta u(r) P_1(\nu) \qquad \text{and} \qquad u_\nu = \eta v(r)(1-\nu^2)\frac{dP_1(\nu)}{d\nu}, \tag{4.139}$$

where

$$v = \frac{1}{2}\frac{1}{\rho r^2}\frac{d}{dr}(\rho r^2 u). \tag{4.140}$$

(Compare with Eqs. [4.130]–[4.131].) Retaining turbulent friction in the surface layers, one can also show that the function u satisfies the following equation:

$$\mathcal{L}^{VI} u - \frac{4\pi G^3 m^3 \rho^3}{L p r^4}\frac{n - 3/2}{n+1} u = -\frac{G^2 m^2 \rho^2}{r^4} \kappa\rho \exp(-\tau), \tag{4.141}$$

where n is the effective polytropic index. (Compare with Eq. [4.63].)

Following Section 4.3.1, we shall prescribe the usual radiative-zero boundary condition. For electron-scattering opacity, we have $\kappa = constant$, $n = 3$, $p = p_b z^4$, $\rho = \rho_b z^3$, $T = T_b z$, and $\mu_V = 10^N \mu_b z$, where $z = R - r$. Letting next

$$x = \frac{R-r}{\delta} \qquad \text{and} \qquad y = \frac{u}{u_1}, \tag{4.142}$$

one can rewrite Eq. (4.141) in the form

$$\sum_{k=0}^{6} \frac{a_k}{x^k}\frac{d^{6-k} y}{dx^{6-k}} - x^4 y = -x^4 \exp(-\alpha_1 x^4), \tag{4.143}$$

where

$$u_1 = \frac{2\kappa L}{3\pi G M} \qquad \text{and} \qquad \alpha_1 = \frac{1}{4}\kappa\rho_b\,\delta^4. \tag{4.144}$$

Equation (4.75), with $n = 3$, defines the boundary-layer thickness δ. (Compare Eq. [4.143] with Eq. [4.76].) Of course, the solutions of Eq. (4.143) must satisfy the boundary conditions (4.77)–(4.79). However, because the motions generated by the reflection effect must vanish at some depth from the surface, condition (4.80) must be replaced by the following condition:

$$y \to 0, \qquad \text{as} \qquad x \to \infty. \tag{4.145}$$

Equations (4.143), (4.145), and (4.77)–(4.79) form the basic equations of the problem.

Table 4.4. *The functions u and rv in the surface boundary layer.*

	$N = 5$		$N = 6$	
x	u	rv	u	rv
0.0	0.	3.6299E+5	0.	5.2076E+4
0.2	1.3084E+2	−3.6179E+5	2.3604E+1	−5.1793E+4
0.4	2.5672E+2	−3.4923E+5	4.5883E+1	−4.9218E+4
0.6	3.6435E+2	−3.1731E+5	6.4100E+1	−4.3798E+4
0.8	4.4216E+2	−2.7198E+5	7.6742E+1	−3.7005E+4
1.0	4.8658E+2	−2.2146E+5	8.3607E+1	−2.9828E+4
1.2	4.9889E+2	−1.7091E+5	8.5073E+1	−2.2837E+4
1.4	4.8297E+2	−1.2372E+5	8.1863E+1	−1.6411E+4
1.6	4.4429E+2	−8.2151E+4	7.4931E+1	−1.0814E+4
1.8	3.8927E+2	−4.7645E+4	6.5365E+1	−6.2057E+3
2.0	3.2453E+2	−2.0864E+4	5.4292E+1	−2.6548E+3
2.2	2.5680E+2	−1.7779E+3	4.2781E+1	−1.4281E+2
2.4	1.9138E+2	1.0236E+4	3.1753E+1	1.4236E+3
2.6	1.3276E+2	1.6263E+4	2.1925E+1	2.1955E+3
2.8	8.3842E+1	1.7674E+4	1.3763E+1	2.3593E+3
3.0	4.5990E+1	1.5942E+4	7.4771E+0	2.1127E+3
3.2	1.9177E+1	1.2467E+4	3.0481E+0	1.6416E+3
3.4	2.2549E+1	8.4233E+3	2.7242E−1	1.1010E+3
3.6	−6.6589E+0	4.6728E+3	−1.1723E+0	6.0390E+2
3.8	−9.7740E+0	1.7310E+3	−1.6591E+0	2.1675E+2

Source: Tassoul, J. L., and Tassoul, M., *Astrophys. J.*, **261**, 273, 1982.

In Table 4.4 we list the functions u and rv (in cm s^{-1}) for a $3M_\odot$ Cowling point-source model, with $n = 3$, $\kappa = 0.34$ cm^2 g^{-1}, and $u_1 = 6.45 \times 10^7$ cm s^{-1}. The values are listed for $N = 5$ ($\alpha_1 = 20$ and $\delta/R = 3.6 \times 10^{-3}$) and $N = 6$ ($\alpha_1 = 50$ and $\delta/R = 4.6 \times 10^{-3}$). Figure 4.9 illustrates the function $y(x)$ when $\alpha_1 = 20$. It is apparent from Table 4.4 and Eq. (4.139) that the axially symmetric circulation pattern consists of a main cell (or gyre) within the boundary layer ($0 \leq R - r \lesssim 0.01R$) and secondary cells at lower depths ($R - r \gtrsim 0.01R$). Because the flow speed decreases exponentially with optical depth, the dominant mass flow takes place within the outermost external layer of the absorbing star, however. The circulatory currents are symmetrical with respect to the line joining the centers of gravity, with rising motions in the "hot spot" ($\nu = 1$) and sinking motions at the antipode ($\nu = -1$). There is thus a mean steady current that is flowing away from the "hot spot" on the stellar surface and a mean steady countercurrent that is flowing away from the antipode at a somewhat lower level. The whole flow, in fact, takes place within a very thin superficial shell ($0.99 \lesssim r/R \leq 1$). Typically, with $\eta = 10^{-2}$ and $N = 6$, Table 4.4 indicates that $|u_r| \leq 0.85$ cm s^{-1} and $|u_\theta| \leq 520$ cm s^{-1}. Even though there are still uncertainties about these maxima (again because u and rv are quite sensitive to the values of N), these speed estimates are far removed from the various evaluations based on frictionless solutions that can be found in the literature. All these evaluations are utterly inadequate because they do not satisfy the kinematic boundary condition (4.38).

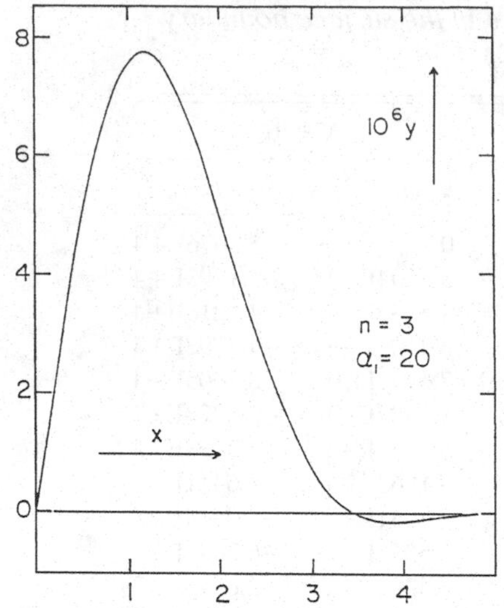

Fig. 4.9. Function $y(x)$ in the surface boundary layer, when $n = 3$ and $\alpha_1 = 20$. *Source:* Tassoul, J. L., and Tassoul, M., *Astrophys. J.*, **261**, 273, 1982.

4.7 Meridional circulation in a magnetic star

In Section 4.3.1 we have obtained a self-consistent description of meridional streaming and concomitant differential rotation in the chemically homogeneous envelope of an early-type, nonmagnetic star. Since these matters have been largely clarified by now, here we shall go a step further and discuss the role of a prescribed magnetic field in an early-type star. For the sake of simplicity, we shall assume that the large-scale field is not maintained by a contemporary dynamo operating in the convective core, but rather that it is the slowly decaying relic of the field present in the gas from which the star formed.

4.7.1 *The magnetically driven currents*

In an inertial frame of reference, the momentum equation for the large-scale flow becomes

$$\frac{D\mathbf{v}}{Dt} = -\operatorname{grad} V - \frac{1}{\rho}\operatorname{grad} p + \frac{1}{\rho}\mathbf{F}(\mathbf{v}) + \frac{1}{4\pi\rho}\operatorname{curl}\mathbf{H}\times\mathbf{H}, \tag{4.146}$$

where **H** denotes the mean magnetic field and **F** is the turbulent viscous force per unit volume (see Section 3.6). We also have

$$\operatorname{div}\mathbf{H} = 0 \tag{4.147}$$

and

$$\frac{\partial\mathbf{H}}{\partial t} = \operatorname{curl}(\mathbf{v}\times\mathbf{H}) - \operatorname{curl}(\beta\operatorname{curl}\mathbf{H}), \tag{4.148}$$

where β is to be interpreted as the coefficient of *magnetic eddy diffusivity* in the turbulent radiative envelope. For the sake of simplicity, we shall assume that $\beta = \bar{\beta}T^{-\nu}$, where

$\bar{\beta}$ and ν (>0) are two constants. As usual, these equations must be combined with Eqs. (4.3)–(4.6).

Because we are considering a rotating magnetic star that does not greatly depart from spherical symmetry, the large-scale meridional flow is the linear superposition of rotationally driven currents and magnetically driven currents. To calculate these currents, we shall prescribe an axially symmetric *dipolar* field. Neglecting the circulation velocity and letting $\mathbf{H} = \mathbf{P}$ in Eq. (4.148), we thus have, in spherical polar coordinates (r, $\mu = \cos\theta, \varphi$),

$$\mathbf{P} = \bar{H}\left[\mathcal{P}(r,t)P_1(\mu)\mathbf{1}_r + \mathcal{Q}(r,t)(1-\mu^2)\frac{dP_1(\mu)}{d\mu}\mathbf{1}_\mu\right], \tag{4.149}$$

where $\bar{H}$ is a constant and $P_1(\mu) = \mu$. We have

$$\mathcal{P} = p_m(r)\exp(-\sigma t) \tag{4.150}$$

and

$$\mathcal{Q} = \frac{1}{2}\frac{1}{r^2}\frac{d}{dr}\left(r^2 p_m\right)\exp(-\sigma t), \tag{4.151}$$

with $p_m(R) = 1$ so that $\bar{H}$ is the initial polar field strength. The constant σ is the lower eigenvalue of

$$p_m'' + \frac{4}{r}p_m' + \frac{\sigma}{\bar{\beta}}T^\nu p_m = 0, \tag{4.152}$$

with $Rp_m' + 3p_m = 0$ at $r = R$, and p_m finite at $r = 0$. Given this large-scale magnetic field, we shall now expand about hydrostatic equilibrium in powers of the nondimensional parameter

$$\lambda = \frac{\bar{H}^2 R^4}{GM^2}. \tag{4.153}$$

To the decaying dipolar field $\mathbf{P}$ corresponds the following meridional velocity:

$$\mathbf{u} = \lambda u(r,t)P_2(\mu)\mathbf{1}_r + \lambda v(r,t)(1-\mu^2)\frac{dP_2(\mu)}{d\mu}\mathbf{1}_\mu, \tag{4.154}$$

where

$$u = u_m(r)\exp(-2\sigma t) \quad \text{and} \quad v = v_m(r)\exp(-2\sigma t). \tag{4.155}$$

As usual, we also have

$$v_m = \frac{1}{6}\frac{1}{\rho r^2}\frac{d}{dr}\left(\rho r^2 u_m\right) \tag{4.156}$$

and $2P_2(\mu) = 3\mu^2 - 1$. (Recall that $u_\theta = -ru_\mu/\sin\theta$.)

Following Section 4.3.1, boundary-layer theory was used to calculate the functions u_m and v_m. Extensive numerical results have been obtained for a Cowling point-source model embedded into a vacuum ($M = 3M_\odot$, $R = 1.75R_\odot$, $L = 93L_\odot$), with electron-scattering opacity and with $\mu_V = 10^5\mu_{\rm rad}$ in the boundary layers. Equations (4.9) and (4.153) become, therefore,

$$\epsilon = 3 \times 10^{-6}\, v_{\rm eq}^2 \qquad \text{and} \qquad \lambda = 9 \times 10^{-17}\, \bar{H}^2, \tag{4.157}$$

where the equatorial velocity $v_{\rm eq}$ ($= \Omega_0 R$) and the constant $\bar{H}$ are measured in km s^{-1} and gauss, respectively. (Letting $v_{\rm eq} = 60$ km s^{-1} and $\bar{H} = 10^3$ G, one has $\epsilon \approx 10^{-2}$ and $\lambda \approx 10^{-10}$.) Now, solving Eq. (4.152) for the decay time τ_p ($= \sigma^{-1}$), one finds that $\bar{\beta}\tau_p = 4.2 \times 10^{23}$ (when $\nu = 1.5$) and $\bar{\beta}\tau_p = 5.4 \times 10^{37}$ (when $\nu = 3.5$). For example, if we neglect turbulence altogether, β becomes equal to its ideal value $\nu_m = 10^{13}\, T^{-3/2}$ cm^2 s^{-1} so that one has $\tau_p = 4 \times 10^{10}$ yr. Obviously, shorter decay times can be obtained by choosing other values for the free parameters $\bar{\beta}$ and ν; these times must be compared with the main-sequence lifetime $t_{\rm ms}$ of a $3M_\odot$ star, which is of the order of 2×10^8 yr.

As was already noted, correct to the orders ϵ and λ, the large-scale meridional flow is the vectorial sum of rotationally driven currents and magnetically driven currents. Henceforth we shall call them the "Ω-currents" and the "H-currents," respectively. Table 4.5 lists the functions u, rv, u_m, and rv_m, in cgs units, when $\nu = 1.5$ and $\nu = 3.5$. All entries in the last four columns must be multiplied by the exponential factor $\exp(-2t/\tau_p)$; they do not depend on $\bar{\beta}$. Even when $\bar{H}$ is as large as 10^3–10^4 G, one readily sees that the steady Ω-currents are much faster than the slowly decaying H-currents in the bulk of the radiative envelope. (At $r = 0.6R$, one has $|u_r| \approx 10^{-6}$ cm s^{-1} for the Ω-currents, whereas $|u_r| \approx 4 \times 10^{-10}$ cm s^{-1} at $t = 0$ for the H-currents). Just below the surface, however, λu_m may become larger than ϵu, in spite of the fact that both u_m and u vanish at the top of the boundary layer (i.e., at $r = R$). The presence of sizable H-currents in the outermost surface layers of a magnetic star is not at all unexpected, since it is only in the low-density surface regions that the Lorentz force can generate sizable departures from spherical symmetry. Figure 4.10 illustrates the complex circulation pattern of the H-currents; correct to $\mathcal{O}(\lambda)$, it does not depend on the polar field strength. Figure 4.10 must be compared with Figure 4.3, which depicts the corresponding Ω-currents.

From Table 4.5, one readily sees that the values of u_m and v_m are quite sensitive to the exponent ν, that is to say, to the magnitude of the coefficient of magnetic eddy diffusivity. Thus, even though it is the eddy viscosity that ultimately prevents unwanted singularities in the circulation velocities at the surface, the role of the magnetic eddy diffusivity is nevertheless an essential one in the sense that it considerably reduces the magnitude of these velocities near the surface. Unless one makes the unrealistic demand that the motions be strictly laminar in a chemically homogeneous, fully ionized radiative envelope, there is no reason to select the value $\nu = 1.5$, however. This should be especially true because hydrogen is only partially ionized at the surface of many magnetic stars, thus increasing the diffusion coefficient β.

Now, because the Ω-currents and the H-currents are neatly separated to the orders ϵ and λ, these solutions can be used also to obtain the circulation pattern when the axis of the basic dipolar field is inclined at an angle χ to the rotational axis. Because we already know that the H-currents play a negligible role in the bulk of a radiative envelope, we shall

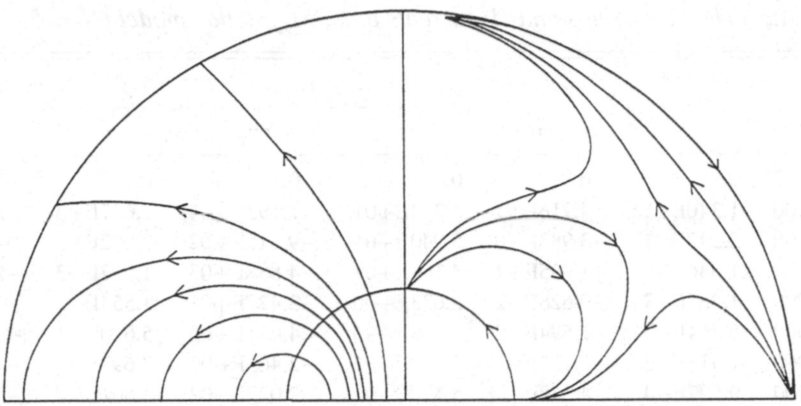

Fig. 4.10. Lines of force of the dipolar magnetic field (*left*) and streamlines of the corresponding quadrupolar circulation (*right*), when $\nu = 3.5$. The shape of these curves does not depend on the polar field strength. Recall that the dipolar field decreases as $\exp(-t/\tau_p)$ whereas the meridional currents decrease as $\exp(-2t/\tau_p)$. Note also the accumulation of streamlines near the core boundary and near the free surface. *Source:* Tassoul, J. L., and Tassoul, M., *Astrophys. J.*, **310**, 786, 1986.

merely follow in time the distortions of an initially dipolar field that may be caused by the Ω-currents alone. Three-dimensional calculations show beyond any doubt that the slow but inexorable Ω-currents will indeed convert an initially inclined dipolar field into a more complex field that has a larger inclination over the rotation axis. Figure 4.11 illustrates the evolution of an initially dipolar field, with initial $\chi = 45°$, $\epsilon = 10^{-3}$, and $\tau_p = 2 \times 10^8$ yr. (The rotation axis is set to be vertical.) Accordingly, assuming a modest increase of the coefficient β over its ideal value ν_m ($\beta/\nu_m \approx 10^2$) and choosing a rotation that is typical for a magnetic star ($\epsilon \approx 10^{-3}$), we have shown that the Ω-currents are by far too inefficient to produce a perpendicular rotator over the main-sequence lifetime of a $3M_\odot$ star. In other words, because the field lines can more easily diffuse through a less-than-ideal body, one has random orientation of the axes, whereas in an idealized stellar model (with $\beta \equiv \nu_m$) one has an excess of perpendicular rotators. Since the observed distribution of the obliquities seems to be at most a marginal nonrandom one, these calculations corroborate the idea that small-scale, eddylike motions comprise an essential ingredient of the many theoretical problems that are raised by rotation and magnetism in a stellar radiative envelope.

4.7.2 Circulation, rotation, and magnetic fields

In Section 4.7.1 we have calculated the meridional flow in the radiative envelope of a rotating magnetic star, assuming that departures from spherical symmetry are not too large. As we know, these rotationally and magnetically driven currents advect angular momentum and, hence, interact with the azimuthal motion. In the absence of a magnetic field, the transport of specific angular momentum can be made to balance the viscous force arising from differential rotation so that the mean state of motion is a steady one (see Eqs. [4.55]–[4.57]). When a large-scale magnetic field pervades the system, however, this balance is modified by the presence of the toroidal component of the Lorentz force. This is the reason why the claim has often been made that there exists a weak, axially

Table 4.5. *The Ω-currents and H-currents in a 3$M_\odot$ stellar model (N = 5).*

			$\nu = 1.5$		$\nu = 3.5$	
r/R	u	rv	u_m	rv_m	u_m	rv_m
0.283182	0.	0.	0.	0.	0.	0.
0.283400	1.340E−1	4.716E+1	3.384E+01	1.193E+04	1.637E+3	5.754E+5
0.283600	2.347E−1	−3.983E+0	5.940E+01	−9.815E+02	2.862E+3	−4.984E+4
0.283800	1.640E−1	−1.935E+1	4.160E+01	−4.888E+03	1.997E+3	−2.364E+5
0.290000	1.355E−2	−9.628E−2	3.633E+00	−2.427E+01	1.554E+2	−1.178E+3
0.300000	5.414E−3	−1.594E−2	1.579E+00	−4.031E+00	5.620E+1	−1.949E+2
0.350000	1.414E−3	−1.125E−3	5.852E−01	−2.463E−01	8.697E+0	−1.218E+1
0.400000	9.577E−4	−4.370E−4	5.428E−01	−2.027E−02	3.601E+0	−3.838E+0
0.450000	8.798E−4	−2.739E−4	6.869E−01	1.244E−01	2.185E+0	−1.893E+0
0.500000	9.566E−4	−2.347E−4	1.050E+00	3.644E−01	1.720E+0	−1.213E+0
0.550000	1.154E−3	−2.717E−4	1.843E+00	9.096E−01	1.652E+0	−9.324E−1
0.600000	1.484E−3	−4.149E−4	3.633E+00	2.328E+00	1.867E+0	−8.139E−1
0.650000	1.983E−3	−7.627E−4	8.022E+00	6.485E+00	2.435E+0	−7.586E−1
0.700000	2.699E−3	−1.533E−3	2.014E+01	2.060E+01	3.656E+0	−6.545E−1
0.750000	3.695E−3	−3.187E−3	5.970E+01	7.924E+01	6.414E+0	−1.207E−1
0.800000	5.044E−3	−6.745E−3	2.249E+02	4.060E+02	1.373E+1	2.833E+0
0.850000	6.824E−3	−1.472E−2	1.249E+03	3.306E+03	3.953E+1	2.293E+1
0.900000	9.116E−3	−3.480E−2	1.435E+04	6.315E+04	1.928E+2	2.604E+2
0.950000	1.200E−2	−1.062E−1	1.030E+06	1.014E+07	3.419E+3	1.288E+4
0.986000	1.461E−2	−5.019E−1	4.058E+09	1.052E+12	1.097E+6	4.772E+7
0.990000	1.418E−2	−7.772E−1	8.503E+10	1.040E+12	4.637E+6	−5.647E+7
0.995000	8.944E−3	−1.154E+0	1.937E+11	−2.036E+13	6.421E+6	−7.520E+8
0.999000	1.872E−3	−1.246E+0	5.452E+10	−3.622E+13	1.482E+6	−9.882E+8
1.000000	0.	−1.255E+0	0.	−3.646E+13	0.	−9.838E+8

Source: Tassoul, J. L., and Tassoul, M., *Astrophys. J.*, **310**, 786, 1986.

symmetric magnetic field that can offset the advection of specific angular momentum and so keeps the rotation effectively uniform in space, with little or no turbulent motions in the radiative envelope. It is the purpose of this section to conduct an examination of the ways a large-scale magnetic field can indeed maintain almost uniform rotation in the radiative envelope of an early-type star (see also Section 5.4.2).

Let us first assume that the magnetic field is symmetric about the rotation axis. Expressing the mean velocity **v** and the mean magnetic field **H** as the sum of poloidal and toroidal parts,

$$\mathbf{v} = \mathbf{u} + \Omega\varpi\mathbf{1}_\varphi \qquad \text{and} \qquad \mathbf{H} = \mathbf{P} + \mathcal{T}\varpi\mathbf{1}_\varphi, \tag{4.158}$$

we can thus write the φ components of Eqs. (4.146) and (4.148) in the forms

$$\rho\frac{\partial}{\partial t}(\Omega\varpi^2) + \rho\mathbf{u}\cdot\operatorname{grad}(\Omega\varpi^2) = \operatorname{div}(\rho\nu\varpi^2\operatorname{grad}\Omega) - \frac{1}{4\pi}\mathbf{P}\times\operatorname{curl}(\mathcal{T}\varpi\mathbf{1}_\varphi) \tag{4.159}$$

Fig. 4.11. Evolution of an initially dipolar magnetic field, with $\chi = 45°$, $\epsilon = 10^{-3}$, and $\tau_p = 2 \times 10^8$ yr. The rotation axis is vertical. The lines of force, which do penetrate the convective core, are depicted in the radiative envelope only: at $t = 5 \times 10^4$ yr, at $t = 4.2 \times 10^7$ yr, at $t = \tau_p$, and in the asymptotic limit $t \to \infty$. *Source:* Tassoul, J. L., and Tassoul, M., *Astrophys. J.*, **310**, 805, 1986.

and

$$\frac{\partial \mathcal{T}}{\partial t} + \text{div}(\mathcal{T}\mathbf{u}) = \mathbf{P} \cdot \text{grad}\ \Omega - \frac{1}{\varpi}\{\text{curl}[\beta\, \text{curl}(\mathcal{T}\varpi \mathbf{1}_\varphi)]\}_\varphi. \tag{4.160}$$

In principle, we must calculate the functions **u**, Ω, **P**, and $\mathcal{T}$ from Eqs. (4.146) and (4.148) and the auxiliary equations (4.3)–(4.6). Following standard practice, however, we shall assume some plausible forms for the circulation velocity **u** and the poloidal magnetic field **P**. Accordingly, it is possible to calculate the functions Ω and $\mathcal{T}$ from Eqs. (4.159) and (4.160), with **u** and **P** being two prescribed vectors. As usual, these coupled parabolic equations must be solved with some initial conditions and a prescribed set of boundary conditions at the core boundary $r = R_c$ and at the free surface $r = R$. For the sake of simplicity, all numerical calculations reported below were made for a spherical fluid shell with constant density ρ, constant kinematic viscosity ν $(= \mu_V/\rho)$, and constant magnetic diffusivity β. The convective core is assumed to be maintained in strict uniform rotation. Two extensive sets of numerical calculations have been made.

Tassoul and Tassoul (1989) have considered a quadrupolar magnetic field, vanishing at the core–envelope interface and at the free surface, and a prescribed quadrupolar meridional circulation. Specifying initial states of uniform or almost uniform rotation, they obtained solutions that are characterized by an inexorable approach to *a state of isorotation* (i.e., rotation with constant angular velocity along each field line) with large differential rotation between field lines after about ten Alfvén times, with no apparent trend toward solid-body rotation.*

Moss, Mestel, and Tayler (1990) have considered a dipolar magnetic field, fully anchored into the convective core and threading the free surface, and a quadrupolar meridional circulation. They also introduced a high-viscosity buffer zone above the core–envelope interface, in which, however, they retained low values for the magnetic diffusivity. Starting from a state of almost uniform rotation, they obtained solutions in which there is a periodic, low-amplitude shear reversal about a state of uniform rotation, along with spatially extended latitudinal oscillations in the toroidal magnetic field.

How can one explain the differences between these two independent sets of calculations? Recall first that, in both works, the magnetic field is symmetric about the rotation axis so that the magnetic transport of angular momentum in the radiative envelope takes place along but not across the poloidal field lines. Accordingly, if these lines thread neither the free surface nor the core–envelope interface, viscous friction is the only mechanism that can potentially couple different field lines. Hence, the redistribution of angular momentum takes place along the field lines through the propagation of Alfvén waves. Since ohmic dissipation acts to damp out these waves, it will thus enforce a constant angular velocity along each poloidal field line, although this constant is in general different for each field line. This solution corresponds to the state of isorotation that was obtained by Tassoul and Tassoul (1989). Obviously, this is quite different from the situation in which the poloidal field lines are anchored into the rigidly rotating core. If so, then, there is a coupling between the convective core and the radiative envelope, so that significant mutual coupling of different poloidal field lines will occur. As was noted by Moss, Mestel, and Tayler (1990), it is the anchoring of all poloidal field lines in a rigidly rotating, strongly viscous convective core that is ultimately responsible for the establishment of a state of almost uniform rotation, on a time shorter than the main-sequence lifetime of an early-type star. Of course, if the core is not rotating as a solid body or if some poloidal field lines do not penetrate into the core, the large-scale poloidal magnetic field will not necessarily enforce almost uniform rotation throughout the radiative envelope.

It is appropriate at this juncture to briefly discuss the work of Charbonneau and MacGregor (1992), who have studied the rotational evolution of an initially nonrotating radiative envelope, following an impulsive spin-up of the core to a constant angular velocity. This was accomplished by solving Eqs. (4.159) and (4.160) for a given axially symmetric vector **P**, neglecting all fluid motions other than the azimuthal flow

* The concept of isorotation – as opposed to solid-body rotation – has its roots in the work of Ferraro (1937) and Mestel (1961).

associated with the evolving differential rotation (i.e., letting $\mathbf{u} \equiv \mathbf{0}$ in these equations). For fully core-anchored poloidal field configurations, they found that uniform rotation is always enforced on a time very much shorter than the main-sequence lifetime, yet generally much larger than the core-to-surface Alfvén transit time. However, they also found that the relatively rapid transition toward uniform rotation depends critically on all poloidal field lines having at least one footpoint anchored on the rigidly rotating core. This is well illustrated by their unanchored and partially anchored solutions, which in many cases either do not attain solid-body rotation or do so on a purely viscous time scale.

We can only conclude from these three sets of calculations that the extent to which a weak poloidal magnetic field can produce a state of almost uniform rotation in a stellar radiative envelope depends critically on assumptions regarding the behavior of the field lines at the core–envelope interface. To be specific, *if the convective core is not maintained in strict uniform rotation, or if all field lines are not fully anchored into the core, the configuration does not converge toward a state of almost uniform rotation in the radiative envelope.* Given our almost complete ignorance of the state of motion in a convective core and of whether all poloidal field lines do penetrate into the core of an early-type star, there is thus no reason to claim that there always exists an axially symmetric magnetic field that can enforce almost uniform rotation despite the inexorable advection of angular momentum by the meridional currents. Such a magnetic field may or may not exist, depending on the field-line topology and rotation in the convective core.

Of course, as was noted by Moss (1992) and others, if the poloidal magnetic field is not symmetric about the rotation axis, nonuniform rotation will generate magnetic torques that can interchange angular momentum between poloidal field lines. Preliminary calculations have been made when the magnetic axis is perpendicular to the rotation axis, suggesting that the azimuthal magnetic forces do indeed establish almost uniform rotation in the radiative envelope of a perpendicular rotator. In my opinion, no firm conclusion can be made until independent studies present reliable calculations of a large number of fully three-dimensional models having arbitrary inclinations. This is another way of saying that, contrary to an often held belief, the presence of a large-scale magnetic field does not make the problem of stellar rotation any simpler.

4.8 Discussion

Self-gravitation and self-generated radiation are the two main factors that make most problems of stellar hydrodynamics quite different from those encountered in the geophysical sciences and in laboratory hydrodynamics. Self-gravitation acts as the "container" of a star, making its outer surface free rather than solid. Hence, it is self-gravitation that allows for the small departures from spherical symmetry – regardless of whether their ultimate cause is the centrifugal force, the Lorentz force, or the tidal interaction with a companion. Moreover, as explained in this chapter, it is the transport of self-generated radiation in a nonspherical configuration that causes the slow but inexorable currents and concomitant differential rotation in a stellar radiative zone.

In the case of a slowly rotating, early-type star, the large-scale meridional flow is quadrupolar in structure, with rising motions at the poles and sinking motions at the equator. Typically, the time scale of these thermally driven currents in the bulk of a radiative envelope is equal to the Kelvin–Helmholtz time divided by the ratio of centrifugal force to gravity at the equator (see Eq. [4.37]), which is known as the Eddington–Sweet time.

The complexity of the problem derives from the fact that turbulent friction becomes of paramount importance near the core–envelope interface and the free surface. To be specific, near each of these boundaries, a thin layer exists in which turbulent processes allow the velocity to make the transition from the value required by the nature of the boundary to the value that is appropriate to the interior, frictionless solution. Simultaneously, the frictional force acting on the mean azimuthal flow can be made to balance the transport of angular momentum by the large-scale meridional flow, therefore ensuring that all three components of the momentum equation are properly satisfied.

As far as hydrodynamics is concerned, perhaps the most challenging feature of these motions is that they bear no relation whatsoever to the large-scale circulation encountered in geophysics.

For example, as was seen in Section 2.5.1, for large-scale atmospheric motions away from the Earth's surface the balance is essentially geostrophic (see Eq. [2.79]). On the contrary, one readily sees that Eq. (4.57) defines the balance between the turbulent viscous force acting on the mean azimuthal motion and the inexorable transport of angular momentum by the thermally driven currents. When written in a rotating frame of reference, this equation merely states that the Coriolis force acting on the meridional circulation balances the azimuthal viscous force. In other words, *the concept of geostrophy does not apply to the thermally driven currents in a nonspherical stellar radiative envelope.*

A comparison between the results obtained in Sections 2.5.2 and 4.3.1 also shows that *the boundary layers in a stellar radiative envelope are definitely not of the Ekman type.* To be specific, because the meridional flow is essentially caused by the nonspherical part of the temperature field, these boundary layers are of the mixed *thermo-viscous* type (see Eq. [4.60]). That is to say, whereas turbulent friction plays a dominant role in the energy equation, the mechanical balance is mainly between the pressure-gradient force and the effective gravity; the viscous force is very small in the equations of motion themselves. These boundary layers are also of a singular nature because it is not possible to obtain the boundary-layer solutions by merely adding thermo-viscous corrections to the interior, frictionless solution (see Eqs. [4.71], [4.76], and [4.100]). To the best of my knowledge, there is no equivalent in any other field.

Admittedly, in Section 4.3 we have made use of steady state models to represent the largest scale of motion in a stellar radiative envelope, while applying parametric expressions to describe the effects of all smaller scales. These solutions are basically very similar to the linear and nonlinear solutions that were obtained for oceanic boundary currents: Bryan's nonlinear solution smoothly reduces to Munk's *linear* solution as the eddy viscosity is gradually increased (see Section 2.6.3). More recently, because it has been established that mid-ocean eddies ($\sim$ 50 km) are prevalent, their models have been superseded by high-resolution models that include this eddy field within the large-scale oceanic circulation. In principle, a similar improvement could be made in the case of a stellar radiative zone, taking into account the smaller scales of motion. Unfortunately, very little is known about the intensity, length and time scales, and the spatial distribution of these transient motions. At this writing, it is therefore quite difficult to resolve the eddy field and, at the same time, the large-scale flow in the radiative envelope of a rotating star.

4.9 Bibliographical notes

Sections 4.1 and 4.2. The existence of meridional currents was originally suggested by:

1. Vogt, H., *Astron. Nachr.*, **223**, 229, January 1925.
2. Eddington, A. S., *The Observatory*, **48**, 73, March 1925.

Their time scale was discussed in:

3. Eddington, A. S., *Mon. Not. R. Astron. Soc.*, **90**, 54, 1929.
4. Sweet, P. A., *Mon. Not. R. Astron. Soc.*, **110**, 548, 1950.

Equation (4.37) was properly derived in Reference 4. The expansion method is due to:

5. Milne, E. A., *Mon. Not. R. Astron. Soc.*, **83**, 118, 1923.

A systematic study was made by:

6. Chandrasekhar, S., *Mon. Not. R. Astron. Soc.*, **93**, 390, 1933 (reprinted in *Selected Papers*, **1**, p. 183, Chicago: The University of Chicago Press, 1989).

Other pioneering contributions to the problem of meridional circulation were made by:

7. Gratton, L., *Mem. Soc. Astron. Italiana*, **17**, 5, 1945.
8. Öpik, E. J., *Mon. Not. R. Astron. Soc.*, **111**, 278, 1951.
9. Mestel, L., *Mon. Not. R. Astron. Soc.*, **113**, 716, 1953.
10. Baker, N., and Kippenhahn, R., *Zeit. Astrophys.*, **48**, 140, 1959.
11. Mestel, L., in *Stellar Structure* (Aller, L. H., and McLaughlin, D. B., eds.), p. 465, Chicago: The University of Chicago Press, 1965.
12. Mestel, L., *Zeit. Astrophys.*, **63**, 196, 1966.
13. Smith, R. C., *Mon. Not. R. Astron. Soc.*, **148**, 275, 1970.

Section 4.3. The importance of viscous friction was already noted in Reference 2. This idea was further studied in:

14. Randers, G., *Astrophys. J.*, **94**, 109, 1941.
15. Krogdahl, W., *Astrophys. J.*, **99**, 191, 1944.

The analysis in this section is taken from:

16. Tassoul, J. L., and Tassoul, M., *Astrophys. J. Suppl.*, **49**, 317, 1982.
17. Tassoul, J. L., and Tassoul, M., *Astrophys. J.*, **264**, 298, 1983.
18. Tassoul, M., and Tassoul, J. L., *Astrophys. J.*, **271**, 315, 1983.
19. Tassoul, J. L., and Tassoul, M., *Geophys. Astrophys. Fluid Dyn.*, **36**, 303, 1986.
20. Tassoul, M., *Astrophys. J.*, **427**, 388, 1994.

Adequate boundary-layer analyses, at the core and at the surface, were originally made in Reference 16.

As far as the surface boundary layer is concerned, the claim has been made that one can construct a nonsingular surface solution that satisfies all the boundary conditions, *without requiring any boundary layer*; see:

21. Sakurai, T., *Geophys. Astrophys. Fluid Dyn.*, **36**, 257, 1986.
22. Zahn, J. P., *Astron. Astrophys.*, **265**, 115, 1992.

As was shown in Reference 20, neglecting the viscous force acting on the meridional flow, one can obtain a solution that satisfies all the boundary conditions at the free surface. Unfortunately, this nonsingular solution does *not* satisfy all the basic equations – one component of the momentum equation remains necessarily unfulfilled. As a matter of fact, the Sakurai–Zahn approach is inadequate because it cannot take into account the mathematical singularity of the equations at the free surface (see Eq. [4.100], which has a *pole* at $x = 0$). It is also shown in Reference 20 that the dynamics in a slowly rotating, early-type star demands some frictional forces acting on the meridional flow to be present. This requires the consideration of singular, thermo-viscous boundary layers, both at the core and at the free surface.

Section 4.4. Our most detailed study of meridional circulation will be found in:

23. Tassoul, M., and Tassoul, J. L., *Astrophys. J.*, **440**, 789, 1995.

This paper contains many numerical illustrations, as well as several comments on the current literature.

Section 4.5. See:

24. Tassoul, M., and Tassoul, J. L., *Astrophys. J.*, **267**, 334, 1983.

Section 4.6. See:

25. Tassoul, J. L., and Tassoul, M., *Astrophys. J.*, **261**, 265, 1982; *ibid.*, p. 273.

The reflection effect in a nonsynchronous binary component was considered by:

26. Tassoul, M., and Tassoul, J. L., *Mon. Not. R. Astron. Soc.*, **232**, 481, 1988.

The reference to Hosokawa is to his paper:

27. Hosokawa, Y., *Sci. Rep. Tôhoku Univ., Ser. I*, **43**, 207, 1959.

Section 4.7.1. Magnetically driven currents are discussed in:

28. Tassoul, J. L., and Tassoul, M., *Astrophys. J.*, **310**, 786, 1986.

The effects of meridional streaming on an oblique rotator have been considered by:

29. Tassoul, J. L., and Tassoul, M., *Astrophys. J.*, **310**, 805, 1986.
30. Moss, D., *Mon. Not. R. Astron. Soc.*, **244**, 272, 1990.

Section 4.7.2. Since the early 1950s, rumor had it that there exists a weak poloidal magnetic field that is symmetric about the rotation axis and which can quickly enforce almost uniform rotation in a stellar radiative envelope, with little or no turbulent friction, despite the inexorable advection of angular momentum by the meridional currents. In Reference 28, as a sideline to the actual calculation of these currents, we expressed serious doubts about the existence in general of such a field. Not unexpectedly, we had opened Pandora's box, letting out the following sequel of papers:

31. Mestel, L., Moss, D. L., and Tayler, R. J., *Mon. Not. R. Astron. Soc.*, **231**, 873, 1988.
32. Tassoul, M., and Tassoul, J. L., *Astrophys. J.*, **345**, 472, 1989.
33. Moss, D. L., Mestel, L., and Tayler, R. J., *Mon. Not. R. Astron. Soc.*, **245**, 559, 1990.

Reference 31 presents a set of numerical calculations based on Eqs. (4.159) and (4.160), the results of which are summarized by the following claim: "in both the analytical and numerical treatments above, we have always *calculated* the Ω-field, showing the departures from uniformity to be small" (p. 883). In order to ascertain the universal validity of that statement, similar calculations were made in Reference 32, prescribing in Eqs. (4.159) and (4.160) a poloidal field **P** that is *not* anchored in the convective core. It was found that the configuration tends toward a state of isorotation that has a large gradient in the angular velocity near the rotation axis. Reference 33 presents additional calculations, recognizing the difference that exists between anchored and unanchored poloidal field lines. Since unanchored field lines do not necessarily enforce almost uniform rotation, it is now conjectured that a magnetic field that is *not* symmetric about the rotation axis is most likely to quickly establish almost uniform rotation in a stellar radiative envelope.

Other contributions are due to:

34. Charbonneau, P., and MacGregor, K. B., *Astrophys. J.*, **387**, 639, 1992.
35. Moss, D., *Mon. Not. R. Astron. Soc.*, **257**, 593, 1992.

See especially Reference 34, which also contains a useful comparison between their own results and those presented in References 31–33.

The references to Ferraro and Mestel are to their papers:

36. Ferraro, V. C. A., *Mon. Not. R. Astron. Soc.*, **97**, 458, 1937.
37. Mestel, L., *Mon. Not. R. Astron. Soc.*, **122**, 473, 1961.

5

Solar rotation

5.1 Introduction

Until recently, only surface measurements of the solar rotation rate were available. Since the mid-1980s, with the advent of helioseismology, much has been learned about the internal rotation of the Sun through the inversion of p-mode frequency splittings. As was noted in Section 1.2.2, it now appears that the observed surface pattern of differential rotation with latitude prevails throughout most of the solar convection zone, with equatorial regions moving faster than higher latitudes. In contrast, the underlying radiative core appears to rotate nearly uniformly down to $r \approx 0.1$–$0.2R_\odot$, at a rate that is intermediate between the polar and equatorial rates of the photosphere. Within the central region $r \lesssim 0.2R_\odot$, some measurements suggest that the angular velocity increases with depth, implying rotation at a rate between 2 and 4 times that of the surface; other measurements strongly suggest, however, that the solar inner core rotates rigidly down to the center.

The problem presented by the observed solar differential rotation is one of long standing and many efforts have been made to formulate a plausible flow pattern that reproduces the large-scale motions in the solar atmosphere. Following Lebedinski's (1941) pioneering work, many theories have been proposed to explain how the equatorial acceleration originated and is maintained in the solar convection zone. Broadly speaking, they can be divided into two classes, depending on the mechanism proposed to produce and maintain the equatorial acceleration: (i) the interaction of rotation with *local* turbulent convection and (ii) the interaction of rotation with *global* turbulent convection in a rotating spherical shell. Till the late 1980s, however, the most detailed models invariably predicted rotation profiles that were constant on cylinders concentric to the rotation axis. Obviously, these solutions are at variance with the current observations, which suggest an angular velocity that is constant on radii in the convection zone, at least at mid-latitudes. In Section 5.2 we shall explain how the disparities between the rotation profiles deduced from the helioseismological data and what has been predicted by these early models can be resolved.

Now, a number of recent observations has shown that solar-type stars undergo rotational deceleration as they slowly evolve on the main sequence (see Eq. [1.7]). As we shall see in Section 7.2, this spin-down is presumably the consequence of angular momentum loss via magnetically channeled stellar winds and/or sporadic mass ejections emanating from the surface layers. *The central question is how this inexorable braking of the outer convection zone will affect the rotational state of the radiative interior*. In the case of the Sun, the absence of marked differential rotation in the outer layers of its radiative core implies that angular momentum redistribution within that region must be very efficient indeed. Within the framework of the eddy–mean flow interaction presented in Section 3.6, three distinct mechanisms for angular momentum redistribution might be operative: (i) large-scale meridional currents, (ii) turbulent friction acting on the differential rotation, and (iii) large-scale magnetic fields. In Section 5.3 we discuss the time-dependent meridional flow in the Sun's radiative interior, taking into account the development with age of a gradient of mean molecular weight in the hydrogen-burning core. Section 5.4 presents quantitative studies of the rotational evolution of the Sun's radiative interior, with angular momentum being removed from the convective envelope to simulate the effects of the solar wind and/or episodic mass ejections.

5.2 Differential rotation in the convection zone

The interaction of rotation with convection appears to be the most likely mechanism for the generation of the observed solar differential rotation. Two different approaches have been proposed to explain the maintenance of differential rotation and concomitant meridional circulation. One class of models is based on the appealing assumption that the variations in angular velocity arise mainly from the nonlinear interaction of rotation with the largest scales of convection, when a radial superadiabatic gradient of temperature prevails. These *global-convection models* resolve numerically as many of the large scales as possible in a rotating spherical shell and parameterize, via eddy diffusivities, the transport of momentum and heat by all the smaller unresolved scales. In the other class of models, the role of global convection is assumed to be unimportant. What is essential in these *mean-field models* is the effect of the large-scale azimuthal flow on the local convective motions that are not greatly influenced by the Sun's spherical shape. As usual, following closely the method presented in Section 3.6, the role of this turbulent convection is parameterized by the use of eddy viscosities, which are specified functions of rotation. Unavoidably, this mathematical description of the interaction between turbulent convection and rotation depends on adjustable parameters.

5.2.1 *Mean-field models*

In this approach the large-scale motions in the solar convection zone are described by means of stationary, axially symmetric flow patterns, with turbulent convection giving rise to Reynolds stresses and eddy viscosity coefficients. In spherical polar coordinates (r, θ, φ), the mean velocity $\mathbf{v}$ is of the form

$$\mathbf{v} = \mathbf{u} + \Omega r \sin\theta \mathbf{1}_\varphi, \tag{5.1}$$

where $\mathbf{u}$ is the two-dimensional meridional velocity. Because we have assumed axial symmetry for the mean flow, mass conservation implies that

$$\operatorname{div}(\rho\mathbf{u}) = 0, \tag{5.2}$$

where ρ is the mean density.

For mean steady motions, the φ component of Eq. (3.123) becomes

$$\rho\mathbf{u} \cdot \operatorname{grad}(\Omega\varpi^2) = \frac{\sin\theta}{r^2}\frac{\partial}{\partial r}(r^3\sigma_{r\varphi}) + \frac{1}{\sin\theta}\frac{\partial}{\partial\theta}(\sin^2\theta\,\sigma_{\theta\varphi}), \tag{5.3}$$

where $\varpi = r \sin\theta$. This equation merely expresses the fact that turbulent friction acting on the differential rotation can be made to balance the transport of specific angular momentum, $\Omega\varpi^2$, by the meridional flow. If the influence of rotation and gravity was negligible, the turbulent transport of momentum would occur downward along the gradient of angular velocity, so that the Reynolds stresses $\sigma_{r\varphi}$ and $\sigma_{\theta\varphi}$ would be proportional to $\partial\Omega/\partial r$ and $\partial\Omega/\partial\theta$, respectively. However, because anisotropy prevails in a rotating fluid embedded in a gravitational field, the stresses $\sigma_{r\varphi}$ and $\sigma_{\theta\varphi}$ contain both diffusive and nondiffusive parts. Following Section 3.6, one has

$$\sigma_{r\varphi} = \mu_V r \frac{\partial\Omega}{\partial r} \sin\theta + \lambda_V \Omega \sin\theta \tag{5.4}$$

and

$$\sigma_{\theta\varphi} = s\mu_V \frac{\partial\Omega}{\partial\theta} \sin\theta + \lambda_H \Omega \cos\theta, \tag{5.5}$$

where $s = \mu_H/\mu_V$ is the *anisotropy parameter*. It is immediately apparent that the nondiffusive parts, which are proportional to the mean rotation rate, maintain rather than smooth out differential rotation in the solar convection zone. They depend on two independent parameters, λ_V and λ_H, which define the anisotropies in the vertical (i.e., along the effective gravity) and horizontal directions. Appropriate expansions for these parameters are

$$\lambda_V = \lambda_{V0}(r) + \lambda_{V1}(r) \sin^2\theta + \cdots \tag{5.6}$$

and

$$\lambda_H = \lambda_{H1}(r) \sin^2\theta + \cdots. \tag{5.7}$$

Note that the parameter λ_H, which is related to the anisotropy of turbulence in planes perpendicular to the effective gravity, vanishes at the poles. In principle, the radial functions λ_{V0}, λ_{V1}, and λ_{H1} may be derived from the equations governing the fluctuating part of the instantaneous velocity field (e.g., Rüdiger 1989).

Neglecting the inertial terms $\mathbf{u} \cdot \operatorname{grad} \mathbf{u}$, one can also rewrite Eqs. (3.125) and (3.126) for mean steady motions in the compact form

$$\operatorname{grad} V + \frac{1}{\rho} \operatorname{grad} p - \Omega^2 \varpi \mathbf{1}_\varpi = \frac{1}{\rho} \mathbf{F}_p(\mathbf{u}), \tag{5.8}$$

where $\mathbf{F}_p(\mathbf{u})$ is the poloidal part of the turbulent viscous force per unit volume acting on the meridional flow (see Eq. [3.123]) and $\mathbf{1}_\varpi$ is the radial unit vector in cylindrical polar coordinates (ϖ, φ, z). Taking the curl of Eq. (5.8), one obtains

$$\frac{1}{\rho^2} \operatorname{grad} p \times \operatorname{grad} \rho - \frac{\partial}{\partial z} \left(\Omega^2 \varpi\right) \mathbf{1}_\varphi = \mathbf{R}(\mathbf{u}), \tag{5.9}$$

where, for shortness, $\mathbf{R}(\mathbf{u})$ is the curl of the viscous force. If $\mathbf{R}(\mathbf{u})$ makes a negligible contribution to this equation, one readily sees that any barotropic model for the solar convection zone has the angular velocity constant on cylinders aligned with the rotation axis; that is, $p = p(\rho)$ implies that $\Omega = \Omega(\varpi)$, and conversely. This result is a mere consequence of the Poincaré–Wavre theorem (see Section 3.2.1). It is an important result, however, because we know that the angular velocity is not constant on cylinders within the solar convection zone. Since detailed models for the Sun indicate that $\mathbf{R}(\mathbf{u})$ is indeed negligible in the bulk of that zone, it follows that strict barotropy is most certainly an inadequate approximation for the solar rotation problem.

In Section 3.3.2 we have shown that the anisotropy of turbulent convection due to the preferred direction of gravity can produce differential rotation and meridional circulation in the solar convection zone. Since the early 1970s, a variety of models have been calculated, taking into account in an approximate manner the convective energy transport. To complete Eqs. (5.2), (5.3), and (5.8) we thus let

$$\rho T \mathbf{u} \cdot \operatorname{grad} S + \operatorname{div}(\mathcal{F} + \mathcal{F}_c) = 0. \tag{5.10}$$

The specific entropy is given by

$$S = c_V \log \frac{p}{\rho^{5/3}} + constant, \tag{5.11}$$

where c_V is the specific heat at constant volume. The radiative flux is given by the standard expression

$$\mathcal{F} = -\frac{4ac}{3}\frac{T^3}{\kappa\rho} \operatorname{grad} T \tag{5.12}$$

(see Eqs. [3.5] and [3.6]), and the convective flux is taken to be of the form

$$\mathcal{F}_c = -\kappa_c T \operatorname{grad} S, \tag{5.13}$$

where $\kappa_c(r)$ is the turbulent heat transport coefficient. One also has

$$p = \frac{\mathcal{R}}{\bar{\mu}} \rho T, \tag{5.14}$$

where $\bar{\mu}$ is the mean molecular weight. As usual, this set of equations must be solved with appropriate boundary conditions at the base and at the top of the rotating spherical shell.

Baroclinic models based on the concept of anisotropic eddy viscosity exhibit angular velocity profiles that are not constant on cylinders. They also produce a slow meridional flow, with typical surface velocities of the order of 1 m s^{-1}. Moreover, all these baroclinic models have very small (≈ 1 K) pole–equator temperature differences. Unfortunately, in order to reproduce the observed equatorial acceleration, the anisotropy parameter s ($= \mu_H/\mu_V$) must be larger than one. This is a most surprising result since one expects turbulent convection to provide more transport in the radial than in the horizontal directions. This inadequacy of these solutions strongly suggests that Lebedinski's (1941) anisotropic eddy viscosity might not be the ultimate cause of the Sun's differential rotation.

As was originally pointed out by Weiss (1965), the solar differential rotation could be generated by meridional currents driven by a pole–equator temperature difference. This approach is based on the fact that rotation has a small but significant influence upon turbulent convection, thus resulting in a convective heat transport that depends on heliocentric latitude. This gives rise to an inexorable meridional flow that transports angular momentum toward the equator and thus sustains the differential rotation. Following this idea, several authors have developed models of differentially rotating spherical shells – assuming a latitude-dependent heat transport coefficient $\kappa_c(r, \theta)$ and an isotropic eddy viscosity. Many of these models succeed in maintaining angular velocity profiles that are not constant on cylinders. However, some of them have meridional velocities at the surface that are too large, while others have pole–equator temperature differences that are too large.

More recently, Kitchatinov and Rüdiger (1995) have pointed out that the conflict between mean-field models and solar observations can be resolved by taking into account an anisotropic eddy viscosity as well as an anisotropic turbulent heat transport. Thus, instead of letting $\kappa_c = \kappa_c(r)$ or $\kappa_c = \kappa_c(r, \theta)$ in Eq. (5.13), they prescribe that the convective heat flux has the components, in Cartesian coordinates,

$$\mathcal{F}_{ci} = -\rho c_P \sum_{j=1}^{3} \chi_{ij} \left[\frac{\partial T}{\partial x_j} - \left(\frac{\partial T}{\partial x_j} \right)_{\mathrm{ad}} \right], \tag{5.15}$$

where $(\partial T/\partial x_j)_{\mathrm{ad}}$ is the adiabatic gradient of mean temperature and χ_{ij} is a tensor describing the turbulent heat transport ($i = 1, 2, 3$). As was done for the eddy viscosities and related coefficients, the components of this tensor can be obtained from the equations governing the fluctuating quantities (e.g., Rüdiger 1989). Fortunately, these models involve only one adjustable parameter, which is the ratio of the mixing length to the pressure-scale height. Figure 5.1 illustrates one particular solution. Note that the angular velocity distribution closely fits the helioseismological data reported in Section 1.2.2, with the rotation becoming virtually rigid below the convection zone. This model has a small (≈ 5 K) pole–equator temperature difference, which is consistent with the observations. However, it also predicts a slow equatorward meridional motion on the free surface, which is not observed in the Doppler measurements (see Section 1.2.1). Nonetheless, this is the first mean-field model that satisfies almost all the observational constraints. Given this result, it thus seems highly probable that anisotropy plays a key role in the solar rotation problem, since calculations involving isotropic transport coefficients always yield angular velocities that are constant on cylinders in the models. This effect is illustrated in the bottom part of Figure 5.2, which depicts a model corresponding to an isotropic thermal conductivity.

5.2.2 *Global-convection models*

In the global-convection theories of the Sun's differential rotation the largest convective cells are influenced by rotation, leading to a continuous redistribution of angular momentum, which we observe as a differential rotation. Actually, it is the combined effect of the spherical geometry and the Coriolis force acting on these large-scale convective motions that generates variations with latitude and radius of the angular velocity. Extensive numerical calculations have been made, independently, by Gilman and Glatzmaier in the early 1980s. Their models solve the nonlinear, three-dimensional, time-dependent equations for thermal convection in a rotating spherical shell of compressible fluid. Both sets of models are based on the assumption that the convective velocities are small compared to the local sound speed, thus filtering out the pressure waves. Moreover, because it is not possible to resolve all scales of motion, from the largest to the smallest, it is also assumed that the small unresolved scales give rise to viscous and thermal diffusivities, which are specified functions of the coordinates.

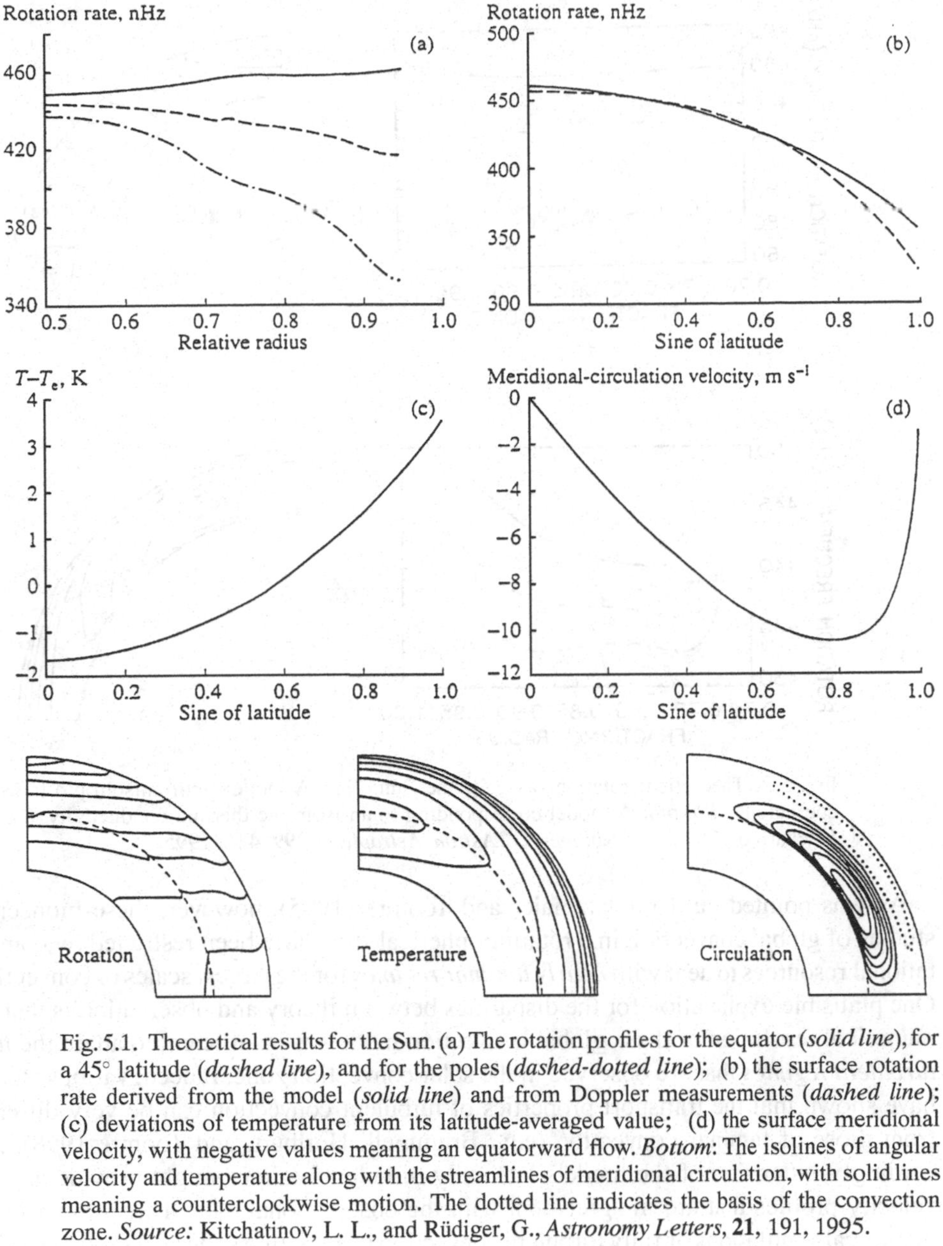

Fig. 5.1. Theoretical results for the Sun. (a) The rotation profiles for the equator (*solid line*), for a 45° latitude (*dashed line*), and for the poles (*dashed-dotted line*); (b) the surface rotation rate derived from the model (*solid line*) and from Doppler measurements (*dashed line*); (c) deviations of temperature from its latitude-averaged value; (d) the surface meridional velocity, with negative values meaning an equatorward flow. *Bottom*: The isolines of angular velocity and temperature along with the streamlines of meridional circulation, with solid lines meaning a counterclockwise motion. The dotted line indicates the basis of the convection zone. *Source:* Kitchatinov, L. L., and Rüdiger, G., *Astronomy Letters*, **21**, 191, 1995.

Although the numerical techniques employed in these models are quite different, the results obtained by Gilman and Glatzmaier are qualitatively the same. In particular, it is found that their simulated global convection in a rotating spherical shell tends to take the form of north–south (banana) rolls, the tilting of which yields Reynolds stresses to drive the zonal flows that maintain differential rotation. Unfortunately, in these early models the simulated angular velocity in the convection zone is constant on cylinders coaxial with the rotation axis, which is not in agreement with the helioseismological data reported in Section 1.2.2.

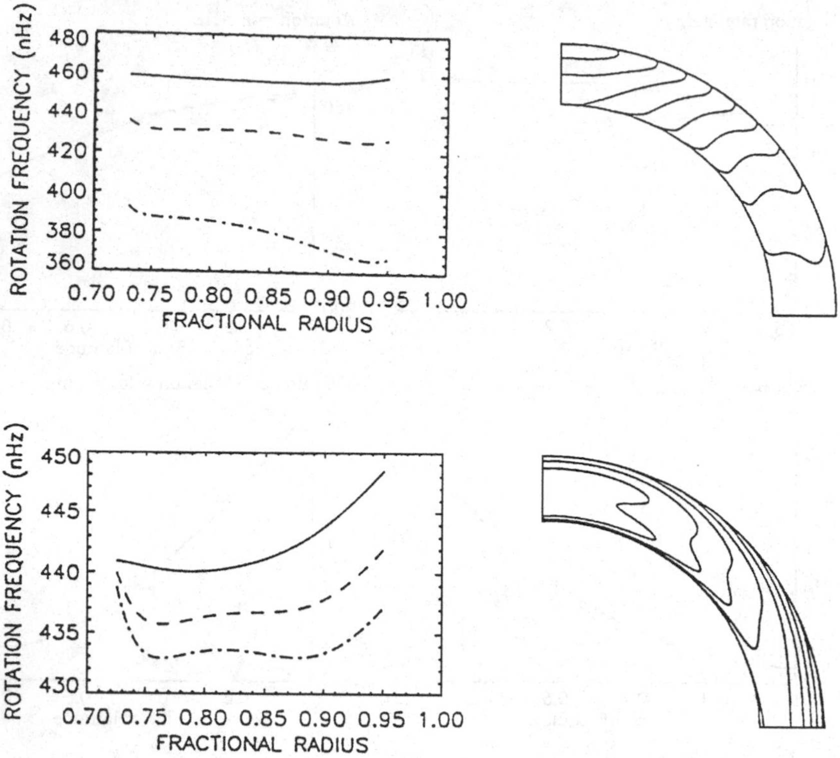

Fig. 5.2. Theoretical rotation rates for the Sun. *Top*: A model with anisotropic transport coefficients. *Bottom*: A model corresponding to an isotropic thermal conductivity. *Source:* Kitchatinov, L. L., and Rüdiger, G., *Astron. Astrophys.*, **299**, 446, 1995.

As was pointed out by Glatzmaier and Toomre (1995), however, these pioneering studies of global convection in a rotating spherical shell have been restricted by computational resources to deal with *nearly laminar regimes* for the largest scales of convection. One plausible explanation for the disparities between theory and observation is that the numerical resolution of these global-convection models is insufficient to attain the *fully turbulent regimes* that are observed in the solar convection zone. Indeed, various studies have shown that the transport properties of turbulent convection can be very different from those of laminar convection (e.g., Brummell, Hurlburt, and Toomre 1998). Accordingly, extension of the models into fully turbulent regimes might provide angular velocity profiles that are in agreement with the observational data. Three-dimensional numerical simulations of fully turbulent convection in a rotating spherical shell have been produced. Advances in computation permit these simulations to have a spatial resolution about tenfold greater in each dimension than those of the earlier studies. In particular, it is found that the north–south roll-like convective cells have broken up with the increased nonlinearity; this orderly convection is replaced by convection dominated by intermittent plumes of matter, with the downflow motions stronger in amplitude than the upflow motions. Although these extensions to fully turbulent regimes are quite promising, it is not yet clear at this writing to what extent the new global-convection models adequately describe the observed rotation profile in the solar convective zone.

5.3 Meridional circulation in the radiative core

Because the conditions in the radiative zone of a rotating star are not spherically symmetric, the transport of radiation will in general tend to heat the polar and equatorial regions unequally, thus causing a large-scale flow of matter in planes passing through the rotation axis. This problem was already discussed in Sections 4.3 and 4.4, where we obtained consistent solutions for the meridional flow and concomitant differential rotation in the radiative envelope of a nonmagnetic, early-type star. This section is concerned with the large-scale circulation generated by the small departures from spherical symmetry in the Sun's radiative core. Not unexpectedly, the development with age of a gradient of mean molecular weight $\bar{\mu}$ in the hydrogen-burning core makes this problem much more intricate since, then, *any model that possesses full internal consistency necessarily becomes time dependent in its mean properties*. To be specific, starting from an initially homogeneous radiative core in an unevolved solar model, we shall discuss the effects of a growing $\bar{\mu}$-gradient on the meridional flow – making allowance for hydrogen-core burning as the model leisurely evolves away from the zero-age main sequence. However, since we are chiefly interested in the interaction between the $\bar{\mu}$-distribution and the rotationally driven currents, we shall "turn off" the solar-wind torque that slows down the outer convective envelope.

In an inertial frame of reference, the large-scale velocity field $\mathbf{v}$ is the combination of a rotation and a meridional flow, as defined in Eq. (5.1). To complete the continuity equation (Eq. [5.2]) and the momentum equation (Eq. [3.123]) we must add the energy equation,

$$\rho T \mathbf{u} \cdot \operatorname{grad} S = \rho \epsilon_{\rm Nuc} + \operatorname{div}(\chi \operatorname{grad} T), \tag{5.16}$$

and Poisson's equation,

$$\nabla^2 V = 4\pi G \rho, \tag{5.17}$$

where $\epsilon_{\rm Nuc}$ is the rate of energy released by the thermonuclear reactions, χ is the coefficient of radiative conductivity, and G is the constant of gravitation. Let us rewrite the equation of state in the form

$$\frac{p}{\rho} = \mathcal{R} \frac{T}{\bar{\mu}}, \tag{5.18}$$

where $\mathcal{R}$ is the universal gas constant. Neglecting diffusion altogether, we must also prescribe that

$$\frac{\partial \bar{\mu}}{\partial t} + \mathbf{u} \cdot \operatorname{grad} \bar{\mu} = \mathcal{S}_{\rm Nuc}, \tag{5.19}$$

where $\mathcal{S}_{\rm Nuc}$ is the rate of variation of mean molecular weight caused by nuclear burning.

Following current practice, we shall expand about hydrostatic equilibrium in powers of the small parameter

$$\epsilon = \frac{\Omega_0^2 R^3}{GM}, \tag{5.20}$$

where Ω_0 is the (constant) overall rotation rate. Hence, we have

$$\mathbf{u} = \epsilon \mathbf{u}_1 + \epsilon^2 \mathbf{u}_2 + \cdots \tag{5.21}$$

and

$$\Omega = \Omega_0(w_0 + \epsilon w_1 + \epsilon^2 w_2 + \cdots). \tag{5.22}$$

As explained in Section 4.3.1, the truncated expansion (5.22) is asymptotically convergent provided one has $t_V < t_{ES}$, where t_V is the viscous time and t_{ES} is the circulation time (see Eqs. [4.37] and [4.54]). Since we are neglecting the continuous removal of angular momentum from the surface convective layers, we shall assume strict solid-body rotation to $\mathcal{O}(\epsilon^{1/2})$, that is, we shall let $w_0 \equiv 1$ in Eq. (5.22).

Correct to $\mathcal{O}(\epsilon)$, one can write

$$\rho = \rho_0 + \epsilon\rho_1 \tag{5.23}$$

and similar expansions for the pressure and the gravitational potential. Now, because we want to recover a spherically symmetric model in the limit $\epsilon \to 0$, Eq. (5.18) implies that, correct to $\mathcal{O}(\epsilon)$, one must write

$$\bar{\mu} = \bar{\mu}_0 + \epsilon\bar{\mu}_1 \tag{5.24}$$

and a similar expansion for the temperature. By assumption, we have $\bar{\mu}_0 = \bar{\mu}_0(r, t)$ and $T_0 = T_0(r, t)$ in the spherical model corresponding to $\epsilon = 0$. Since we want to obtain a solution that possesses full internal consistency, it follows at once from Eqs. (5.19) and (5.21) that the function $\bar{\mu}_0$ must satisfy the following equation:

$$\frac{\partial\bar{\mu}_0}{\partial t} = \mathcal{S}_0, \tag{5.25}$$

where $\mathcal{S}_0(r, t)$ is the (prescribed) rate of variation of the mean molecular weight in the reference spherical model. Here we shall assume that, in spherical polar coordinates (r, θ, φ), one initially has

$$\bar{\mu}(r, \theta, t = 0) \equiv \bar{\mu}_0(r, 0) \equiv \textit{constant}, \tag{5.26}$$

where the values of $\bar{\mu}_0$ are then allowed to change in time in a manner that depends on the given function $\mathcal{S}_0$ and on Eq. (5.25).

In Section 4.2.1 we have shown that the functions ρ_1, p_1, and V_1 can be obtained from Poisson's equation and the poloidal part of the momentum equation, which do not depend on $\bar{\mu}_1$ and T_1. In particular, the continuity of gravity across the outer nonspherical surface implies that

$$\rho_1 = \rho_{1,0}(r, t) + \rho_{1,2}(r, t)P_2(\cos\theta) \tag{5.27}$$

and similar expansions for p_1 and V_1, where $P_2(\cos\theta)$ is the Legendre polynomial of degree two. By virtue of Eq. (5.18), however, the expansions for $\bar{\mu}_1$ and T_1 contain, in principle, an infinite number of additive terms of the form $\bar{\mu}_{1,2k}(r, t)P_{2k}(\cos\theta)$ and $T_{1,2k}(r, t)P_{2k}(\cos\theta)$, with $k = 0, 1, 2, \ldots$. If so, then, the radial component u_{1r} should also contain an infinite number of additive terms of the form $u_{1,2k}(r, t)P_{2k}(\cos\theta)$, with $k = 0, 1, 2, \ldots$. Obviously, these terms essentially depend on the initial $\bar{\mu}$-distribution in the nonspherical model.

Fortunately, *by making use of Eq. (5.26), which is a most plausible initial condition, one can easily show that all terms belonging to* $k = 2, 3, 4, \ldots$ *must identically vanish from* $\bar{\mu}_1$, T_1, *and* u_{1r}. Indeed, since $\rho_{1,2k} \equiv p_{1,2k} \equiv 0$ when $k \geq 2$, it follows from

Eq. (5.18) that one has $T_{1,2k}/T_0 = \bar{\mu}_{1,2k}/\bar{\mu}_0$ ($= a_{2k}$, say) for these values of k. Hence, for each k (≥ 2) Eq. (5.16) implies that $u_{1,2k}$ is a linear and homogeneous function of a_{2k} and its derivatives. Next, linearizing Eq. (5.19) and eliminating $u_{1,2k}$, one obtains a *homogeneous* differential equation for each a_{2k}, when $k \geq 2$. Now, it readily follows from Eq. (5.26) that $a_{2k}(r, 0) \equiv 0$ since, by assumption, our initial model is chemically homogeneous. One can also let $a_{2k}(0, t) = a_{2k}(R_n, t) = 0$ for all t (≥ 0), where R_n is the radius of the sphere outside which (at the prescribed level of numerical accuracy) nuclear burning and the $\bar{\mu}$-gradient may be neglected. Since for each k (≥ 2) the function a_{2k} is the solution of a linear and homogeneous differential equation, these initial and homogeneous boundary conditions imply that one has $a_{2k}(r, t) \equiv 0$ for all t (≥ 0), when $k = 2, 3, 4, \ldots$.

In other words, starting from an initially homogeneous core, we can rightfully write

$$\bar{\mu}_1 = \bar{\mu}_{1,0}(r, t) + \bar{\mu}_{1,2}(r, t) P_2(\cos\theta) \tag{5.28}$$

and a similar expansion for T_1. The corresponding meridional velocity is, therefore,

$$\mathbf{u}_1 = u(r, t) P_2(\cos\theta) \mathbf{1}_r - r v(r, t) \sin\theta \frac{d P_2(\cos\theta)}{d \cos\theta} \mathbf{1}_\theta. \tag{5.29}$$

Equation (5.2) provides the link between the functions u and v. One finds that

$$v = \frac{1}{6} \frac{1}{\rho r^2} \frac{d}{dr}(\rho r^2 u), \tag{5.30}$$

where we have omitted the subscript "0" from the density in the spherical model.

Correct to $\mathcal{O}(\epsilon^{3/2})$, the back reaction of the first-order part of the meridional flow on the constant overall rotation is

$$w_1 = \beta_1(r, t) \frac{d P_1(\cos\theta)}{d \cos\theta} + \beta_3(r, t) \frac{d P_3(\cos\theta)}{d \cos\theta}. \tag{5.31}$$

The functions β_1 and β_3 satisfy two equations that are quite similar to Eqs. (4.82) and (4.83), with $\partial\beta_1/\partial t$ and $\partial\beta_3/\partial t$ being retained since u and v depend on time.

Now, it is immediately apparent that the functions $p_{1,2}$ and $\rho_{1,2}$ can be obtained from Eqs. (4.24) and (4.25). By virtue of Eq. (5.18), however, Eq. (4.27) must be replaced by

$$\frac{T_{1,2}}{T} = \left(\frac{\rho'}{p'} - \frac{\rho}{p} \right) h + a, \tag{5.32}$$

where the function h can be obtained from Eq. (4.23). For shortness, we have also let

$$a = \frac{\bar{\mu}_{1,2}}{\bar{\mu}}. \tag{5.33}$$

As usual, we have omitted the subscript "0" from the functions in the spherical model corresponding to $\epsilon = 0$. A prime denotes a derivative with respect to the radial variable r.

Inserting next these solutions into the energy equation, one finds that the radial function u can be written in the form

$$u = u_\Omega(r, t) + u_{\bar{\mu}}[a(r, t), r, t], \tag{5.34}$$

thus indicating that the large-scale meridional flow is the sum of "Ω-currents" and "$\bar{\mu}$-currents". After collecting and rearranging terms, we obtain

$$u_\Omega = \frac{2lr^4}{G^2m^3}\frac{n+1}{n-3/2}(\alpha_0 h' + \alpha_1 h), \tag{5.35}$$

where the functions $\alpha_0(r,t)$ and $\alpha_1(r,t)$ depend on the reference spherical model, and n is the effective polytropic index (see Eq. [4.33]). The function l is the net amount of energy crossing the spherical surface of radius r per second, that is,

$$l = -4\pi r^2 \chi T'. \tag{5.36}$$

Parenthetically note that Eq. (5.35) merely reduces to Sweet's function (4.32) in the outer parts of the Sun's radiative core, where one has $l \equiv L$, $\epsilon_{\rm Nuc} \equiv 0$, and $\bar{\mu} \equiv constant$. Similarly, one can show that the function $u_{\bar{\mu}}$ has the form

$$u_{\bar{\mu}} = -\frac{l}{4\pi Gm\rho}\frac{n+1}{n-3/2}\frac{T}{T'}\mathcal{D}''a, \tag{5.37}$$

where $\mathcal{D}''a$ is a second-order differential operator acting on the function a, that is,

$$\mathcal{D}''a = \frac{\partial^2 a}{\partial r^2} + A_0\frac{\partial a}{\partial r} + A_1 a, \tag{5.38}$$

where the functions $A_0(r,t)$ and $A_1(r,t)$ depend on the reference spherical model. Equations (5.37) and (5.38) were originally obtained by Mestel (1953).

Making use of Eqs. (5.19) and (5.25), one also has

$$\frac{\partial}{\partial t}(\bar{\mu}a) + \bar{\mu}'u = \mathcal{S}_{1,2}, \tag{5.39}$$

where $\mathcal{S}_{1,2}$ depends on the choice that is made for the function $\mathcal{S}_{\rm Nuc}$. Substituting for u in accordance with Eq. (5.34), one can calculate the function a from Eq. (5.39), which is parabolic in structure. Thence, the radial function u can be obtained from Eqs. (5.34)–(5.38).

Now, one readily sees that $n \to 3/2$ near the top of the radiative core, thus implying the existence of a mathematical singularity in our frictionless solution. As explained in Section 4.3.1, this major inadequacy can be resolved by making use of the thermoviscous boundary-layer solution depicted in Figure 4.1, letting $x = (R_c - r)/\delta_c$ in Eq. (4.70) since we are now approaching the singularity from below the inner boundary. This modification is not essential for the subsequent discussion, however, because the interaction between the $\bar{\mu}$-distribution and the rotationally driven currents takes place in the bulk of the Sun's radiative core, away from the core–envelope interface.

Numerical calculations have been performed by making use of an evolutionary sequence of a standard 1 $M_\odot$ model. Figures 5.3 and 5.4 illustrate at selected instants the functions u and a in the chemically homogeneous part of the Sun's radiative core (see Eqs. [5.34] and [5.35]). It is worth noting that the function a always remains much smaller than unity. To order ϵ, it is thus correct to make use of the truncated expansion (5.24) and the linearized equation (5.39).

It is immediately apparent from Figure 5.3 that, almost from the start, the $\bar{\mu}$-currents oppose the Ω-currents – the large-scale circulatory motions die out as the $\bar{\mu}$-gradient

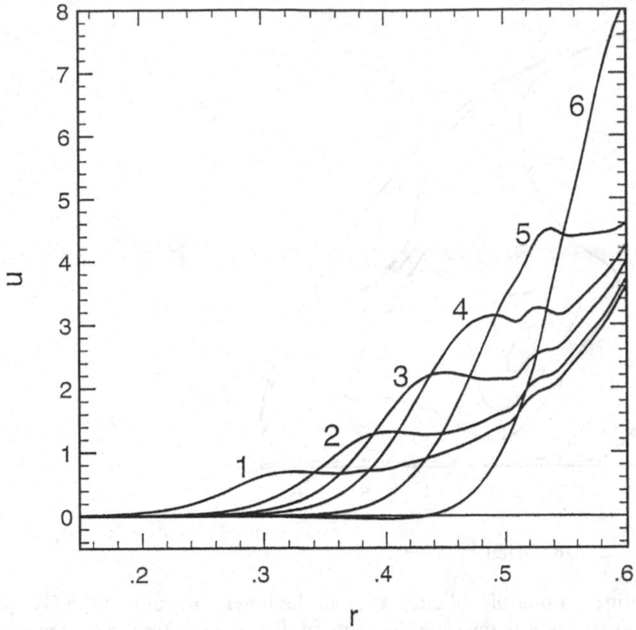

Fig. 5.3. The radial function u, at different instants, as a function of r (measured in units of $R_{\odot}$). (1) $t = 0.4652$ Gyr; (2) $t = 1.015$ Gyr; (3) $t = 1.495$ Gyr; (4) $t = 1.975$ Gyr; (5) $t = 2.935$ Gyr; (6) $t = 4.937$ Gyr. The quantity u is measured in units of 10^{-5} cm s^{-1}. *Source (revised):* Tassoul, M., and Tassoul, J. L., *Astrophys. J.*, **279**, 384, 1984.

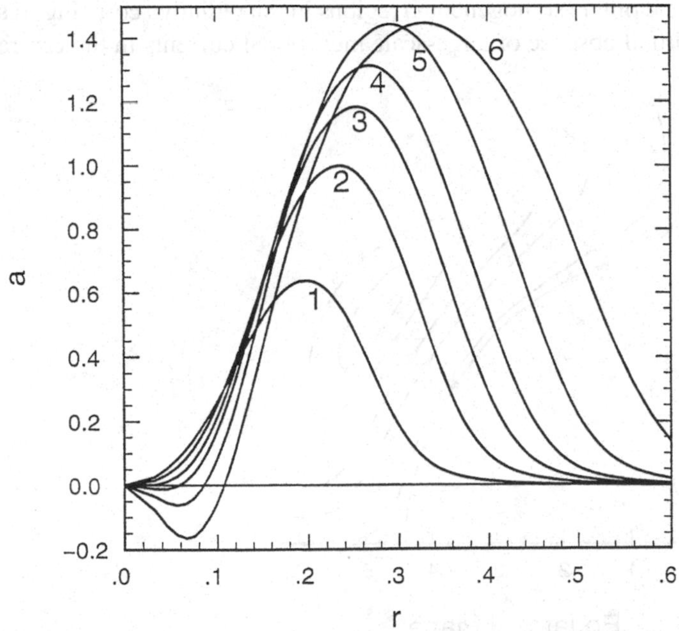

Fig. 5.4. The dimensionless function a, at different instants, as a function of r (measured in units of $R_{\odot}$). See Figure 5.3 for the labeling of the curves. The quantity a is measured in units of 10^{-2}. *Source (revised):* Tassoul, M., and Tassoul, J. L., *Astrophys. J.*, **279**, 384, 1984.

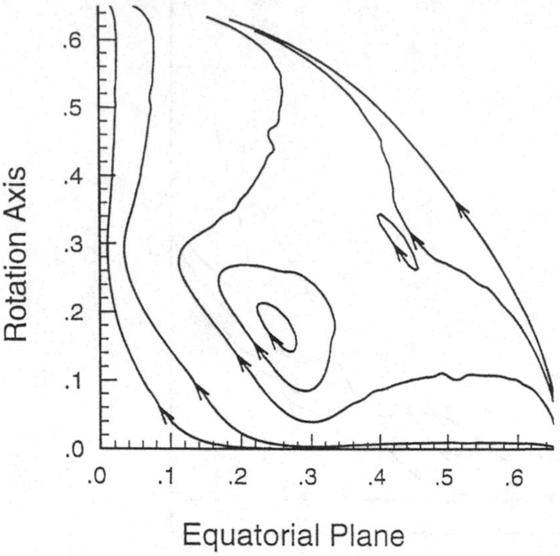

Fig. 5.5. Streamlines of meridional circulation in the inner core, at $t \approx 0.5$ Gyr (curve 1 of Figure 5.3). The variable r is measured in units of $R_\odot$. *Source (revised):* Tassoul, M., and Tassoul, J. L., *Astrophys. J.*, **279**, 384, 1984.

spreads throughout the Sun's radiative core. Within the numerical accuracy of these first-order calculations, thus, a $\bar{\mu}$-gradient virtually kills off meridional streaming, so that no substantial advection of matter by circulation may take place between the inner (inhomogeneous) and outer (homogeneous) regions in the radiative core. Figures 5.5 and 5.6 illustrate the virtual absence of large-scale meridional currents in the central region

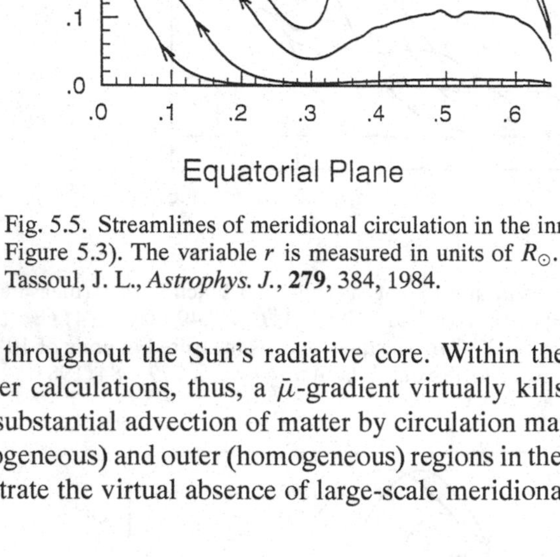

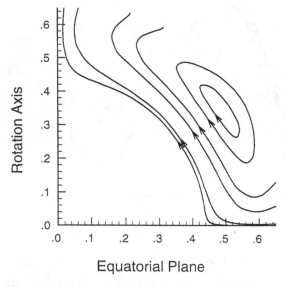

Fig. 5.6. Same as Figure 5.5, but at $t \approx 5$ Gyr (curve 6 of Figure 5.3). Note the virtual disappearance of meridional currents in the domain $r/R_\odot \lesssim 0.40$. *Source (revised):* Tassoul, M., and Tassoul, J. L., *Astrophys. J.*, **279**, 384, 1984.

of an evolved solar model. In the chemically homogeneous part of the radiative core, however, the typical time scale of the rotationally driven currents remains of the order of the Eddington–Sweet time, $t_{\mathrm{ES}} = R/\epsilon|u_\Omega| \approx (GM^2/RL)/\epsilon$ (see Eq. [4.37]). Typically, because $\epsilon \approx 10^{-4}$ and $|u_\Omega| \approx 10^{-5}$ cm s^{-1} in the outer part of the Sun's radiative core, one finds that $t_{\mathrm{ES}} \gtrsim 10^{12}$ yr, which is much larger than the Sun's age.

Now, as was pointed out at the end of Section 4.3.1, ϵw_1 describes the back reaction of the meridional flow on the basic rotation rate w_0 (see Eq. [5.22]). It is a simple matter to show that one has $\epsilon|w_1/w_0| \approx \epsilon|\rho u_\Omega r/\mu_V| \approx t_V/t_{\mathrm{ES}}$, where μ_V is the vertical coefficient of eddy viscosity and t_V is the viscous time scale. If we let $\mu_V = 10^N \mu_m$, where μ_m is the microscopic viscosity and N is a positive number, detailed numerical calculations indicate that one has $\epsilon|w_1/w_0| \approx \epsilon 10^{4-N}$. Since $\epsilon \approx 10^{-4}$ in the Sun, one readily sees that a moderate amount of turbulence ($N \approx$ 2–3, say) is amply sufficient to ensure that, to a first approximation, the viscous friction acting on the mean azimuthal flow dominates over the advection of specific angular momentum by the rotationally driven currents.

5.4 Spin-down of the solar interior

As we shall see more closely in Section 7.2, the Sun and solar-type stars sustain a continuous loss of mass as a result of magnetized stellar winds and/or episodic mass ejections emanating from their outer convection zones. The rotational deceleration of the convective envelope that results from the application of this torque leads to the creation of internal stresses that act to redistribute angular momentum within the radiative core. Yet, as was pointed out in Section 1.2.2, analyses of helioseismological data strongly suggest that from the base of the convection zone down to $r \approx 0.1$–$0.2R_\odot$ the Sun's interior is rotating at a rate close to that of the surface equatorial belt, while the inner core is perhaps rotating more rapidly than the chemically homogeneous parts of the radiative interior. Because angular momentum is continuously transferred away from the surface convection zone to outer space, it follows at once that there must exist a very effective mechanism of angular momentum transport inside the Sun, thus keeping the bulk of the radiative interior rotating approximately uniformly in spite of the inexorable solar-wind torque.

Broadly speaking, if we describe the mean velocity field as the sum of an overall rotation and a large-scale meridional flow, three mechanisms can redistribute angular momentum within the Sun's radiative interior: (i) the advection of specific angular momentum by the meridional currents, (ii) the diffusion of momentum arising from turbulent friction acting on the differential rotation, and (iii) the interaction with a large-scale magnetic field. Specifically, making use of Eq. (5.1), one can write the φ component of the momentum equation in the form

$$\rho \frac{\partial}{\partial t}(\Omega \varpi^2) + \rho \mathbf{u} \cdot \mathrm{grad}(\Omega \varpi^2) = \mathrm{div}(\rho \nu \varpi^2 \,\mathrm{grad}\, \Omega) + \frac{1}{4\pi} \mathbf{H}_p \cdot \mathrm{grad}(\varpi H_\varphi), \tag{5.40}$$

where $\varpi = r \sin\theta$ and ν is the kinematic viscosity (see Eq. [4.146]). The vectors $\mathbf{H}_p$ and $H_\varphi \mathbf{1}_\varphi$ are, respectively, the poloidal and toroidal parts of the magnetic field.

In Section 5.3 we have shown that the typical speed $|\mathbf{u}|$ of the thermally driven meridional currents is so slow that, to a first approximation, the advection of angular momentum by these currents can be neglected in Eq. (5.40). In fact, two categories of models have been proposed. In one of them, angular momentum redistribution is treated as a turbulent diffusion process, with advection by the meridional flow and magnetic fields being neglected altogether. The other group of models is based on the idea that this redistribution is dominated by magnetic stresses arising from the shearing of a preexisting poloidal magnetic field. It is to these two distinct approaches that we now turn.

5.4.1 Rotation and turbulent diffusion

It has long been recognized that standard evolution theory is quite successful in explaining the main properties of stars. Yet, as more data become available, the limits of the standard spherical models have become more apparent. For example, the observed solar lithium abundance is a factor of 200 smaller than that found in meteorites, indicating that some downward particle transport has occurred in the outer parts of the Sun's radiative core. As was noted by Endal and Sofia (1981), rotation might be the ultimate cause of this slow mixing process, since rotationally induced instabilities will generate a wide spectrum of small-scale motions that produce internal mixing of certain chemical species.

Following these authors, the mechanisms that redistribute angular momentum can be divided into two categories, dynamical and thermal, according to the time scales associated with the triggering mechanisms. Hence, whenever the shear-flow and symmetric instabilities arise in their models, the angular velocity gradient is instantaneously readjusted to a state of marginal stability by radial exchange of angular momentum (see Eqs. [3.93] and [3.101]).* However, because of the longer time scales for the thermal instabilities, the overall redistribution of angular momentum and chemical composition are computed using the coupled diffusion equations:

$$\rho r^4 \frac{\partial \Omega}{\partial t} = \frac{\partial}{\partial r}\left(\rho r^4 D \frac{\partial \Omega}{\partial r}\right), \tag{5.41}$$

for the angular velocity $\Omega(r, t)$, and

$$\rho r^2 \frac{\partial X_i}{\partial t} = f \frac{\partial}{\partial r}\left(\rho r^2 D \frac{\partial X_i}{\partial r}\right), \tag{5.42}$$

for the mass fraction $X_i(r, t)$ of chemical species i. The function D, which is sensitive to both angular velocity and chemical composition gradients, is the coefficient of eddy viscosity due to the rotationally induced thermal instabilities. (It was denoted by ν in Eq. [5.40].) Note that these equations may be derived at once from Eqs. (3.133) and (3.134), assuming that the ratio of eddy diffusivity to eddy viscosity is equal to the constant f. As usual, the eddy coefficient D is taken as the product of some typical length L_c and some typical speed V_c, which is assumed to be the sum of velocities generated by the Eddington–Sweet currents and some thermal instabilities. As was noted in Section 3.6, however, *such a formulation is at best phenomenological because it is not yet known how to model the variations of the function D with any confidence.* In fact, because the eddy coefficients cannot be calculated from first principles alone, their

* Parenthetically note that the ever-present barotropic and baroclinic instabilities discussed in Section 3.4.3 are not taken into account in these models.

overall magnitude can be determined only by adjusting the constant f and the empirical formula for the function D to the observational constraints.

Several evolutionary models that include the combined effects of rotationally induced mixing and angular momentum redistribution in the Sun's radiative core have been calculated by Pinsonneault, Kawaler, Sofia, and Demarque (1989). Following current practice, the effects of rotation were treated as small distortions superimposed on spherically symmetric models (see Section 6.2). For some reason, however, Eq. (5.41) was replaced by

$$\rho r^2 \frac{I}{M} \frac{\partial \Omega}{\partial t} = \frac{\partial}{\partial r}\left(\rho r^2 \frac{I}{M} D \frac{\partial \Omega}{\partial r}\right), \tag{5.43}$$

where I is the moment of inertia and M is the mass of the Sun. In some calculations, Eq. (5.42) was also modified to include the combined effects of rotationally induced mixing and microscopic diffusion. Following Chaboyer, Demarque, and Pinsonneault (1995), we thus have

$$\rho r^2 \frac{\partial X_i}{\partial t} = \frac{\partial}{\partial r}\left[\rho r^2 f_m D_{m,1} X_i + \rho r^2 (f_m D_{m,2} + f D) \frac{\partial X_i}{\partial r}\right], \tag{5.44}$$

where $D_{m,1}$ and $D_{m,2}$ are derived from the microscopic diffusion coefficients and multiplied by the adjustable parameter f_m. As usual, these equations must be supplemented by appropriate initial and boundary conditions. In particular, one must prescribe some general expression for the continuous loss of angular momentum due to the magnetically coupled solar wind.

The evolutionary models have been calibrated to match the usual global properties of the present-day Sun, as well as its observed rotation rate. Numerical calculations indicate that the value of f is approximately 0.033. This result is in perfect agreement with the fact that turbulent diffusion of matter is a much less effective process than turbulent diffusion of momentum in a stably stratified system (see Section 3.6). Note also that those models that include rotation and microscopic diffusion have convection zone depths of $0.710 R_\odot$, providing a good match to the observed depth.

As far as rotation is concerned, the models have an oblateness in agreement with the observed upper limit. This is a consequence of a general feature of these models, namely, that they all rotate slowly in the outer layers where the contribution to oblateness is greatest. Angular momentum transport in the models is also remarkably efficient in smoothing out differential rotation in the radiative core. The possible range of rotation profiles for models with angular momentum transport is compared to a model with the same surface rotation velocity but without transport in Figure 5.7. Note that the rotation curve for $r > 0.6 R_\odot$ is almost flat in the models. Inside the radius $r = 0.6 R_\odot$, however, the degree of differential rotation depends on the choice of parameters. Now, as was noted in Section 1.2.2, inversion of the available p-mode oscillation data suggests a nearly flat rotation curve down to $r \approx 0.1$–$0.2 R_\odot$. Accordingly, it appears most likely that a more efficient angular momentum transport mechanism is present in the Sun – one that is not present in the models developed by the Yale group.

At this juncture it is appropriate to mention the work of Schatzman (1996), who pointed out that gravity waves generated by turbulent stresses in the solar convection zone might also contribute to the almost uniform rotation of the Sun's radiative interior. Original calculations by Kumar and Quataert (1997) and others show that there is enough angular

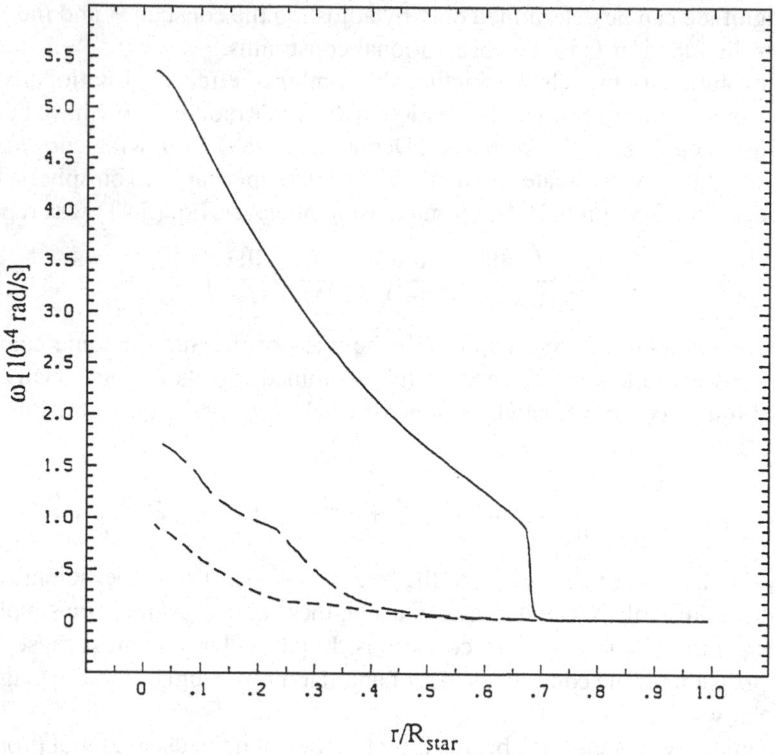

Fig. 5.7. Angular velocity as a function of radius in the present-day Sun for three distinct models of angular momentum transport. (ω is the angular velocity.) The *solid line* is the rotation curve the present-day Sun would have if it started with an average initial angular momentum and evolved to the age of the Sun without any transport of angular momentum from the radiative interior to the surface convection zone. The *long-dashed line* is a model with very inefficient angular momentum transport. The *short-dashed line* is a model with very efficient angular momentum transport. *Source:* Pinsonneault, M. H., Kawaler, S. D., Sofia, S., and Demarque, P., *Astrophys. J.*, **338**, 424, 1989.

momentum in gravity waves generated by convection that they can force the outer parts of the radiative interior into corotation with the base of the convection zone in about 10^7 yr. Even though these results are dependent on the description chosen for the turbulent motions in the solar convection zone, they clearly show that turbulent diffusion due to random gravity waves is a physical process that cannot be ignored.

5.4.2 *Rotation and magnetic fields*

In Section 5.4.1, the models for the evolution of the internal solar rotation have been computed assuming angular momentum transport solely by hydrodynamical means. In this section we shall investigate the rotational deceleration of a solar model containing a large-scale poloidal magnetic field in its radiative core, in response to the torque applied to it by a magnetically coupled wind. The first quantitative study of this problem was made by Charbonneau and MacGregor (1992). Their investigation was conducted using a numerical model that includes treatment of both convection zone braking by the magnetized solar wind and internal angular momentum redistribution by magnetic and viscous stresses.

For the sake of simplicity, we shall neglect the meridional velocity **u** in Eq. (5.40), and we shall assume strict axial symmetry for the large-scale magnetic field. Equation (5.40) then becomes

$$\rho \frac{\partial}{\partial t}(\Omega \varpi^2) = \operatorname{div}(\rho \nu \varpi^2 \operatorname{grad} \Omega) + \frac{1}{4\pi} \mathbf{H}_p \cdot \operatorname{grad}(\varpi H_\varphi), \tag{5.45}$$

where the poloidal magnetic field $\mathbf{H}_p(r, \theta)$ is assumed to be time independent and known a priori. With these simplifications, the spin-down problem reduces to solving Eq. (5.45) and the φ component of the induction equation,

$$\frac{\partial H_\varphi}{\partial t} = \beta \left(\nabla^2 - \frac{1}{\varpi^2} \right) H_\varphi + \varpi \mathbf{H}_p \cdot \operatorname{grad} \Omega, \tag{5.46}$$

where β is the magnetic diffusivity (see Eq. [4.148]). Both ν and β are assumed to be constant throughout the radiative core, with the adopted value for ν being small enough that viscous transport of angular momentum is negligible compared to magnetic transport. Such a formulation is self-consistent because it takes into account (i) the generation of the toroidal component $H_\varphi(r, \theta, t)$ by shearing of the poloidal field and (ii) the back reaction on the angular velocity $\Omega(r, \theta, t)$ due to the nonvanishing Lorentz force associated with the time-varying toroidal component of the magnetic field. When supplemented by some initial and boundary conditions, Eqs. (5.45) and (5.46) describe a two-dimensional problem for the two unknown functions, $\Omega(r, \theta, t)$ and $H_\varphi(r, \theta, t)$, governed by two coupled, linear, quasi-hyperbolic equations.

A large set of calculations have been performed by Charbonneau and MacGregor (1993), starting on the zero-age main sequence from a state of solid-body rotation at 50 times the present solar rate and zero toroidal field. They identify two distinct regions in the interior: a convective envelope, which they assume to rotate as a solid body at all times at the rate $\Omega_{\mathrm{CE}}(t)$, and an underlying radiative core. The solutions were computed for four distinct poloidal field configurations, as shown in Figure 5.8, and for poloidal field strengths B_0 of 0.01, 0.1, 1, and 10 G. Note that the fields D1 and D2 are such that direct magnetic coupling exists between the convective envelope and the radiative core, while for the fields D3 and D4 the envelope is magnetically decoupled from the underlying core.

These spin-down calculations enable us to draw a detailed picture of the magnetic and rotational evolution of an internally magnetized solar-type model, which is acted upon by the torque associated with a magnetically coupled wind. The evolution can be divided into three more or less distinct phases: an initial phase of toroidal field buildup, lasting between a few thousand to a few million years, depending on the topology and strength of the internal poloidal field; a second period in which large-scale toroidal oscillations set up in the radiative core during the first phase are damped; and a third period, lasting from age of about 10^7 yr onward, characterized by a state of dynamical balance between the total stresses (magnetic plus viscous) at the base of the convective envelope and the wind-induced surface torque, leading to a quasi-static internal magnetic and rotational evolution.

The time evolution of internal differential rotation is shown in Figure 5.9. The dimensionless quantity $\Delta\Omega$ is constructed by integrating the difference $\Omega(r, \theta, t) - \Omega_{\mathrm{CE}}(t)$ over the magnetized part of the radiative interior, thus providing a global measure of the difference in angular velocity between the convective envelope and the magnetized part of the

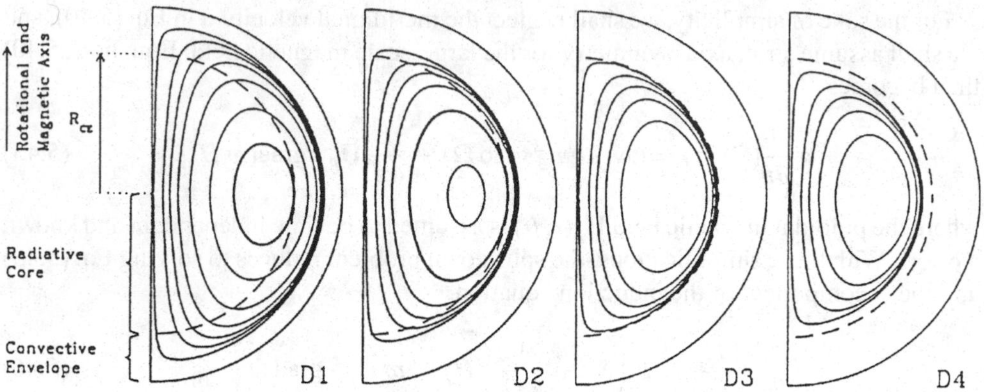

Fig. 5.8. Poloidal magnetic field configurations. The bottom of the convective envelope is located at $R_{CE} = 0.74302 R_{\odot}$. *Source:* Charbonneau, P., and MacGregor, K. B., *Astrophys. J.*, **417**, 762, 1993.

radiative core. Note that in all cases $\Delta\Omega$ initially increases very rapidly before reaching a maximum at about $t \approx 10^8$ yr. At later times, $\Delta\Omega$ declines at a rate nearly independent of poloidal field strength and configuration. An important common property of these solutions is the weak differential rotation that most of them exhibit by the time they have attained the solar age. Except for the D4 configurations, all solutions have $\Delta\Omega < 0.02$

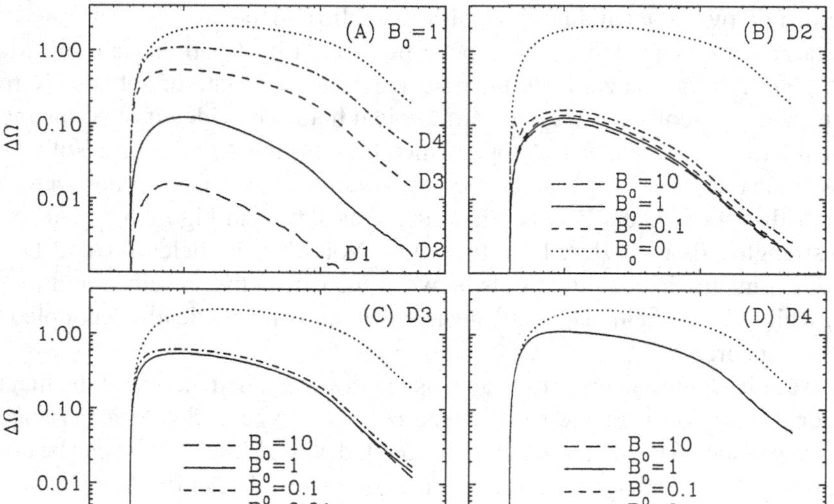

Fig. 5.9. Time evolution of the internal differential rotation, as defined in the text by the quantity $\Delta\Omega$, for various poloidal field configurations and strengths. In (A) are shown solutions for the four poloidal configurations of Figure 5.8, all at a strength of 1 G. In (B), (C), and (D) are shown the effects of varying the poloidal field strength for a given poloidal configuration. *Source:* Charbonneau, P., and MacGregor, K. B., *Astrophys. J.*, **417**, 762, 1993.

at $t = 4.5 \times 10^9$ yr. This is in contrast to the unmagnetized models, which often have significant angular velocity gradients in their radiative cores even at the solar age.

These quantitative studies are important because, for the very first time, they demonstrate the existence of classes of large-scale internal magnetic fields that can accommodate rapid spin-down of the surface layers near the zero-age main sequence and yield a weak internal differential rotation in the radiative core by the solar age. The lack of significant differential rotation from the base of the solar convection zone down to $r \approx 0.1\text{–}0.2R_\odot$ would then exclude from further consideration poloidal magnetic configurations of the D4 type. Within the current observational uncertainties, all D1, D2, and D3 solutions are compatible with the results reported in Section 1.2.2. However, none of these solutions exhibit enhanced angular velocity inside the radius $r = 0.2R_\odot$, as some helioseismological observations have suggested. Following Charbonneau and MacGregor (1992), this can be achieved by choosing poloidal magnetic fields such that the inner core remains magnetically decoupled from the surrounding regions. Admittedly, there is no firm justification for such a choice, but it seems to be the only way to have a rapidly rotating inner core (if any) in the present-day Sun.

More recently, Rüdiger and Kitchatinov (1996) have performed a large set of spin-down calculations, making allowance for the differential rotation in the convective envelope (see Section 5.2.1). Their work thus combines differential rotation at the base of the solar convection zone, rotational braking due to a magnetically coupled solar wind, and an axially symmetric magnetic field in the Sun's radiative interior. A reasonable picture emerges only if the following two conditions are met: (1) viscosity is strongly enhanced compared to its microscopic value, and (2) the internal magnetic field does not penetrate into the outer convection zone. As was shown in Section 4.7.2, an axially symmetric poloidal magnetic field makes the rotation uniform along each field line, although the constant angular velocity is in general different for each field line. If the internal poloidal field was anchored into the differentially rotating convective envelope, the latitudinal rotation inhomogeneity would thus penetrate deep into the radiative interior, which is not observed. With the magnetic field fully embedded into the core, however, their models do reproduce the thin layer where a transition from differential to rigid-body rotation occurs at the bottom of the solar convection zone. (This transition layer is known as the *solar tachocline*.) The problem is then presented by the "dead zone" permeated by the field lines that never come close to the base of the convective envelope (see the D3 and D4 configurations in Figure 5.8). This is the reason why a sizable amount of eddy viscosity is needed to link this region to the base of the solar convection zone across the magnetic field. To be specific, the models of Rüdiger and Kitchatinov (1996) require an eddy viscosity of the order of $10^4 \mu_m$, where μ_m is the microscopic viscosity (see also the end of Section 5.3). As they noted, however, there is no contradiction at this point with the models of Charbonneau and MacGregor (1993), since these solutions also require an amplification factor of the order of 10^4 in the coefficient of viscosity.

In summary, two independent sets of spin-down calculations have been made. They differ in one important respect, however. In the Charbonneau–MacGregor models the convective envelope is assumed to rotate uniformly at all times, whereas the latitudinal differential rotation of that zone is properly retained in the Rüdiger–Kitchatinov models. In both sets of models, it is found that there exist large-scale magnetic fields that yield a weak internal differential rotation by the solar age. In the former case, however, the poloidal field lines may or may not penetrate into the convective envelope (see Figure 5.8). By contrast, in the latter case, the helioseismological observations are reproduced only with the poloidal magnetic field fully contained within the radiative core. In both cases, the models are quite insensitive to the magnitude of the internal magnetic field, provided the poloidal field strength B_0 is larger than 10^{-3} G. However, despite the high efficiency of these magnetic fields in transporting angular momentum, *turbulent* friction is always needed to enforce almost uniform rotation in the radiative interior by the solar age.

5.5 Discussion

In Sections 4.3 and 4.4 we have presented a simple but adequate description of the mean state of motion in a nonmagnetic early-type star that consists of a uniformly rotating convective core and a surrounding radiative envelope. Assuming no mass loss from the star's surface, we have shown that there exists a mean steady solution for the large-scale motion in the radiative envelope, which is the combination of an overall differential rotation and slow circulatory currents in planes passing through the rotation axis. As was pointed out, however, the major impediment to the complete resolution of this problem is the lack of quantitative observational data about the velocity field in the surface layers of an early-type star. This is in contrast to the late-type stars, as a variety of recent observational results have shed important new light on both the internal rotation of the Sun and the rotational evolution of solar-type stars. It is therefore appropriate at this juncture to critically review the degree of development of the main theories of solar rotation.

It is generally believed that the interaction of rotation with convection plays an essential role in the generation and maintenance of differential rotation and concomitant meridional circulation in the solar convection zone. Unfortunately, although it has been suggested that rotation may be interacting with either local turbulent convection or global turbulent convection, no scheme has yet been generally accepted as being basically correct. In fact, because a general theory of turbulent convection still lies in the distant future, in all likelihood further progress will result from a balanced approach that involves increasingly reliable helioseismological observations combined with more and more sophisticated numerical simulations.

Considerable progress has been made in determining the processes that affect the internal rotation of the Sun. In Section 5.3 we have shown that thermally driven meridional currents inexorably advect angular momentum in the chemically homogeneous parts of the Sun's radiative core, thus tending to induce small departures from solid-body rotation in these regions. For the rotation rate of the present-day Sun, this large-scale advection of angular momentum is probably negligible, although in a more detailed study it might effectively contribute to the angular momentum redistribution within the outer parts of the Sun's radiative core.

Two very efficient mechanisms for angular momentum redistribution in the solar interior have been thoroughly investigated: *turbulent friction* acting on the differential rotation and large-scale *magnetic fields*. As was shown in Section 5.4, both of them provide the means by which the solar-wind torque is communicated to the interior, while enforcing almost uniform rotation in the radiative core by the solar age.

In the turbulent models illustrated in Figure 5.7 the angular momentum redistribution within the Sun's radiative core is treated diffusively. As was repeatedly pointed out, this approach is, at best, a semiquantitative one (see, e.g., Section 3.6). Indeed, by making use of the crude concept of eddy viscosity, one necessarily relegates all eddy and/or wave events to a passive means of dissipating the large-scale flow, thus implying an ill-defined energy cascade from the largest to the smallest scales of motion. And because one cannot calculate the eddy coefficients from first principles alone, it follows that one must integrate Eqs. (5.41) and (5.42) under widely different conditions, thence guessing the form and values of the empirical formula for the function D that best fit the global properties of the present-day Sun. Note also that these turbulent models, which often have significant angular velocity gradients in their radiative cores even at the solar age, are generally characterized by the presence of a small, rapidly rotating central core. Such a behavior is attributable to the fact that the development with age of a gradient of mean molecular weight in the hydrogen-burning core leads to a much reduced eddy viscosity in these parts of the solar interior, thus preventing them from participating to the overall redistribution of angular momentum.

In Section 5.4.2 we have shown that a more efficient means for transporting angular momentum in the Sun's radiative core is through the intermediary of a large-scale internal magnetic field. Detailed numerical simulations demonstrate the existence of classes of poloidal fields allowing rapid surface spin-down at early epochs, while producing almost uniform rotation throughout the Sun's radiative core by the solar age. However, these calculations show that a certain amount of a turbulent friction is always required to couple the field lines. They also indicate that the observed surface rotation rate is a rather poor indicator of the strength and geometry of hypothetical large-scale magnetic fields pervading the solar radiative regions. As far as the internal rotation is concerned, the most important property of these models is the weak overall differential rotation that most of them exhibit by the time they have attained the solar age. This is in contrast to the diffusive models presented in Section 5.4.1, which exhibit enhanced angular velocity in their central regions $r \lesssim 0.2R_\odot$. Since the actual rotation rate inside this radius is still very uncertain, we are therefore led to the conclusion that the relative importance of the two basic mechanisms for angular momentum redistribution deep inside the Sun is also an open question.

5.6 Bibliographical notes

Comprehensive introductions to the Sun are:

1. Stix, M., *The Sun*, Berlin: Springer-Verlag, 1989.
2. Foukal, P., *Solar Astrophysics*, New York: John Wiley and Sons, 1990.

Sections 5.1 and 5.2. The concept of anisotropic eddy viscosity was originally applied to the solar rotation problem by Lebedinski (Reference 31 of Chapter 3). The reference to Weiss is to his paper:

3. Weiss, N. O., *The Observatory*, **85**, 37, 1965.

The first *detailed* mean-field models are due to:

4. Köhler, H., *Solar Physics*, **13**, 3, 1970.
5. Durney, B. R., and Roxburgh, I. W., *Solar Physics*, **16**, 3, 1971.

Subsequent mean-field models are reviewed in:

6. Stix, M., in *The Internal Solar Angular Velocity* (Durney, B. R., and Sofia, S., eds.), p. 392, Dordrecht: Reidel, 1987.

7. Rüdiger, G., *Differential Rotation and Stellar Convection*, New York: Gordon and Breach, 1989.

More details about the Kitchatinov–Rüdiger models will be found in:

8. Kitchatinov, L. L., and Rüdiger, G., *Astron. Astrophys.*, **299**, 446, 1995.

Observational constraints on the solar rotation theories have been discussed at length in:

9. Durney, B. R., *Astrophys. J.*, **378**, 378, 1991; *ibid.*, **407**, 367, 1993.
10. Chiu, H. Y., and Paternò, L., *Astron. Astrophys.*, **260**, 441, 1992.
11. Durney, B. R., *Solar Phys.*, **169**, 1, 1996.

The main properties of the global-convection models devised by Gilman and Glatzmaier are summarized in:

12. Gilman, P. A., and Miller, J., *Astrophys. J. Suppl.*, **61**, 585, 1986.
13. Glatzmaier, G. A., in *The Internal Solar Angular Velocity* (Durney, B. R., and Sofia, S., eds.), p. 263, Dordrecht: Reidel, 1987.

Fully turbulent regimes are considered in:

14. Glatzmaier, G. A., and Toomre, J., in *Gong 94: Helio- and Astero-Seismology* (Ulrich, R. K., Rhodes, E. J., Jr., and Däppen, W., eds.), *A.S.P. Conference Series*, **76**, 200, 1995.

Papers of related interest are:

15. Pulkkinen, P., Tuominen, I., Brandenburg, A., Nordlund, A., and Stein, R. F., *Astron. Astrophys.*, **267**, 265, 1993.
16. Canuto, V. M., Minotti, F. O., Schilling, O., *Astrophys. J.*, **425**, 303, 1994.
17. Brummell, N. H., Xie, X., and Toomre, J., in *Gong 94: Helio- and Astero-Seismology* (Ulrich, R. K., Rhodes, E. J., Jr., and Däppen, W., eds.), *A.S.P. Conference Series*, **76**, 192, 1995.
18. Brummell, N. H., Hurlburt, N. E., and Toomre, J., *Astrophys. J.*, **473**, 494, 1996; *ibid.*, **493**, 955, 1998.
19. Vandakurov, Yu. V., *Astronomy Letters*, **23**, 55, 1997.

Very little is known about the solar *tachocline*, that is, the thin velocity boundary layer below the convection zone, where there exists an unresolved transition to almost uniform rotation. This and related matters are treated in:

20. Spiegel, E. A., and Zahn, J. P., *Astron. Astrophys.*, **265**, 106, 1992.
21. Gilman, P. A., and Fox, P. A., *Astrophys. J.*, **484**, 439, 1997.
22. Rüdiger, G., and Kitchatinov, L. L., *Astron. Nachr.*, **318**, 273, 1997.

Other papers may be traced to Reference 23 (p. 216) of Chapter 1; see also:

23. Basu, S., *Mon. Not. R. Astron. Soc.*, **288**, 572, 1997.

Section 5.3. The inhibiting role of a $\bar{\mu}$-gradient was first pointed out by Mestel (Reference 9 of Chapter 4). Quantitative studies will be found in:

24. McDonald, B. E., *Astrophys. Space Sci.*, **19**, 309, 1972.

25. Huppert, H. E., and Spiegel, E. A., *Astrophys. J.*, **213**, 157, 1977.

The time-dependent models reported in this section were originally obtained by:

26. Tassoul, M., and Tassoul, J. L., *Astrophys. J.*, **279**, 384, 1984.

The effect of a stellar-wind torque on the meridional flow was further discussed in:

27. Tassoul, M., and Tassoul, J. L., *Astrophys. J.*, **286**, 350, 1984.

Section 5.4.1. These diffusive models have their roots in the work of Endal and Sofia:

28. Endal, A. S., and Sofia, S., *Astrophys. J.*, **243**, 625, 1981.

Detailed evolutionary models have been reported in:

29. Pinsonneault, M. H., Kawaler, S. D., Sofia, S., and Demarque, P., *Astrophys. J.*, **338**, 424, 1989.
30. Chaboyer, B., Demarque, P., and Pinsonneault, M. H., *Astrophys. J.*, **441**, 865, 1995.

An illustration of their empirical coefficient D will be found in Reference 40 (Figure 16, p. 548) of Chapter 3. Compare with the results obtained by:

31. Tassoul, J. L., and Tassoul, M., *Astron. Astrophys.*, **213**, 397, 1989.

Angular momentum transport by gravity waves has been discussed in:

32. Schatzman, E., *J. Fluid Mech.*, **322**, 355, 1996.
33. Kumar, P., and Quataert, E. J., *Astrophys. J. Letters*, **475**, L143, 1997.

The efficiency of this transport mechanism has been confirmed independently by:

34. Zahn, J. P., Talon, S., and Matias, J., *Astron. Astrophys.*, **322**, 320, 1997.

Section 5.4.2. See:

35. Charbonneau, P., and MacGregor, K. B., *Astrophys. J. Letters*, **397**, L63, 1992.
36. Charbonneau, P., and MacGregor, K. B., *Astrophys. J.*, **417**, 762, 1993.

Their original results have received confirmation in the following work:

37. Kitamaya, O., Sakurai, T., and Ma, J., *Geophys. Astrophys. Fluid Dyn.*, **83**, 307, 1996.

Solar spin-down models that include differential rotation in the convective envelope are due to:

38. Rüdiger, G., and Kitchatinov, L. L., *Astrophys. J.*, **466**, 1078, 1996.

Section 5.5. A detailed comparison between theory and helioseismological observations of the Sun's internal angular velocity profile will be found in:

39. Charbonneau, P., Tomczyk, S., Schou, J., and Thompson, M. J., *Astrophys. J.*, **496**, 1015, 1998.

6

The early-type stars

6.1 Introduction

An inspection of Figure 1.6 shows that the mean projected equatorial velocity of main-sequence stars increases slowly with spectral type, reaching a maximum of about 200 km s^{-1} in the late B-type stars. Thence, the mean velocity $\langle v \sin i \rangle$ decreases slowly for later spectral types until about F0, where it starts dropping precipitously through the F-star region. As is well known, this rapid transition to very small rotational velocities occurs at approximately the spectral type where subphotospheric convection zones become suddenly much deeper on the main sequence. Accordingly, because Sun-like stars are most likely to develop episodic mass ejections and magnetically channeled stellar winds, it is generally thought that these stars are losing mass – and, hence, angular momentum – as they slowly evolve on the main sequence. Postponing to Chapter 7 the study of these low-mass stars ($M \lesssim 1.5M_\odot$), in this chapter we shall consider stars more massive than the Sun ($M \gtrsim 1.5M_\odot$) that are in radiative equilibrium in their surface layers.

In Chapter 4 we have already discussed the large-scale meridional currents and concomitant differential rotation in the radiative envelope of an early-type star, when the departures from spherical symmetry are not too large. Admittedly, the aim of that chapter was to develop a clear understanding of the many hydrodynamical phenomena that arise in a rotating star. In the following sections of this chapter we shall instead examine a selection of practical topics dealing with rotation, meridional circulation, and turbulence in the early-type stars. The chapter is organized as follows. The modifications brought by axial rotation on the overall structure of a main-sequence star are discussed in Section 6.2.1. Section 6.2.2 is devoted to the effects of rotation on the observable parameters, which depend on the inclination of the rotation axis to the line of sight. Section 6.3 presents a detailed study of axial rotation along the upper main sequence. In Section 6.4, which is of direct relevance to the study of chemically peculiar stars, we consider the interaction between microscopic diffusion and rotationally driven motions in a stellar radiative envelope. We conclude the chapter with a brief discussion of the changes in rotation as an early-type star evolves off the main sequence.

6.2 Main-sequence models

The main objective of this section is the construction of reliable numerical models of rotating stars consisting of a convective core, in which hydrogen burning is taking place, and a chemically homogeneous radiative envelope. In fact, very little is known about the interaction between rotation and convection in the core of an early-type star. For mathematical simplicity, it is often assumed that convective cores rotate uniformly; as was correctly pointed out by Tayler (1973), however, there is still considerable uncertainty about this point. The state of motion in the outer envelope of an early-type star has received comparatively much greater attention. Unfortunately, the study of a stellar radiative zone is complicated by the necessity to come to terms with a whole spectrum of eddylike motions that continuously interact with the mean flow, that is, the overall rotation and the slow but inexorable meridional currents. Following Section 3.6, we shall explicitly resolve these large-scale motions, while parameterizing the smaller-scale transient eddies through the use of Reynolds stresses and eddy viscosities.

In cylindrical polar coordinates (ϖ, φ, z), the mean velocity $\mathbf{v}$ becomes

$$\mathbf{v} = \mathbf{u} + \Omega\varpi\mathbf{1}_\varphi, \tag{6.1}$$

where $\mathbf{u}$ is the two-dimensional meridional velocity. Since we are considering an axially symmetric configuration, mass conservation implies that

$$\operatorname{div}(\rho\mathbf{u}) = 0, \tag{6.2}$$

where ρ is the mean density. Neglecting the acceleration and inertia of the meridional flow, we can rewrite the poloidal part of Eq. (3.123) in the form

$$\frac{1}{\rho}\operatorname{grad} p = -\operatorname{grad} V + \Omega^2\varpi\mathbf{1}_\varpi + \frac{1}{\rho}\mathbf{F}_p(\mathbf{u}), \tag{6.3}$$

where p is the pressure, V is the gravitational potential, and $\mathbf{F}_p(\mathbf{u})$ is the poloidal part of the turbulent viscous force per unit volume acting on the circulation. Similarly, by use of Eq. (6.1), one can show that the φ component of Eq. (3.123) has the form

$$\rho\frac{\partial}{\partial t}(\Omega\varpi^2) + \rho\mathbf{u}\cdot\operatorname{grad}(\Omega\varpi^2) = F_\varphi(\Omega), \tag{6.4}$$

where $F_\varphi(\Omega)$ is the azimuthal component of the turbulent viscous force per unit volume acting on the differential rotation (see Eq. [3.133]). To complete these equations we must add Poisson's equation,

$$\nabla^2 V = 4\pi G\rho, \tag{6.5}$$

an equation of state,

$$p = \frac{\mathcal{R}}{\bar{\mu}}\rho T + \frac{1}{3}aT^4, \tag{6.6}$$

and the energy equation,

$$\rho T\mathbf{u}\cdot\operatorname{grad} S = \rho\epsilon_{\text{Nuc}} - \operatorname{div}\mathcal{F}_t, \tag{6.7}$$

where S is the specific entropy and $\mathcal{F}_t$ is the total (radiative and convective) flux vector (see Eqs. [5.11]–[5.13]). Remaining symbols have their standard meanings.

The above set of partial differential equations provides seven scalar relations among the seven unknown functions Ω, $\mathbf{u}$, p, ρ, T, and V. Thus, in principle, the internal structure of a rotating star with meridional circulation is entirely determined by these equations, together with some initial conditions and the usual set of boundary conditions (see Section 2.2.2). The main difficulty of the problem lies in the fact that neither the internal stratification of a rotating star nor the shape of its free surface are known in advance. Another difficulty arises because we know very little about the transport of specific angular momentum, $\Omega\varpi^2$, in a stellar interior. In principle, the angular velocity Ω can be calculated from Eq. (6.4), which merely expresses that the advection of specific angular momentum by the meridional currents must balance the effects of turbulent friction acting on the mean azimuthal flow. In practice, because the coefficients of eddy viscosity cannot be calculated from first principles alone, the actual dependence of the angular velocity on the coordinates and time remains quite uncertain. As was pointed out in Section 4.8, *the precise determination of the rotation law in a stellar radiative envelope must await the development of numerical models that resolve the transient eddylike motions in sufficient detail to reproduce their transport properties adequately.* Parenthetically note that the presence of a weak poloidal magnetic field does not solve the problem either since, as was shown in Section 4.7.2, such a field does not necessarily maintain almost uniform rotation throughout the radiative envelope of an early-type star.

With the advent of high-speed computers in the 1960s, significant advances have been made in the study of the internal structure of rotating stars. However, because the actual distribution of angular momentum within a star is still largely unknown, in all numerical models proposed to date the rotation law is always specified in an ad hoc manner. In this section we shall thus assume that there are no internal motions other than rotation, and we shall merely replace Eq. (6.4) by some prescribed rotation law, either $\Omega = constant$ or some function $\Omega = \Omega(\varpi)$ that satisfies the essential stability condition defined in Eq. (3.98). If so, then, Eq. (6.3) simplifies to the usual condition of mechanical equilibrium for a barotrope,

$$\frac{1}{\rho}\,\text{grad}\,p = -\,\text{grad}\,\Phi, \tag{6.8}$$

where

$$\Phi = V(\varpi, z) - \int^{\varpi} \Omega^2(\varpi')\varpi' d\varpi' \tag{6.9}$$

(see Section 3.2.1). Given these simplifications, one readily sees that the basic equations are quite similar in structure to those for nonrotating stars, except that Eq. (6.5) must be solved in two dimensions with an outer boundary that is itself an unknown. Another difficulty stems from the fact that Eq. (6.8) is incompatible with the energy equation in a circulation-free barotrope (see Section 3.3.1). Accordingly, it is also assumed that, though radiative equilibrium does not hold at every point, it does hold on average (i.e., averaged over each level surface $\Phi = constant$).

A great number of techniques have been devised to determine the equilibrium structure of rotating polytropes and barotropic stars. To the best of my knowledge, Milne (1923) was the first to construct barotropic models for slowly rotating stars, using a first-order perturbation technique and treating the effects of uniform rotation as a small distortion superimposed on a known spherical model (see Eqs. [4.9]–[4.25]). As was originally shown by Takeda (1934), however, fairly accurate results can be obtained by means of a double-approximation technique. In the central regions, where the rotational distortion is small, a first-order expansion is used. This solution is then matched to a solution in the low-density surface layers, where the gravitational field arises mainly from the matter present in the slightly oblate inner core. Since, in general, the domains of validity of the two approximation regimes overlap, self-consistent solutions may readily be constructed. More recently, Kippenhahn and Thomas (1970) have shown that the use of two zones is unnecessary for the same degree of accuracy can be obtained in choosing an appropriate geometrical representation for the level surfaces. Their technique has been widely used because, without much trouble, rotation can be incorporated into the usual programs of stellar evolution (see, e.g., Section 5.4.1). Unfortunately, although it provides satisfactory results for quasi-spherical models in slow uniform rotation, other methods must preferably be used when the level surfaces greatly deviate from concentric spheres.

Progress in the study of rapidly rotating barotropes has been made by using full numerical solutions of all the relevant structure equations. Notably, Ostriker and Mark (1968) have developed the *self-consistent-field method*, which was especially designed to relax altogether the restrictive assumption of quasi-sphericity. In this method, Eq. (6.5) is replaced by its integral solution,

$$V = -G \int_{\mathcal{V}} \frac{\rho(\mathbf{r}')}{|\mathbf{r} - \mathbf{r}'|} \, dv', \tag{6.10}$$

where the triple integral must be evaluated over the volume $\mathcal{V}$ of the configuration. Given an angular momentum distribution, an iterative procedure is established in which an approximate expression for the total potential Φ is derived from a trial density distribution $\rho_0(\varpi, z)$. A new density distribution $\rho_1(\varpi, z)$ is then obtained from the equilibrium equations. For convenience, the external boundary condition on the gravitational potential is applied on a sphere exterior to the model. This is the basis of the self-consistent-field method, in which Poisson's equation and the equilibrium equations are solved alternately. This iterative scheme works remarkably well for the more massive stars, but it fails to converge even for a nonrotating main-sequence model if its mass is less than about $9 M_\odot$ (i.e., if its central mass concentration is sufficiently high). This is the reason why Clement (1978) has presented a two-dimensional, finite-difference technique for solving Poisson's equation simultaneously with the equilibrium equations. The method does not appear to be limited by the large central concentrations that characterize intermediate mass stars and those with high angular momentum. Rapidly rotating main-sequence models in the mass range that is not accessible to the self-consistent-field method have been computed with this two-dimensional numerical technique.

6.2.1 *Uniform rotation versus differential rotation*

As was originally shown by Milne (1923), uniform rotation has two general effects on the structure of a star. It leads to (i) a global expansion of the star due to the local centrifugal force and (ii) a departure from sphericity due to the nonspherical part of the effective gravity. To be specific, because the centrifugal force takes over from the pressure part of the burden of supporting the weight of the overlying layers in the energy-producing regions, the global-expansion effect causes a reduction in the total luminosity of the star when it is compared to its nonrotating counterpart having the same mass. Moreover, because a uniformly rotating star is slightly oblate, in its equatorial belt part of the mass is supported by the centrifugal force whereas this is not the case in the polar regions. Accordingly, the pressure and hence the net outward flux of energy

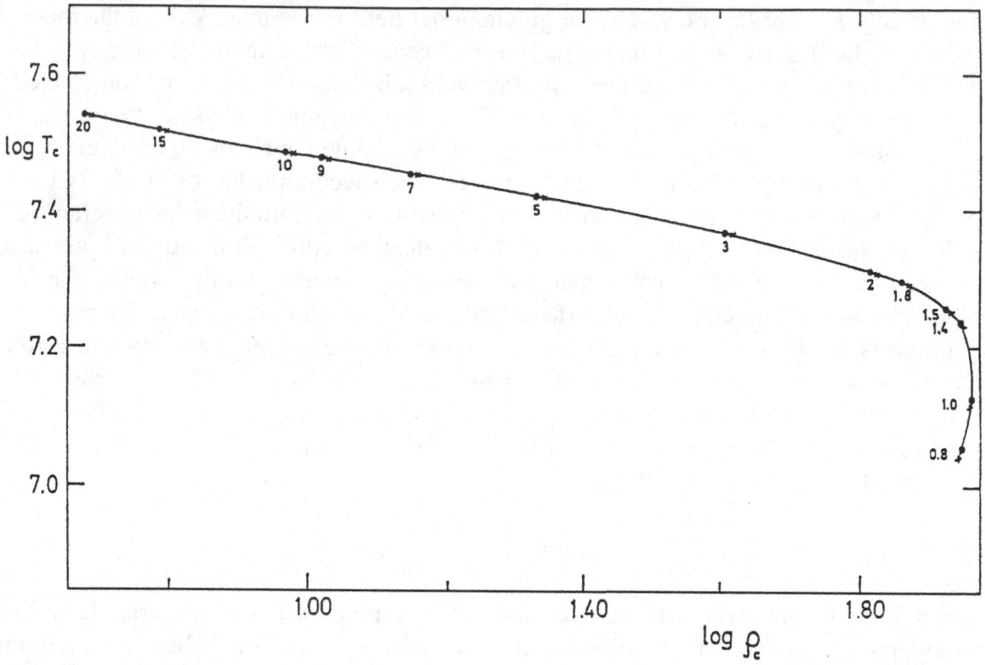

Fig. 6.1. The central temperature T_c as a function of the central density ρ_c for main-sequence stars. The curve is drawn through the data referring to nonrotating stars (*dots*). The crosses refer to critically rotating stars. The numerals along the curve define the mass of the models. *Source:* Sackmann, I. J., *Astron. Astrophys.*, **8**, 76, 1970.

must also be smaller at the equator than at the poles. In other words, the nonsphericity effect induces a dependence of effective temperature on latitude, with the polar regions appearing hotter than the equatorial belt.

To illustrate these results, I shall summarize the numerical work of Sackmann (1970), who, by making use of Takeda's double-approximation technique, has constructed a large set of models for main-sequence stars in the mass range 0.8–20$M_\odot$. Her calculations show that for each mass along the main sequence it is possible to construct a series of uniformly rotating models, with each series terminating with a model for which the effective gravity vanishes at the equator. The maximum luminosity change caused by solid-body rotation is about 7% for high-mass stars and somewhat smaller for low-mass stars with a radiative envelope. (For stars with masses below 1.5$M_\odot$, this change becomes much larger, though less certain.) Figure 6.1 demonstrates that a uniformly rotating star of mass M has similar central properties as a nonrotating star with mass $M - \Delta M$, where $\Delta M > 0$. We observe that the values of T_c and ρ_c for rotating stars on the verge of equatorial breakup fall exactly along the curve for nonrotating stars, with their positions being somewhat shifted in the direction of the lower masses. Note also that the largest deviation between the values for critically rotating stars and nonrotating stars is as small as 0.001 in $\log_{10} T_c$ and 0.004 in $\log_{10} \rho_c$! Following Sackmann, one has

$$\frac{\Delta M}{M} = \frac{3}{2}\bar{\epsilon}, \tag{6.11}$$

Table 6.1. *The percentage decrease in mass necessary to make the central pressure of critically rotating models equal to that of a nonrotating model.*

$M/M_\odot$	$\Delta M/M$ (%)	$M/M_\odot$	$\Delta M/M$ (%)
0.8	3.0	3	2.2
1.0	4.1	5	2.0
1.4	0.7	7	2.7
1.5	0.0	9	2.3
1.8	1.2	10	2.0
2.0	1.4	20	2.8

Source: Sackmann, I. J., *Astron. Astrophys.*, **8**, 76, 1970.

where $\bar{\epsilon}$ is the pressure-weighted average of the ratio of centrifugal force to gravity over the whole star. Table 6.1 illustrates this mass-lowering effect at breakup rotation along the main sequence. For the sake of completeness, in Figure 6.2 we also depict the critical equatorial velocity v_c at the point of equatorial breakup. Note that the velocity v_c steadily decreases as one passes down the main sequence from $20M_\odot$ to $1.4M_\odot$ and that it rises again as the mass is decreased below $1.4M_\odot$.

The above results strongly suggest that solid-body rotation can be considered as a small perturbation superimposed on the structure of a nonrotating star. For differentially

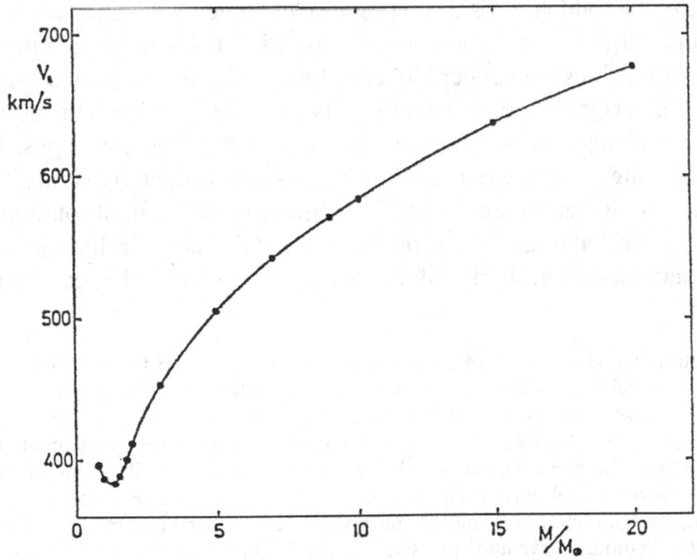

Fig. 6.2. The critical equatorial velocity v_c as a function of mass along the main sequence. *Source:* Sackmann, I. J., *Astron. Astrophys.*, **8**, 76, 1970.

rotating configurations, however, the situation is quite different because these systems can store a much higher total angular momentum than a uniformly rotating model with the same ratio of centrifugal force to gravity at the equator (cf. Section 2.8.3). Accordingly, we surmise that sequences of stellar models in nonuniform rotation do not terminate, therefore allowing for much larger observable effects than in a uniformly rotating model on the verge of equatorial breakup. That this is indeed the case was properly demonstrated by Bodenheimer (1971) and Clement (1979).

Several series of differentially rotating models have been constructed, each with fixed mass M and fixed angular momentum distribution $\Omega\varpi^2$, but with increasing values for the total angular momentum J. The rotational characteristics of three $30M_\odot$ models are illustrated in Figure 6.3. Note that considerable polar flattening occurs, with the ratio of equatorial to polar radii ranging up to about 4. Yet, none of these models approaches the limit of zero effective gravity at the equator. Not unexpectedly, in contrast to the case of solid-body rotation, conditions in the central regions now show large changes caused by differential rotation. This is illustrated in Figure 6.4, which shows that the effect of an increase in J is to shift the configuration closely parallel to and downward along the curve corresponding to nonrotating stars. A similar mass-lowering effect was found by Clement, who enlarged Bodenheimer's analysis by constructing sequences of differentially rotating models in the whole mass range 1.5–$30M_\odot$.

As mentioned, the problem is complicated by the fact that we have no direct knowledge of the angular momentum distribution within a star. Fortunately, the Bodenheimer–Clement calculations indicate that, given a mass M and a total angular momentum J, the changes in central temperature and density and in total luminosity are not strongly dependent on the interior angular velocity gradient. In view of the rather arbitrary nature of the assumed rotation laws, this is a most useful result.

In summary, uniform rotation has a mass-lowering effect on the internal structure of a main-sequence star, which gives a rotating model some of the characteristics of a nonrotating model of lower mass. Thus, uniform rotation leads to lower interior temperatures, lower luminosities, and either higher or lower interior densities depending on whether the star's mass is greater or smaller than about $1.5M_\odot$, which is the point where main-sequence stars change from convective cores to convective envelopes. Detailed calculations strongly suggest that this mass-lowering effect is generally valid since it applies to solid-body rotation as well as to various degrees of differential rotation. This is consistent with the view that rotating stars on the upper main sequence have less massive convective cores and, therefore, shorter lifetimes than their nonrotating counterparts.*

* Recall that all barotropic models presented in this section have rotation laws that satisfy the constraint imposed by dynamical stability with respect to axisymmetric motions; that is, their specific angular momentum $\Omega\varpi^2$ increases outward so that their angular velocity falls off more slowly than ϖ^{-2}, where ϖ is the distance from the rotation axis (see Eq. [3.98]). More recently, Clement (1994) has probed the limiting case $\Omega\varpi^2 = \textit{constant}$, which corresponds to a marginally stable configuration. Accurate two-dimensional models have been computed, assuming that one has $\Omega \propto \varpi^{-2}$ outside the cylinder containing the convective core and a solid-body rotation inside that cylinder. Calculations show that these extreme models have more massive convective cores than their nonrotating or rigidly rotating counterparts, at least for stars with masses below $12M_\odot$. In more massive configurations, however, the convective cores always decrease in mass fraction for any distribution of specific angular momentum.

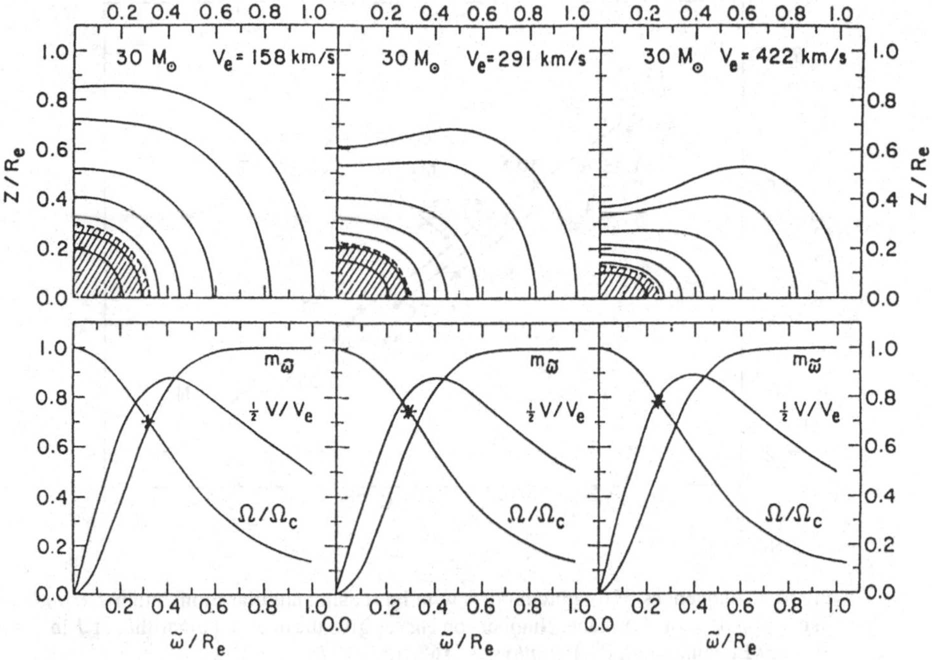

Fig. 6.3. Detailed structure of three models for $30M_\odot$. R_e is the total equatorial radius and v_e is the equatorial velocity. The shaded area indicates the convective core. The upper portions show isopycnic contours enclosing mass fractions 0.2, 0.4, 0.6, 0.8, 0.95, 0.999, and 1.0. The lower portions give the ratio of the angular velocity Ω to the central value Ω_c, the fraction m_ϖ of the total mass interior to the corresponding cylindrical surface about the rotation axis, and the ratio of the circular velocity v to the surface value v_e. The boundary of the convective core is indicated by an asterisk. *Source:* Bodenheimer, P., *Astrophys. J.*, **167**, 153, 1971.

6.2.2 *Effects of rotation on the observable parameters*

The most conspicuous effect of rotation is to distort a star into an oblate configuration. This is well illustrated in Figure 6.3, although it is not known whether such high degrees of differential rotation are present in real stars. Yet, it is these departures from sphericity and the luminosity changes that are of paramount importance for the observable effects of rotation on the radiation emanating from a star. As we recall from Section 3.3.1, a barotropic model with a radiative envelope has an emergent flux $|\mathcal{F}|$ that varies in proportion to the surface effective gravity g (see Eq. [3.41]). Since this quantity is smaller at the equator than at the poles, both the local effective temperature and surface brightness are, therefore, lower at the equator than at the poles. This implies in turn that the various magnitudes and color indices of a rotating star will be functions of the aspect angle i between the line of sight and the rotation axis.

The theoretical problem divides naturally into three parts: (1) Building an interior model so that the effective temperature and gravity become known as functions of latitude on its free surface, (2) computing the energy spectrum of radiation as a function of aspect angle when a suitably realistic model atmosphere is fitted at each point of the free surface, and (3) integrating the emergent flux to obtain the usual photometric parameters for

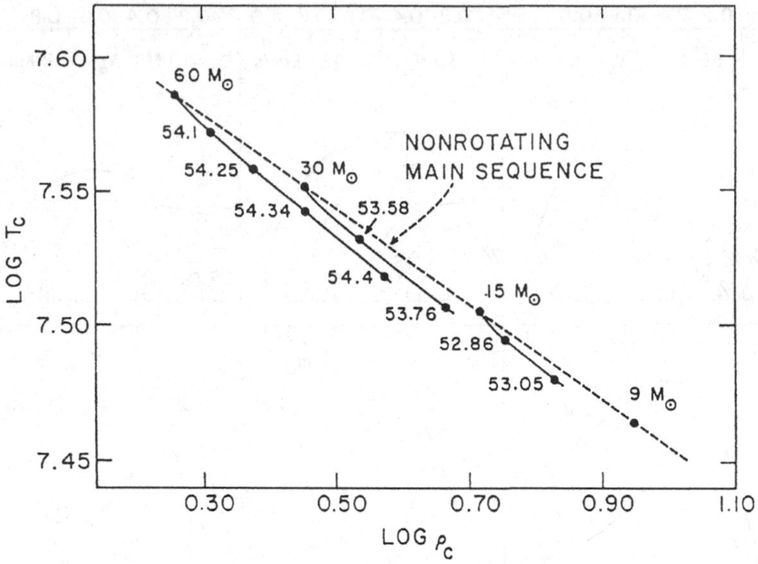

Fig. 6.4. Sequences of rotating models with increasing angular momentum J (*solid curves*) in the ($\log \rho_c - \log T_c$)-plane. Numbers on curves give the decimal logarithm of J in cgs units. *Source:* Bodenheimer, P., *Astrophys. J.*, **167**, 153, 1971.

each aspect angle. Figure 6.5 illustrates the results obtained by Maeder and Peytremann (1970), who have computed the energy spectrum of radiation for uniformly rotating stars of $5M_\odot$, $2M_\odot$, and $1.4M_\odot$. Each rotational track represents configurations ranging from the nonrotating model to the uniformly rotating model for which $\Omega/\Omega_c = 0.99$, where Ω_c is the angular velocity at breakup rotation. For each mass, different values of the inclination i have been considered, with the aspect angle increasing from $i = 0°$ ("pole-on" stars) to $i = 90°$ ("equator-on" stars). For the $2M_\odot$ models, the percentage of stars under the random-orientation hypothesis is also indicated. (This is of course valid for all masses.) We observe that *a pole-on star appears brighter than a nonrotating star of the same mass, but has almost the same color*. This is so because one is directly facing the brighter polar regions as well as a larger projected area resulting from the star's oblateness. Figure 6.5 also shows that *an equator-on star appears fainter and considerably redder than a nonrotating star of the same mass*. The reason lies in the fact that limb darkening reduces the brightness of the polar regions while gravity darkening makes the equatorial belt cooler.

How do these theoretical results compare with the available observational data for normal main-sequence stars? By comparing their uniformly rotating models with various observed quantities, Maeder and Peytremann (1970) found that there was agreement with observation for stars earlier than about spectral type A7 but that later types showed effects at least two times larger than predicted by solid-body rotation. If so, then, what rotation law do upper-main-sequence stars actually follow? The problem has been considered by Smith (1971), who made a statistical study of the data available for rotating stars in the Praesepe and Hyades clusters. In agreement with other works, it is found that these stars seem not to be rotating uniformly. Unfortunately, a

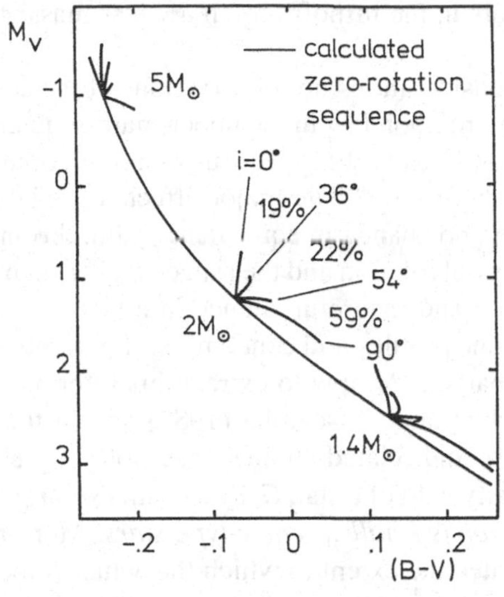

Fig. 6.5. Color–magnitude diagram with rotational tracks for $5M_{\odot}$, $2M_{\odot}$, and $1.4M_{\odot}$, and various angles i. The termination point are for $\Omega/\Omega_c = 0.99$. *Source (revised):* Maeder, A., and Peytremann, E., *Astron. Astrophys.*, **7**, 120, 1970. (Courtesy of Dr. A. Maeder.)

detailed study of the errors involved also shows the uncertainties to be such that the observations cannot be said to support any particular law of nonuniform rotation. More recently, Collins and Smith (1985) have made use of detailed stellar atmosphere models to compute the photometric effects of differential as well as rigid rotation in the A-type stars. Their analysis confirms the known qualitative result that differential rotation produces a larger scatter in the color–magnitude diagram than does uniform rotation. As was shown by these authors, however, photometry alone can only put rather weak constraints on the angular momentum distribution of the upper-main-sequence stars. This precludes any more definite conclusion about the nature of the rotation law in these stars.

Let us next consider the modifications brought by rotation on the age estimates of open star clusters. As we know, the age of a cluster is obtained from its color–magnitude diagram by fitting the observed sequence in the turnoff region with isochronous lines derived from nonrotating stellar models. The effects of rotation on age estimates are essentially of two kinds: (i) *aspect effects* on the color and magnitude of each star belonging to the cluster and (ii) *structural effects* on the models that are used to draw the theoretical isochronic lines. Both effects have been considered by Maeder (1971) under the assumption of uniform rotation on and above the main sequence. His analysis indicates that the structural effects of uniform rotation on age estimates are negligible in comparison with the aspect effects. However, because the displacement of a rotating star to the right of the main sequence can mimic the displacement due to evolution, neglecting the aspect effects leads to an overestimate in age that may reach up to 70% for clusters with the most rapidly rotating stars. In fact, Maeder has estimated that the age overestimates caused by the neglect of rotation reach about 60–70% for α Persei and the Pleiades. By contrast, the ages of the older clusters undergo very little changes, approximately 10–20%, because the stars in the turnoff region are less massive and so are rotating more slowly.

It is evident that neither theoretical considerations nor observations of the continuum can give a clear expectation for the actual rotation law in the upper-main-sequence stars. To what extent can the study of spectral lines yield useful information about the degree of surface differential rotation in these stars? The major effects of axial rotation on spectral lines is to broaden them, with no change in equivalent width; the amount of broadening depends upon the degree of axial rotation and the aspect angle i. In principle, the extent of surface differential rotation and macroturbulence in a star can be determined from the departures of observed line profiles and concomitant Fourier transforms from their standard theoretical counterparts. Attempts to extract this information from line profiles have been made by Stoeckley and Buscombe (1987) and in the Fourier domain by Gray (1977). Although these and related studies have not yet yielded any definite information on the surface velocity field of a star, *Gray's results strongly suggest that differential rotation does not exist or is small in early-type stars.* More recently, Collins and Truax (1995) have investigated the extent to which the actual velocity field of these stars can be determined by the information contained within a spectral line profile or its Fourier transform. It is found that one may use the classical model of a rotating star to determine projected rotational speeds as long as one does not expect accuracies greater than 10% under ideal conditions, with significantly larger errors for stars exhibiting extreme rotation. Accordingly, the use of the classical model as a probe of surface differential rotation and macroturbulence in a star remains problematic at best.

6.3 Axial rotation along the upper main sequence

In Section 1.3 we summarized the mean rotational properties of single stars. It is the purpose of this section to provide further information about the rotation patterns in specific groups of early-type, main-sequence stars.

6.3.1 Rotation in open clusters

Figure 1.6 provides a comparison between the average rotational velocities of cluster and field stars. It is immediately apparent that the $\langle v \sin i \rangle$ values of the, generally younger, cluster stars are similar to those of the field stars, except that for spectral types later than F0 the cluster stars rotate more rapidly than the field stars. A somewhat different picture emerges when one compares the $\langle v \sin i \rangle$ values for members of individual cluster and field stars. That this is indeed the case is illustrated in Figure 6.6, which shows that open clusters and associations often differ in their mean projected rotational velocities.

The question immediately arises whether the $\langle v \sin i \rangle$ values of a given cluster are unusual because of high or low equatorial velocities, v, or because of preferential inclination angles, i, of the rotation axis. Unfortunately, we do not yet know whether the rotation axes are oriented at random in space or whether there exists a preferential direction in some (if not all) clusters. Hereafter we shall assume that alignment of axes does not contribute appreciably to the unusual projected rotational velocities that are observed in some clusters. With regard to the causes of the differences between clusters,

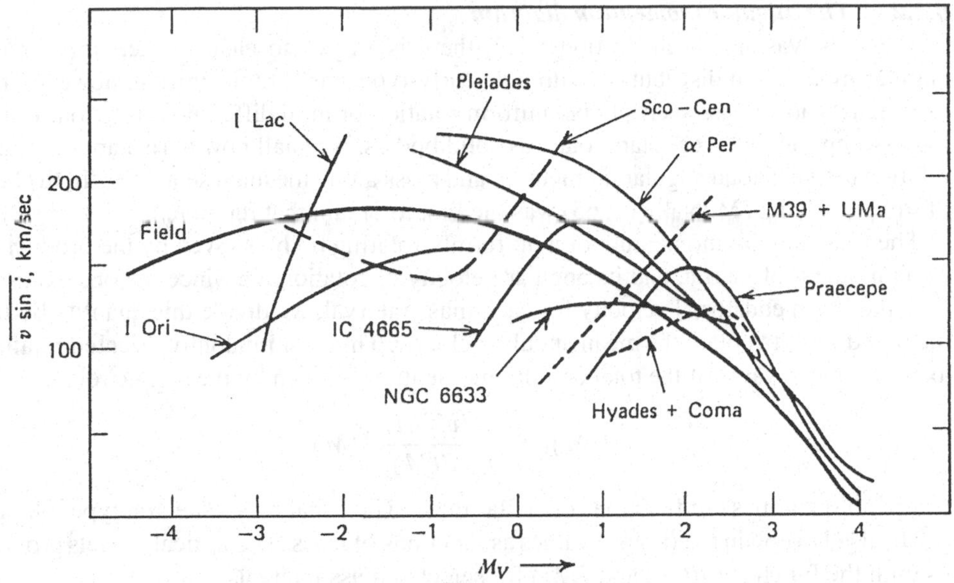

Fig. 6.6. Mean projected equatorial velocities for several open star clusters compared with field main-sequence stars. Adapted from Kraft (1970). *Source:* Gray, D. F., *The Observation and Analysis of Stellar Photospheres*, Cambridge: Cambridge University Press, 1992.

three likely explanations have been considered, namely, evolutionary expansion effects, the proportion of binaries, and the proportion of peculiar stars.

When a star leaves the zero-age main sequence and expands, its rotational velocity decreases. Since the brightest stars in a cluster evolve faster than the less luminous ones, such an evolutionary effect could possibly explain the low rotational velocities of the brightest stars in, for example, IC 4665 (see Figure 6.6). However, the fact that evolutionary expansion is not the main cause of this "turn-down" effect in clusters is well illustrated by the α Persei cluster, where the evolved stars have larger, rather than smaller, mean rotational velocities than field stars! As we shall see in Section 6.3.5, there are at least two ways in which the initial rotational velocities of stars may be gradually modified: by *tidal interaction* in closely spaced binaries (e.g., the Am stars) and by *magnetic braking* in magnetic stars (e.g., the Ap stars). Thus, if some clusters differ in their number of spectroscopic binaries or peculiar stars, we might expect that their $\langle v \sin i \rangle$ values will also depart significantly from the mean rotational velocities of field stars. Detailed studies have shown that clusters with rapidly rotating stars have far fewer binaries and Ap stars than clusters with stars having normal or low rotational velocities (e.g., Levato and Garcia 1984). Hence, we conclude that tidal interaction and magnetic braking are quite effective in reducing rotational velocities, so that a large part of the differences between clusters in their $\langle v \sin i \rangle$ values can be assigned to different frequencies of binaries and Ap stars. But then, as was correctly pointed out by Abt (1970), we have succeeded only in shifting the problem from trying to explain the various mean rotation rates in clusters to trying to explain these frequency differences.

6.3.2 *The angular momentum diagram*

As was shown in Section 6.2.2, there is as yet no clear expectation for the angular momentum distribution within an early-type star. At this writing, however, the most reasonable guess seems to be uniform rotation or mild differential rotation. Using the assumption that these stars rotate as solid bodies, we shall now derive an important relation between total angular momentum and mass along the main sequence. To the best of my knowledge, McNally (1965) was the first to obtain that relation.

The total angular momentum of a uniformly rotating body is given by the product of I, its moment of inertia, and its angular velocity of rotation, Ω. Since the observations give the mean equatorial velocity for each mass interval, we divide this quantity by the mean radius R to obtain the mean angular velocity. Thus, for randomly oriented rotation axes, the mean value of the total angular momentum is given by the simple relation

$$\langle J(M)\rangle = \frac{4}{\pi}\frac{\langle v \sin i\rangle}{R(M)} I(M), \tag{6.12}$$

where all quantities are functions of stellar mass. The usual mass–spectral type relation can be used to obtain the $\langle v \sin i\rangle$ values as functions of mass. Theoretical models provide us with the functions $R(M)$ and $I(M)$ for selected mass intervals.

Updating Kraft's (1970) analysis, Kawaler (1987) has re-derived the mean angular momentum $\langle J(M)\rangle$ along the main sequence using current stellar models and rotational velocities. In Figure 6.7 the circles represent a sample of normal single stars, whereas data indicated by crosses include Am and Be stars in the sample. For comparison, also shown is the line $\langle J(M)\rangle$ that corresponds to rotation at breakup velocity $v_{\rm crit}$, that is, where

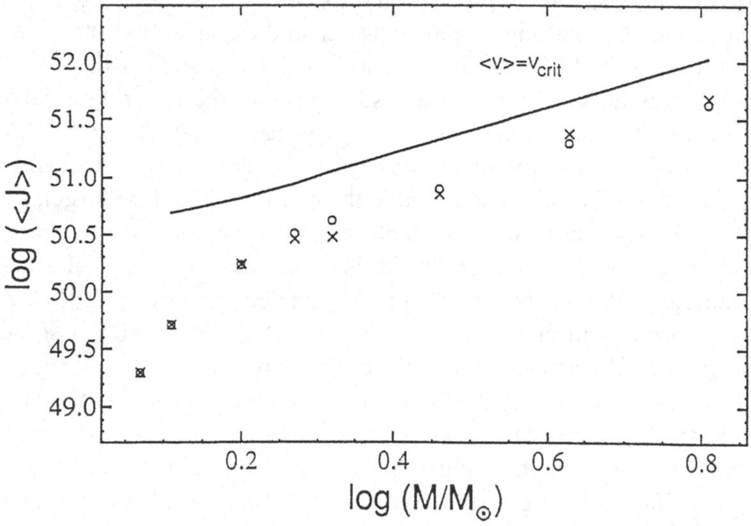

Fig. 6.7. Mean angular momentum as a function of stellar mass, assuming solid-body rotation at the surface rate. The circles represent the sample of normal single stars of Fukuda (1982); the crosses represent the same sample, but include Am and Be stars. The solid line represents the angular momentum for main-sequence models rotating at breakup velocity. *Source:* Kawaler, S. D., *Publ. Astron. Soc. Pacific*, **99**, 1322, 1987. (Courtesy of the Astronomical Society of the Pacific.)

surface gravity and centrifugal force are equal with $v_{crit} = (GM/R)^{1/2}$. These results are consistent with the $\langle v \sin i \rangle$ values being the same fraction of the critical velocity v_{crit} for all main-sequence stars more massive than $1.5M_\odot$.

For normal single stars earlier than spectral type F0, the relation between mean angular momentum per unit mass, $\langle j \rangle = \langle J \rangle / M$, and stellar mass is well represented by a power-law relation of the form $\langle j \rangle \propto M^\alpha$ with $\alpha = 1.09$. (When the Am and Be stars are included in the sample, however, one finds that $\alpha = 1.13$.)* The low-mass stars ($M \lesssim 1.5M_\odot$) deviate from this simple power-law relation, as evidenced by their slow rotational velocities in Figure 1.6. As we shall see in Section 7.2, this sharp break at mass $1.5M_\odot$ can be attributed to angular momentum loss by magnetically controlled winds or episodic mass ejections from stars with outer convection zones. Accordingly, since the high-mass stars ($M \gtrsim 1.5M_\odot$) have no appreciable convective envelopes that could support winds or mass ejections, it is generally believed that these stars have retained most of their initial angular momentum. Hence, it seems likely that the simple power law $\langle J \rangle \propto M^{\alpha+1}$ expresses a fundamental relation between the angular momentum content of an early-type star and its mass, where stars are given, on the average, an amount of angular momentum in proportion to their masses.†

6.3.3 *The rotational velocity distributions*

Figure 1.6 is a plot of the $\langle v \sin i \rangle$ values against spectral type for single, main-sequence stars. In this section we shall briefly discuss the distribution of $v \sin i$ at a given spectral type. Extensive surveys of projected rotational velocities have been assembled by Wolff, Edwards, and Preston (1982). Figure 6.8 illustrates the observed distributions of $v \sin i$ for a number of spectral type ranges.

It is immediately apparent that these distributions are all strikingly similar: They peak at low values of $v \sin i$ and decrease slowly with increasing rotational velocity, with a maximum of about 350 km s^{-1} at all spectral types. Note that the early B-type stars are unique only in having a larger percentage of stars with $v \sin i$ smaller than 40 km s^{-1}. The decrease in rotational velocity for the late A-type stars is also worth noticing since it indicates that the braking mechanism that spins down the stars of later spectral type is already partially operative in the A-type stars. Note also that these observational results rule out simple Maxwellian distributions for the $v \sin i$s along the upper main sequence.

The similarity of the observed distributions strongly suggests that the same physical mechanisms are involved in determining the rotational velocities of all upper-main-sequence stars. Unfortunately, without a clear understanding of the star-formation process and early stellar evolution, we are still unable to explain why slow rotation (i.e., $v \sin i <$ 100 km s^{-1}) is so prevalent among these stars.

6.3.4 *Rotation of Be and shell stars*

If one excepts remnants such as neutron stars and pulsars, the stars of most rapid rotation are the emission-line B stars (i.e., the *Be stars*). There is now widespread

* The values originally obtained by McNally and Kraft were $\alpha = 0.80$ and $\alpha = 0.57$, respectively.

† As was shown by Brosche (1963) and others, the $\langle J \rangle \propto M^2$ rule is closely obeyed over the mass range 10^{18}–10^{48} g, from asteroids up to clusters of galaxies. Explanations have been presented by Wesson (1979) and by Carrasco, Roth, and Serrano (1982).

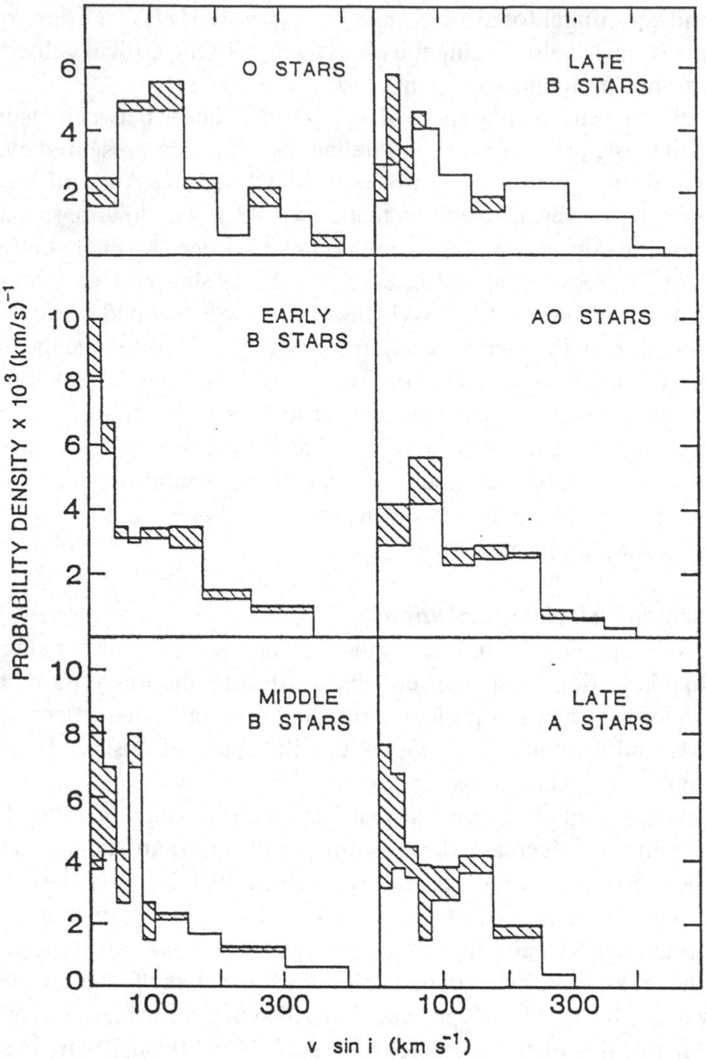

Fig. 6.8. The observed distribution of projected equatorial velocities as a function of spectral type. Hatched areas show spectroscopic binaries discovered to date among the stars within each group. *Source:* Wolff, S. C., Edwards, S., and Preston, G. W., *Astrophys. J.*, **252**, 322, 1982.

agreement that matter is leaving the Be stars at their equator, with the resultant equatorial disk giving the emission seen in the hydrogen lines. Some Be stars also develop, from time to time, a network of deep and narrow absorption lines and they are then called *shell stars*. They are also characterized by extremely broad absorption lines, which, when interpreted as due to axial rotation, makes them as a class the most rapidly rotating Be stars. As was pointed out by Slettebak (1979), this suggests that the shell stars are edge-on normal Be stars: The difference in spectra is due to differences in inclination of the rotation axes.

Mean values of the observed $v \sin i$s range between about 200 and 250 km s^{-1}, with the largest $v \sin i$s being in the neighborhood of 400 km s^{-1}. This raises at once the following question: Do the Be and shell stars rotate at their critical velocity at which centrifugal force balances gravity at the equator? The answer to that question is flatly no. Indeed, as can be seen in Figure 6.2, the theoretical breakup velocities are much larger than 400 km s^{-1} in the mass range 3–15$M_{\odot}$, which corresponds to the masses of normal B-type stars and probably to those of Be type objects as well.

In order to gain further insight into the problem, Porter (1996) has made a detailed statistical study of the projected rotational velocities of these stars. In his discussion the fundamental parameter is not $v \sin i$, however, but the equatorial velocity of the star as a fraction of the breakup velocity, $w = v/v_{crit}$, where v_{crit} is the critical equatorial velocity of the star at breakup rotation. The distribution functions of normal Be stars and shell stars as functions of $w \sin i$ are shown in Figure 6.9. One readily sees that the projected equatorial velocities for shell stars are significantly larger than those for normal Be stars. Statistical tests further indicate that shell stars and normal Be stars are simply related by inclination. This, taken along with theoretical shell line profiles generated in edge-on disks, leads to the following conclusions: (i) shell stars are normal Be stars viewed edge-on and (ii) the shell star distribution with $i = 90°$ is a good representation of the distribution of

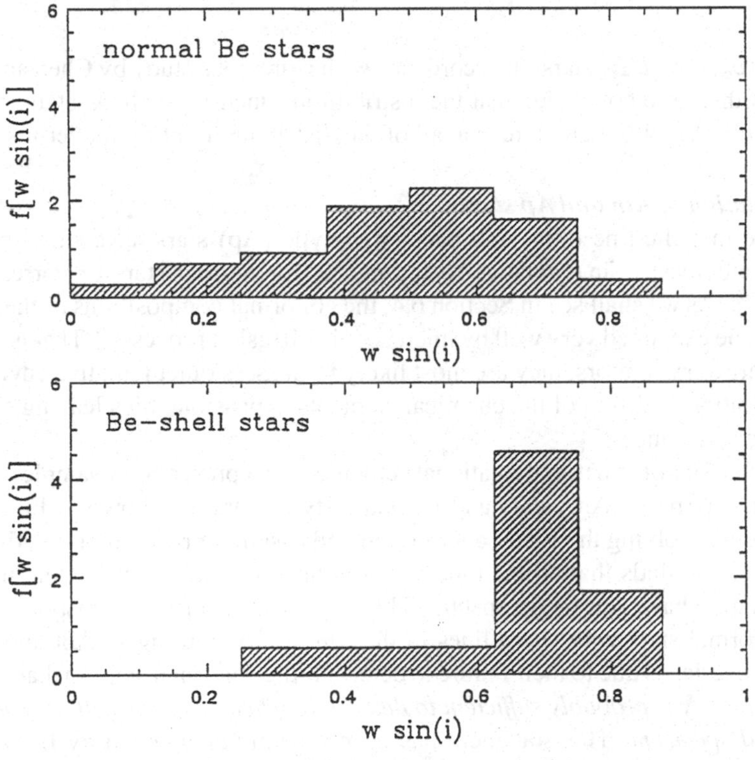

Fig. 6.9. Distribution functions of normal Be stars (*top*) and Be-shell stars (*bottom*) as functions of $w \sin i$. *Source:* Porter, J. M., *Mon. Not. R. Astron. Soc.*, **280**, L31, 1996. (Courtesy of Blackwell Science Ltd.)

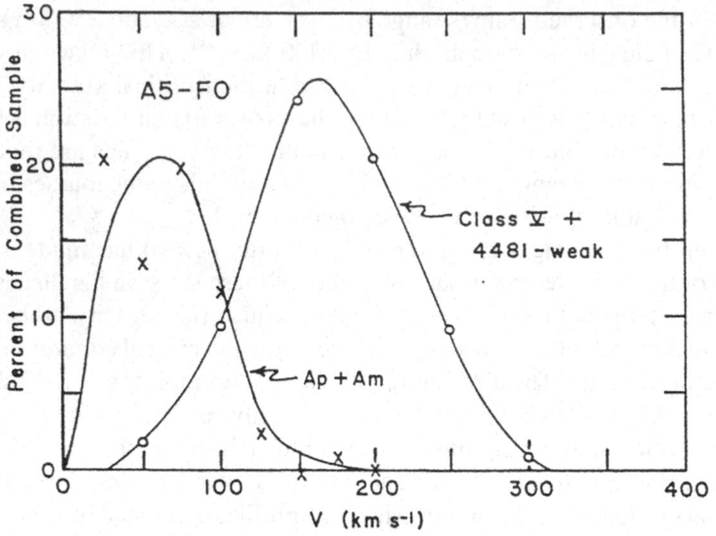

Fig. 6.10. Distributions of equatorial rotational velocities for two samples of A5–F0 stars. The right distribution is for 234 normal class V stars plus 23 stars with weak λ4481 lines; the left distribution is for 133 Ap+Am stars. *Source:* Abt, H. A., and Morrell, N. I., *Astrophys. J. Suppl.*, **99**, 135, 1995.

the ratios $v/v_{\rm crit}$ for *all* Be stars. In accordance with a previous study by Chen and Huang (1987), it is therefore concluded that the distribution function of all Be stars is sharply peaked at $w = 0.7$, although there is a tail of the distribution to the smaller ws.

6.3.5 *Rotation of Am and Ap stars*

The metallic-line (Am) and peculiar A-type (Ap) stars have small projected rotational velocities, $v \sin i$, relative to the means for normal stars of corresponding spectral types. As we shall see in Section 6.4, the abnormal compositions of the Am and Ap stars can be explained very well by microscopic diffusion processes. That is, because these stars are slow rotators, they are most likely to possess quiet radiative envelopes in which gravitational sorting of the chemical elements is possible, thus leaving abnormal atmospheric abundances.

The distributions of equatorial rotational velocities for representative samples of chemically peculiar (Ap and Am) stars and normal A-type stars are shown in Figure 6.10. Thus, after deconvolving the $v \sin i$ distribution and assuming random orientation of the rotation axes, one finds that all the rapid rotators have normal spectra while nearly all the slow rotators have Am or Ap spectra. There is a 10% overlap, corresponding to 39 too many normal stars with sharp lines in the sample. According to Abt and Morrell (1995), this overlap is due to their failure to detect all the abnormal stars so that *a specific rotational velocity is probably sufficient to determine whether a star will have a normal or abnormal spectrum*. This statement has been recently challenged by Budaj (1996, 1997), however.

It has been known for some time that all, or most, Am stars are spectroscopic binaries. A detailed study of the frequency of Am stars among those binaries has been made by Ginestet et al. (1982) and Budaj (1996). Their analyses indicate that the orbital period distribution of the Am stars has a prominent peak in the period range 2–15 days, which is also the region where synchronization is observed. As they showed, this period range coincides with the largest gap in the orbital period distribution of nonpeculiar spectroscopic binaries of spectral types A4–F1, IV and V. To be specific, in the period range 2–100 days, it is found that about 85% of the binaries are Am stars. However, although the Am stars are also observed at larger orbital periods, there is a conspicuous gap in the period range 180–800 days. In Section 8.4.4 we shall explain how *tidal interaction* in binaries with period smaller than 100–200 days can effectively cause their components to have low rotational velocities and thus become Am stars.

In contrast to the Am stars, however, the slow rotation of the Ap stars does not appear to be due to tidal interaction in close binaries. What is, then, the mechanism responsible for the abnormally low rotation rates of the Ap stars, when compared to normal stars of corresponding temperature and luminosity? Unfortunately, whereas their slow rotation is generally attributed to some kind of *magnetic braking*, there remains considerable controversy as to whether most of their angular momentum is lost before or during the main-sequence phase. Observations of Ap stars in open clusters and associations of varying ages can answer that question. According to Wolff (1981), measurements of $v \sin i$ values strongly suggest that the Si-type Ap stars lose angular momentum after they reach the main sequence, while those of the Sr–Cr–Eu group might do so prior to the main-sequence phase. As was noted by North (1984) and Borra et al. (1985), however, her conclusions are based on line-broadening measurements, which are affected not only by the $\sin i$ projection factor but also by the magnetic field strength via Zeeman broadening. This is the reason why they have determined accurate photometric rotation periods of magnetic Ap stars belonging to open clusters and associations. Both studies show that the young cluster stars have essentially the same rotation periods as the older field stars, indicating either that the Ap stars have lost most of their angular momentum before they reach the main sequence or that they are intrinsically slow rotators from their formation on. This result has been recently confirmed by North (1998), who found no evidence for any loss of angular momentum on the main sequence, thus confirming earlier results based on less reliable estimates of surface gravity.

6.4 Circulation, rotation, and diffusion

It is generally thought that diffusion processes are responsible for most of the peculiar abundances observed in the chemically peculiar stars. As was originally noticed by Michaud (1970), abundance anomalies appear, on the main sequence, in the atmospheres of stars most likely to have stable envelopes and atmospheres. These stars are slow rotators and so have less meridional circulation, they often have magnetic fields, and they have an effective temperature for which stellar envelope models give the weakest convection. In its simplest form, the diffusion model assumes that the region below the superficial convection layer of a chemically peculiar star is stable enough so that microscopic diffusion processes can separate the light elements from the heavy ones, that is to say, those chemical elements absorbing more of the outward going radiative flux per atom move to the surface, while those absorbing less sink into the interior. Because those stars where the thermally driven currents are expected to be the slowest also have the largest abundance anomalies, it is evident that a detailed understanding, from first principles, of the interaction between diffusion and meridional circulation becomes essential if we are to understand stellar abundances.

As was pointed out in Section 4.1, because strict radiative equilibrium is impossible for a uniformly rotating star, a state of thermal equilibrium can only be maintained with the help of energy transport by circulatory currents in meridian planes passing through the rotation axis. In the case of a slowly rotating, early-type star, this large-scale meridional flow is quadrupolar in structure, with rising motions at the poles and sinking motions at the equator (see Figure 4.3). In spherical polar coordinates (r, θ, φ), we can thus write

$$\mathbf{u} = \epsilon \left[u(r) P_2(\cos\theta) \mathbf{1}_r - r v(r) \sin\theta \, \frac{d P_2(\cos\theta)}{d \cos\theta} \, \mathbf{1}_\theta \right], \tag{6.13}$$

where P_2 is the Legendre polynomial of degree two and ϵ is the ratio of centrifugal force to gravity at the equator,

$$\epsilon = \frac{v_{\text{eq}}^2 R}{GM}. \tag{6.14}$$

By virtue of Eq. (6.2), one also has

$$v = \frac{1}{6} \frac{1}{\rho r^2} \frac{d}{dr} (\rho r^2 u), \tag{6.15}$$

so that the meridional velocity $\mathbf{u}$ depends on the radial function u only.

In Table 4.1 we list Sweet's (1950) frictionless solution for a Cowling point-source model. One readily sees that this solution, which becomes infinite at the free surface, does not satisfy the essential boundary conditions (4.38). The situation is even worse when the prescribed rotation law is nonuniform since, for then *both* components of the meridional velocity become infinite at the free surface (see Eq. [4.42]). In fact, no further progress has been made until it was realized, in 1982, that turbulent friction acting in the outmost surface layers is an essential ingredient of the problem (see Sections 4.3 and 4.4). That is to say, unless one makes allowance for a *thermo-viscous boundary layer* near the upper boundary of the radiative zone, it is impossible to calculate a meridional flow that satisfies all the boundary conditions and all the basic equations of the problem.* Table 4.2 lists some of the self-consistent solutions obtained by Tassoul and Tassoul (1982, 1995).

Many characteristics of the chemically peculiar stars can be explained on the basis of microscopic diffusion in the presence of meridional circulation in their outer radiative envelopes. Indeed, when this large-scale flow is rapid enough to obliterate the settling of the diffusion of helium, no underabundance of this element is possible and the superficial He convection zone remains important, making the appearance of some of the abnormal abundances impossible. Comparing the meridional circulation velocities for solid-body rotation (i.e., $\alpha = 0$ in Table 4.2) to diffusion velocities of helium below the He convection zone, Michaud (1982) has shown that this zone disappears only in stars with equatorial velocities smaller than about 90 km s^{-1}. This is in agreement with the

* There has been much confusion in the literature about the existence of thermo-viscous boundary layers in rotating stars. This is discussed in the Bibliographical notes for Section 4.3.

cutoff velocity observed for the HgMn stars. Given this encouraging result, detailed two-dimensional diffusion calculations have been carried out by Charbonneau and Michaud (1988) to determine with greater accuracy the maximum rotational velocity allowing the gravitational settling of helium.

In order to couple microscopic diffusion and meridional circulation, one writes the continuity equation in the form

$$\rho \frac{\partial c}{\partial t} + \operatorname{div}\left[\rho(\mathbf{u} + \mathbf{U})c\right] = \operatorname{div}(\rho D_{12} \operatorname{grad} c), \tag{6.16}$$

where $c(r, \theta, t)$ is the concentration of the contaminant, measured with respect to hydrogen, and $D_{12}(r)$ is the coefficient of diffusion of helium in hydrogen. The velocity field $\mathbf{u}(r, \theta)$ corresponds to the meridional circulation, while $\mathbf{U}(r, \theta, t)$ describes the advective part of the diffusion velocity (e.g., Charbonneau and Michaud 1988, pp. 810–811). Equation (6.16) is a parabolic equation that must be solved with appropriate initial and boundary conditions. Calculations have been performed in both $3M_{\odot}$ and $1.8M_{\odot}$ stellar models appropriate, respectively, for HgMn and FmAm stars. The upper limits to the equatorial velocities allowing the chemical separation of helium are found to be 75 and 100 km s^{-1}, respectively, for these stars. Given the various approximations that had to be made in averaging over convection zones and the uncertainties in the meridional circulation velocities near the surface, the agreement with observations is quite satisfactory. This parameter-free model is not so successful, however, in reproducing quantitatively the anomalies of a given star in detail. Mass loss has been suggested as an important ingredient in the FmAm phenomenon.

Now, because turbulent particle transport can also have drastic effects on chemical separation, Charbonneau and Michaud (1991) have performed additional calculations that retain both meridional circulation and anisotropic turbulence. Equation (6.16) was thus replaced by

$$\rho \frac{\partial c}{\partial t} + \operatorname{div}[\rho(\mathbf{u} + \mathbf{U})c] = \operatorname{div}(\rho \mathbf{D} \operatorname{grad} c), \tag{6.17}$$

where the total diffusivity tensor can be written in the form

$$\mathbf{D} = \begin{pmatrix} D_{12} + D_V & 0 \\ 0 & D_{12} + D_H \end{pmatrix}. \tag{6.18}$$

(Compare with Eq. [3.134].) The functions D_V and D_H are the vertical and horizontal coefficients of eddy diffusivity due to the rotationally induced instabilities. Unfortunately, as was explained in Section 3.6, there is as yet no reliable theory that could provide firm analytical expressions or numerical values for these two coefficients. They are essentially free quantities that must be chosen, by trial and error, using the observed abundance anomalies to determine their values. Thus, given some parametric expressions for the eddy diffusivities, the problem can be treated as a two-dimensional initial-boundary value problem. Numerical calculations show that the diffusion model for FmAm stars is particularly constraining regarding the introduction of anisotropic turbulence. In the presence of meridional circulation, it is found that the maximum D_V/D_{12} ratio tolerable with the diffusion model is of the order of 10; otherwise, helium settling is overly impeded in stars rotating below the observed equatorial velocity cutoff. This sets extremely tight constraints on turbulence in early-type stars having equatorial velocities of 100 km s^{-1}

or less. Similar calculations show that the maximum D_H/D_{12} ratio tolerable with the diffusion model for FmAm stars is of the order of 10^6; otherwise, helium settling remains possible in stars rotating above the observed equatorial velocity cutoff. As was pointed by Charbonneau and Michaud (1991), however, this seems to be a prohibitively large value of the ratio D_H/D_{12}.

In summary, the above calculations show that microscopic diffusion in the presence of large-scale meridional currents does explain in a natural way the appearance of the HgMn and FmAm phenomenon in slowly rotating, nonmagnetic stars, without introducing any strong dependence on arbitrary parameters. These calculations also demonstrate that the smaller-scale, eddylike motions cannot be ignored altogether because they, too, can impede the gravitational settling of helium. In principle, given some solution for the meridional circulation, one can integrate Eq. (6.17) to derive upper limits on the coefficients D_V and D_H. As was pointed out in Section 4.4, however, the topology of the meridional flow in the surface layers of an early-type star is quite dependent on the gradient of angular velocity in these regions. Since this uncertainty on the circulation pattern should somewhat reflect on the determination of upper limits on D_V and D_H, it follows that the relative importance of meridional circulation and anisotropic turbulence in reducing chemical separation remains uncertain.

6.5 Rotation of evolved stars

Among the many problems that beset the theory of rotating stars, the redistribution of angular momentum in stellar interiors during evolution is by far the least understood. As we know, the post–main-sequence evolution of a star is accompanied by a strong contraction of its helium-rich core and by a corresponding expansion of the surrounding envelope. Unless there exists a very efficient transport of angular momentum from the core to the envelope, it is evident that the former has to spin up appreciably while at the same time the latter must spin down. The decrease in surface rotation as a star evolves away from the main sequence has been known for several decades. Unfortunately, to compute the gross changes caused by rotation in evolving stars, we are faced with two largely unresolved questions: Is the total angular momentum J of a star conserved or lost during its post–main-sequence evolution? And is there an effective means to redistribute the specific angular momentum $\Omega\varpi^2$ during evolution? The most reliable calculations are those of Endal and Sofia (1979), who have considered different cases of angular momentum redistribution, assuming in all cases conservation of total angular momentum. At the time, their theoretical surface rotation velocities for red-giant models were in agreement with the observed rotation rates for the K giants, so that there was apparently no need to invoke angular momentum losses among these stars.

A different picture emerged, however, when Gray's (1982) Fourier analysis of high signal-to-noise ratio data showed the existence of a discontinuity in rotation for luminosity class III giants.* This sudden drop takes place between spectral types G0 III and G3 III. A similar rotational discontinuity was also seen by Gray and Nagar (1985) in a sample of luminosity class IV subgiants. Near G0 IV, a sudden drop in rotation was

* This discontinuity, which was initially reported at spectral type G5 III, has been confirmed by Gray (1989) but was found to be near G0 III rather than G5 III. This change results primarily from improved spectral types.

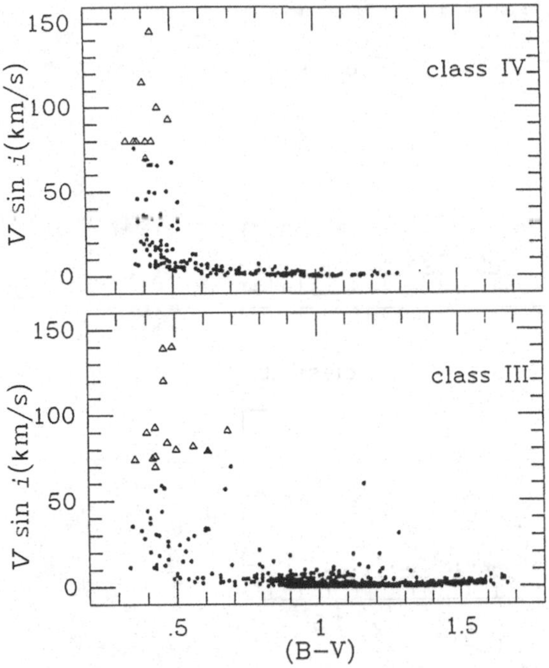

Fig. 6.11. Projected equatorial velocities as functions of $(B - V)$ color. Triangles refer to values taken from "The Bright Star Catalogue." *Source:* de Medeiros, J. R., and Mayor, M., in *Angular Momentum Evolution of Young Stars* (Catalano, S., and Stauffer, J. R., eds.), p. 201, Dordrecht: Kluwer, 1991. (By permission. Copyright 1991 by Kluwer Academic Publishers.)

observed with advancing spectral type, in complete analogy to the drop seen at G0 III in the giants. More recently, a systematic survey of about 2,000 evolved stars was carried out by de Medeiros and Mayor (1991), covering the spectral range from middle F to middle K of luminosity classes IV, III, II, and Ib. Figures 6.11 and 6.12 illustrate the $v \sin i$ measurements of their sample of stars as a function of the $(B - V)$ color. The cutoff in the distribution of rotational velocity for each luminosity class is located at F8 IV, G0 III, F9 II, and near F9 Ib; this corresponds to the $(B - V)$ colors 0.55, 0.70, 0.65, and about 0.70, respectively.

Note the wide range of $v \sin i$ values on the left side of the discontinuity for all luminosity classes. This large spread seems to reflect the broad distributions of rotation rates along the main sequence, as illustrated in Figure 6.8. Note also that the spread in $v \sin i$ values on the left of the cutoff decreases with increasing luminosity. In fact, the supergiant stars show no sudden decrease in rotation, and there is still a large fraction of slow rotators to the left of the discontinuity. This result strongly suggests that the origin of the rotational discontinuity is not the same for all classes.

As was originally suggested by Gray in the 1980s, the rotational discontinuity for the subgiant and giant stars can be interpreted as a result of a strong magnetic braking due to the deepening of their outer convective envelopes at some point in their evolution. To be specific, since the evolution of these stars carries them from hotter to cooler spectral types, a plot of rotation versus $(B - V)$ color delineates the time sequence of their

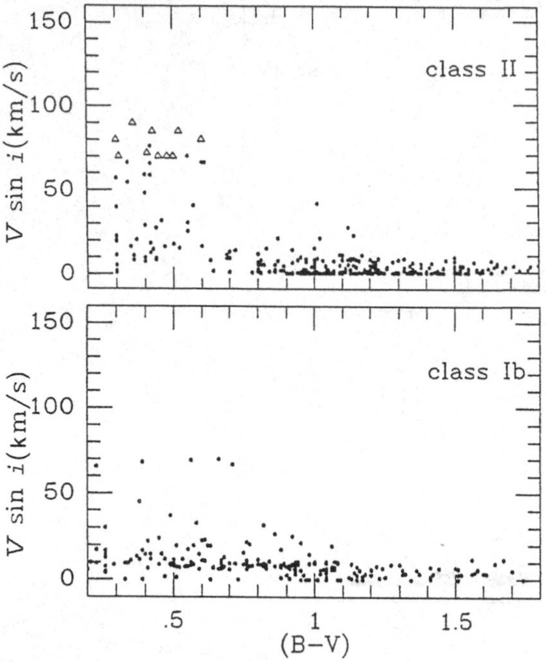

Fig. 6.12. Projected equatorial velocities as functions of $(B - V)$ color. Triangles refer to values taken from "The Bright Star Catalogue." *Source:* de Medeiros, J. R., and Mayor, M., in *Angular Momentum Evolution of Young Stars* (Catalano, S., and Stauffer, J. R., eds.), p. 201, Dordrecht: Kluwer, 1991. (By permission. Copyright 1991 by Kluwer Academic Publishers.)

rotational changes. As their progenitors evolve off the main sequence, the evolutionary increase in moment of inertia slowly reduces the rotation to the values attained on the left of the discontinuities in Figure 6.11. Sudden changes seen near spectral types G0 IV and G0 III occur because the evolutionary deepening of the convective envelope has become sufficient to sustain dynamo activity. Thence, a small amount of material escaping from the star's surface is caught in the open field lines of the dynamo-generated magnetic field, so that large amounts of angular momentum can be carried away by the escaping material (see Section 7.2). In short, the star develops an external magnetic brake that rapidly decelerates the rotation of at least its outer convective envelope.

An important piece of evidence in support of Gray's mechanism comes from the work of Simon and Drake (1989), who have shown that subgiant and giant stars undergo a sudden decrease in chromospheric activity at spectral types G0 IV and G0 III, which correspond to the $(B - V)$ colors 0.6 and 0.7, respectively. The fact that in both cases the observed decline in UV emission coincides with the sharp decrease in surface rotation rates strongly suggests that Gray's mechanism is indeed operative in these stars. As they noted, this joint decay in activity and rotation marks a transformation from acoustic heating in the early F-type stars to a magnetically controlled activity in the cooler stars, thus inducing a strong rotational braking action by means of stellar winds. Detailed calculations by Schrijver and Pols (1993) further indicate that the decrease in the observed rotational velocities of subgiants and giants is stronger than expected from the increase in moment of inertia alone, so that loss of angular momentum through magnetically channeled stellar winds must be substantial between the onset of convection and just beyond the upturn onto the giant branch. For the most luminous classes, however, the discontinuity in rotational velocities is probably the result of another evolutionary effect.

6.6 Bibliographical notes

Because the theory of early-type stars has developed quite slowly over the past twenty years, my first book is still a useful reference for these objects. See:

1. Tassoul, J. L., *Theory of Rotating Stars*, pp. 95–115 and pp. 305–361, Princeton: Princeton University Press, 1978.

The following general references may also be noted:

2. Moss, D., and Smith, R. C., *Rep. Prog. Phys.*, **44**, 831, 1981.
3. Smith, R. C., in *Physics of Be Stars* (Slettebak, A., and Snow, T. P., eds.), p. 123 and p. 486, Cambridge: Cambridge University Press, 1987.
4. Kippenhahn, R., and Weigert, A., *Stellar Structure and Evolution*, pp. 427–453, Berlin: Springer-Verlag, 1990.

Section 6.2. Almost everything we know about the state of rotation in a convective core has been reviewed in:

5. Tayler, R. J., *Mon. Not. R. Astron. Soc.*, **165**, 39, 1973.

The straightforward expansion method is due to Milne (Reference 5 of Chapter 4). The double-approximation technique was first discussed in:

6. Takeda, S., *Mem. College Sci. Kyoto Univ.*, *A*, **17**, 197, 1934.

The following key references may also be noted:

7. Ostriker, J. P., and Mark, J. W.-K., *Astrophys. J.*, **151**, 1075, 1968.
8. Kippenhahn, R., and Thomas, H. C., in *Stellar Rotation* (Slettebak, A., ed.), p. 20, New York: Gordon and Breach, 1970.
9. Clement, M. J., *Astrophys. J.*, **222**, 967, 1978.

Other technical papers may be traced to Reference 1 (pp. 112–115). Subsequent contributions are due to:

10. Gingold, R. A., and Monaghan, J. J., *Mon. Not. R. Astron. Soc.*, **181**, 375, 1977.
11. Smith, B. L., *Astrophys. Space Sci.*, **47**, 61, 1977.
12. Simon, S. A., *Astrophys. J.*, **228**, 357, 1979.
13. Wolfe, R. H., Jr., and Kern, J. W., *Astrophys. Space Sci.*, **64**, 443, 1979.
14. Kopal, Z., *Astrophys. Space Sci.*, **93**, 149, 1983.
15. Eriguchi, Y., and Müller, E., *Astron. Astrophys.*, **146**, 260, 1985; *ibid.*, **147**, 161, 1985.
16. Hachisu, I., *Astrophys. J. Suppl.*, **61**, 479, 1986; *ibid.*, **62**, 461, 1986.
17. Geroyannis, V. S., *Astrophys. J.*, **327**, 273, 1988; *ibid.*, **350**, 355, 1990.

See also:

18. Smith, R. C., and Collins, G. W., *Mon. Not. R. Astron. Soc.*, **257**, 340, 1992.
19. Aksenov, A. G., and Blinnikov, S. I., *Astron. Astrophys.*, **290**, 674, 1994.
20. Uryŭ, K., and Eriguchi, Y., *Mon. Not. R. Astron. Soc.*, **269**, 24, 1994; *ibid.*, **277**, 1411, 1995.

Related contributions are quoted in Reference 18.

Section 6.2.1. The presentation in the text follows:

21. Sackmann, I. J., *Astron. Astrophys.*, **8**, 76, 1970.
22. Bodenheimer, P., *Astrophys. J.*, **167**, 153, 1971.
23. Clement, M. J., *Astrophys. J.*, **230**, 230, 1979; *ibid.*, **420**, 797, 1994.

See also:

24. Shindo, M., Hashimoto, M., Eriguchi, Y., and Müller, E., *Astron. Astrophys.*, **326**, 177, 1997.

Section 6.2.2. The analysis in this section is taken from:

25. Maeder, A., and Peytremann, E., *Astron. Astrophys.*, **7**, 120, 1970; *ibid.*, **21**, 279, 1972.
26. Maeder, A., *Astron. Astrophys.*, **10**, 354, 1971.
27. Smith, R. C., *Mon. Not. R. Astron. Soc.*, **151**, 463, 1971.
28. Collins, G. W., and Smith, R. C., *Mon. Not. R. Astron. Soc.*, **213**, 519, 1985.

The effects of rotation on line profiles are considered in:

29. Gray, D. F., *Astrophys. J.*, **211**, 198, 1977; *ibid.*, **258**, 201, 1982.
30. Stoeckley, T. R., and Buscombe, W., *Mon. Not. R. Astron. Soc.*, **227**, 801, 1987.
31. Collins, G. W., and Truax, R. J., *Astrophys. J.*, **439**, 860, 1995.

Section 6.3.1. The following review paper may be noted:

32. Abt, H. A., in *Stellar Rotation* (Slettebak, A., ed.), p. 193, New York: Gordon and Breach, 1970.

Recent contributions are by:

33. Levato, H., and Garcia, B., *Astrophys. Letters*, **24**, 49, 1984; *ibid.*, p. 161.
34. Glaspey, J. W., *Publ. Astron. Soc. Pacific*, **99**, 1089, 1987.

Section 6.3.2. The reference to McNally is to his paper:

35. McNally, D., *The Observatory*, **85**, 166, 1965.

See also Kraft's discussion (Reference 28 of Chapter 1). The presentation in the text follows:

36. Kawaler, S. D., *Publ. Astron. Soc. Pacific*, **99**, 1322, 1987.

The relationship between total angular momentum and mass for a large variety of astronomical objects has been discussed in:

37. Brosche, P., *Zeit. Astrophys.*, **57**, 143, 1963.
38. Wesson, P., *Astron. Astrophys.*, **80**, 296, 1979.
39. Carrasco, L., Roth, M., and Serrano, A., *Astron. Astrophys.*, **106**, 89, 1982.

Section 6.3.3. See:

40. Wolff, S. C., Edwards, S., and Preston, G. W., *Astrophys. J.*, **252**, 322, 1982.

Section 6.3.4. The following review is particularly worth noting:

41. Slettebak, A., *Space Sci. Review*, **23**, 541, 1979.

Statistical studies of Be and shell stars have been made by:

42. Chen, H. Q., and Huang, L., *Chinese Astron. Astrophys.*, **11**, 10, 1987.
43. Porter, J. M., *Mon. Not. R. Astron. Soc.*, **280**, L31, 1996.

Section 6.3.5. Rotation of the Am and Ap stars is discussed in:

44. Ginestet, N., Jaschek, M., Carquillat, J. M., and Pédoussaut, A., *Astron. Astrophys.*, **107**, 215, 1982.
45. Abt, H. A., and Morrell, N. I., *Astrophys. J. Suppl.*, **99**, 135, 1995.
46. Budaj, J., *Astron. Astrophys.*, **313**, 523, 1996; *ibid.*, **326**, 655, 1997.

Conflicting results about the braking of Ap stars will be found in:

47. Wolff, S. C., *Astrophys. J.*, **202**, 101, 1975; *ibid.*, **244**, 221, 1981.
48. Hartoog, M. R., *Astrophys. J.*, **212**, 723, 1977.
49. Abt, H. A., *Astrophys. J.*, **230**, 485, 1979.
50. North, P., *Astron. Astrophys.*, **141**, 328, 1984; *ibid.*, **334**, 181, 1998; *ibid.*, **336**, 1072, 1998.
51. Borra, E. F., Beaulieu, A., Brousseau, D., and Shelton, I., *Astron. Astrophys.*, **149**, 266, 1985.

Section 6.4. The following key reference may be noted:

52. Michaud, G., *Astrophys. J.*, **160**, 641, 1970.

The meridional circulation solutions are those of Tassoul and Tassoul (References 16, 23, and 25 of Chapter 4). The interaction between microscopic diffusion and meridional circulation was originally considered by:

53. Michaud, G., *Astrophys. J.*, **258**, 349, 1982.
54. Michaud, G., Tarasick, D., Charland, Y., and Pelletier, C., *Astrophys. J.*, **269**, 239, 1983.

Two-dimensional calculations will be found in:

55. Charbonneau, P., and Michaud, G., *Astrophys. J.*, **327**, 809, 1988; *ibid.*, **334**, 746, 1988; *ibid.*, **370**, 693, 1991.
56. Charbonneau, P., Michaud, G., and Proffitt, C. R., *Astrophys. J.*, **347**, 821, 1989.
57. Charbonneau, P., *Astrophys. J.*, **405**, 720, 1993.
58. Turcotte, S., and Charbonneau, P., *Astrophys. J.*, **413**, 376, 1993.

Papers of related interest are:

59. Chaboyer, B., and Zahn, J. P., *Astron. Astrophys.*, **253**, 173, 1992.
60. Charbonneau, P., *Astron. Astrophys.*, **259**, 134, 1992.

See especially Charbonneau's comprehensive discussion.

Section 6.5. A cutoff in the distribution of rotational velocity for evolved stars was originally noticed by Herbig and Spalding. As they pointed out, if a level of 20 km s^{-1} is chosen as reference, later than spectral type G0 giants and subgiants with appreciable rotation rates are very rare. See:

61. Herbig, G. H., and Spalding, J. F., Jr., *Astrophys. J.*, **121**, 118, 1955.

More accurate determinations of the sudden drop in rotation for these stars have been made by:

62. Gray, D. F., *Astrophys. J.*, **251**, 155, 1981; *ibid.*, **262**, 682, 1982; *ibid.*, **347**, 1021, 1989.
63. Gray, D. F., and Nagar, P., *Astrophys. J.*, **298**, 756, 1985.

Detailed discussions of Gray's mechanism are contained in:

64. de Medeiros, J. R., and Mayor, M., in *Angular Momentum Evolution of Young Stars* (Catalano, S., and Stauffer, J. R., eds.), p. 201, Dordrecht: Kluwer, 1991.
65. Gray, D. F., in *Angular Momentum Evolution of Young Stars* (Catalano, S., and Stauffer, J. R., eds.), p. 183, Dordrecht: Kluwer, 1991.
66. Böhm-Vitense, E., *Astron. J.*, **103**, 608, 1992.
67. Schrijver, C. J., and Pols, O. R., *Astron. Astrophys.*, **278**, 51, 1993; *ibid.*, **293**, 640, 1995.

Reference is also made to:

68. Simon, T., and Drake, S. A., *Astrophys. J.*, **346**, 303, 1989.

Their discussion of Gray's discontinuity for the giant stars should be modified, however, since it has been subsequently recognized that the sudden drop in rotation is located near spectral type G0 III and not at G5 III. (This remark does not affect the contents of their paper.)

Several discussions of post–main-sequence evolution may be traced to Reference 1 (pp. 358–361). Subsequent contributions are due to:

69. Endal, A. S., and Sofia, S., *Astrophys. J.*, **210**, 184, 1976; *ibid.*, **220**, 279, 1978; *ibid.*, **232**, 531, 1979.

See especially their third paper. Among the many recent papers relating to post–main-sequence evolution, reference may be made to:

70. Wiita, P. J., *J. Astrophys. Astron.*, **2**, 387, 1981.
71. Sreenivasan, S. R., and Wilson, W. J. F., *Astrophys. J.*, **254**, 287, 1982; *ibid.*, **290**, 653, 1985; *ibid.*, **292**, 506, 1985.
72. Deupree, R. G., *Astrophys. J.*, **357**, 175, 1990; *ibid.*, **439**, 357, 1995; *ibid.*, **499**, 340, 1998.
73. MacGregor, K. B., Friend, D. B., and Gilliland, R. L., *Astron. Astrophys.*, **256**, 141, 1992.
74. Maheswaran, M., and Cassinelli, J., *Astrophys. J.*, **421**, 718, 1994.
75. Sofia, S., Howard, J. M., and Demarque, P., in *Pulsation, Rotation and Mass Loss in Early-Type Stars* (Balona, L. A., Henrichs, H. F., and Le Contel, J. M., eds.), I.A.U. Symposium No 162, p. 131, Dordrecht: Kluwer, 1994.

The results presented in Reference 75 are clearly indicative that rotationally induced chemical mixing plays an essential role in the evolution of massive stars. Unfortunately, as was pointed out at the end of Section 3.6, turbulent mixing in stellar radiative zones is difficult to model with any confidence. Other papers dealing with this poorly understood mechanism may be traced to:

76. Langer, N., *Astron. Astrophys.*, **329**, 551, 1998.

See also my comments in the epilogue.

7

The late-type stars

7.1 Introduction

On the main sequence, it has long been known that large mean rotational velocities are common among the early-type stars and that these velocities decline steeply in the F-star region, from 150 km s^{-1} to less than 10 km s^{-1} in the cooler stars (see Figure 1.6). As was shown in Section 6.3.2, the observed projected velocities indicate that the mean value of the total angular momentum $\langle J \rangle$ closely follows the simple power law $\langle J \rangle \propto M^2$ for stars earlier than spectral type F0, which corresponds to about $1.5 M_\odot$ (see Figure 6.7). The difficulty is not to account for such a relation, which probably reflects the initial distribution of angular momentum, but to explain why it does not apply throughout the main sequence. It has been suggested that the break in the mean rotational velocities beginning at about spectral type F0 might be due to the systematic occurrence of planets around the low-mass stars ($M \lesssim 1.5 M_\odot$), with most of the initial angular momentum being then transferred to the planets. Although this explanation has retained its attractiveness well into the 1960s, there is now ample evidence that it is not the most likely cause of the remarkable decline of rotation in the F-star region along the main sequence. Indeed, following Schatzman's (1962) original suggestion, there is now widespread agreement that *this break in the rotation curve can be attributed to angular momentum loss through magnetized winds and/or sporadic mass ejections from stars with deep surface convection zones*. This interaction between rotation and surface activity, which is the basis for understanding much of the evolution of low-mass stars, will be considered in Section 7.2.

Now, as was shown by Wilson (1963), the average intensity of Ca II emission in a late-type dwarf and, hence, the general degree of its chromospheric activity bear an inverse relationship to its age. A similar trend was found by Kraft (1967) in the rotational velocities of late-F and early-G dwarfs. From a detailed examination of these data, Skumanich (1972) has shown that both rotational velocities and Ca II emission decline with advancing age according to a $t^{-1/2}$ law (see Eq. [1.7]). This coincidence strongly suggests that *there exists a deep physical connection between rotation and surface activity among the low-mass stars*. Further complexity was added to the problem when van Leeuwen and Alphenaar (1982) announced the discovery of a number of rapidly rotating G- and K-dwarfs in the Pleiades, with equatorial velocities up to 170 km s^{-1}. This important result led to a flurry of interest in the rotational evolution of these low-mass stars, which spin down faster than predicted by Skumanich's empirical law shortly upon arriving on the main sequence. In Section 7.3 we shall briefly review the new rotational velocity data for T Tauri stars and late-type dwarfs in young open clusters (see also Section 1.3). The major theoretical models developed to clarify these new findings will be considered in Section 7.4.

7.2 Schatzman's braking mechanism

The relevance of magnetic braking for stars having deep surface convection zones was first recognized by Schatzman (1962). Very briefly, it is assumed that these stars generate episodic mass ejections that act as an expanding plasma in a large-scale magnetic field. As material is ejected from the activity zones, the magnetic field can enforce approximate corotation until the gas has moved out to distances much larger than the star's radius. Beyond this region, because the magnetic stresses become less and less important, the outflowing material can thus leave the star, with each mass element carrying away its angular momentum. As we shall see below, if the gas is kept corotating with the star, a quite small amount of mass loss yields proportionally a much greater loss of angular momentum than matter retaining the angular momentum of the star's surface. Given the efficiency of this mechanism for extracting angular momentum from stars with outer convection zones, the break in the main-sequence rotational velocities can be explained as follows. Since high-mass stars spend relatively little time in the convective phase, magnetic braking is therefore virtually inoperative for these stars. Hence, they suffer very little loss of angular momentum during their pre–main-sequence contraction. In contrast, low-mass stars have a more important convective phase since they retain an outer convection zone all the way to the main sequence. Magnetic braking can thus operate during their entire pre–main-sequence contraction and during their much longer stay on the main sequence. Since the rapid drop in rotational velocity is seen at approximately the point where main-sequence stars develop subphotospheric convection zones, it follows that angular momentum loss preferably occurs in the low-mass stars, thus causing the observed rotational discontinuity in the F-star region. As was shown by Wolff and Simon (1997), recent data strongly suggest that this sharp decrease in mean equatorial velocity along the main sequence, from about $1.6M_\odot$ down to about $1.3M_\odot$, has already been imposed during the pre–main-sequence phase of stellar evolution. For masses less than about $1.3M_\odot$, however, their analysis indicates that further loss of angular momentum occurs rapidly during main-sequence evolution so that, by the age of the Hyades ($\sim$ 600 Myr), mean equatorial velocities for stars in the spectral range F8 V–K5 V are remarkably uniform at any given mass and decline from about 11 km s^{-1} at F8 V to about 4 km s^{-1} at K5 V (see Figure 1.8).

The strongest support for this rotation–activity connection comes from Wilson's (1966) finding that there is a sudden appearance of Ca II emission in the F-star region along the main sequence, whereas it is never observed among the more massive stars. Obviously, the close agreement between the onset of large rotational velocities and the termination of chromospheres is very suggestive of Schatzman's braking mechanism. More recently, Cameron and Robinson (1989) have found another piece of evidence in support of angular momentum loss via discrete mass ejection. They have obtained time series of high-resolution spectra of the Hα profile in the active, rapidly rotating G8–K0 dwarf AB Doradus. Their spectra show transient absorption features that move through the Hα emission profile on rapid time scales. These features strongly suggest the existence of cool, dense clouds embedded in and corotating with the hot extended corona out to several stellar radii from the rotation axis. Their calculations indicate that angular momentum loss could account for rotational braking on a time scale of no more than 100 Myr. If so, these observations might provide an important clue as to how low-mass stars lose the bulk of their angular momentum upon their arrival on the main sequence.

Another mechanism by which stars with convective envelopes can dispose of a considerable fraction of their initial angular momentum is provided by *stellar winds*. Following Mestel (1968), it is subphotospheric convection that is again the essential feature of the mechanism. Waves generated in the outer convection zone are dissipated above the photosphere, thus supplying the heat responsible for the formation of a chromosphere and a corona. When the coronal temperatures are too low to generate a *thermal wind*, however, large centrifugal forces acting on the corotating material can generate an outwardly moving flow (i.e., a *centrifugal wind*). In both cases, the wind motion accelerates outward from very low values at the bottom of the corona to supersonic values far away from the star's surface. Detailed studies have shown that the angular momentum loss rate is equivalent to that carried by a wind kept strictly corotating with the star out to a radius r_A in the circumstellar envelope (e.g., Mestel 1968). By definition, the corotating radius r_A is the mean radius of the *Alfvén surface* defined by

$$v_A = \frac{H_A}{(4\pi\rho_A)^{1/2}}, \tag{7.1}$$

where the indices "A" indicate that the wind speed v, the poloidal field strength H, and the density ρ are evaluated at $r = r_A$. In the simple model developed by Weber and Davis (1967), where the magnetic field in the thermally driven wind is approximately radial in the corotating frame of reference, the effective corotation prescription gives the following expression for the angular momentum loss rate:

$$\frac{dJ}{dt} \approx -\frac{2}{3}\frac{dM}{dt} R^2\Omega \left(\frac{r_A}{R}\right)^2, \tag{7.2}$$

where R is the star's radius and Ω is the angular velocity of rotation. The importance of the large-scale magnetic field can be seen on the following example. From solar-wind data, one finds that $r_A \approx 30R_\odot$ for the Sun; hence, by virtue of Eq. (7.2), the rather weak solar magnetic field increases the angular momentum loss by three orders of magnitude over its value calculated without magnetic field.

Now, from Eq. (7.1) and the definition of the mass flux at $r = r_A$,

$$\frac{dM}{dt} = -4\pi\rho_A\, v_A\, r_A^2\,, \tag{7.3}$$

Eq. (7.2) can be rewritten in the form

$$\frac{dJ}{dt} \approx -\frac{2}{3}\frac{\Omega}{v_A}\left(H_A\, r_A^2\right)^2. \tag{7.4}$$

Since the conservation of magnetic flux implies that $H_A\, r_A^2 = H_0\, R^2$ in the case of a purely radial field, Eq. (7.4) becomes

$$\frac{dJ}{dt} \approx -\frac{2}{3}\frac{\Omega}{v_A}\left(H_0\, R^2\right)^2, \tag{7.5}$$

where H_0 is the average surface magnetic field. If a linear relationship of the form $H_0 \propto \Omega$ is assumed for the dynamo-generated magnetic field, with $J \propto MR^2\Omega$ Eq. (7.5) yields

$$\frac{d\Omega}{dt} \propto -\Omega^3. \tag{7.6}$$

After integrating Eq. (7.6), one obtains

$$\Omega \propto t^{-1/2}, \tag{7.7}$$

which is identical to Skumanich's empirical law (see Eq. [1.7]). This is a most fortunate coincidence since it implies that a simple formulation of angular momentum loss via magnetically channeled stellar winds is adequate to describe the rotational evolution of solar-type stars on the main sequence. As we shall see in Section 7.4.2, however, such a formulation does not describe adequately the spin-down of the very rapidly rotating low-mass stars in young open clusters. In fact, there is now clear indication that the angular momentum loss-rate saturates for surface rotational velocities in excess of 10–20 km s^{-1}.

To the best of my knowledge, there is as yet no complete theory that explains the existence of a dynamo saturation in the most rapid rotators. However, there is increasingly convincing observational evidence to support the idea that the dynamo activity of a late-type star scales with its rotation rate. Dynamo saturation was originally inferred by Vilhu (1984) from the observation that the chromospheric and coronal emission fluxes depend only weakly on rotation at high angular velocities. More recently, Patten and Simon (1996) have undertaken a program to measure photometric rotation periods and X-ray luminosities for late-type stars in the young open cluster IC 2391 (age ~ 30 Myr). In Figure 7.1 we plot the X-ray luminosity L_X against the rotation period $P_{\rm rot}$ for solar-type

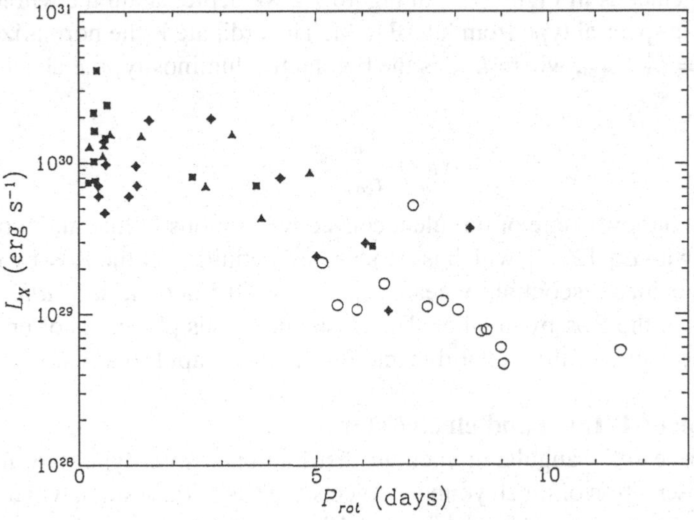

Fig. 7.1. The X-ray luminosity L_X as a function of rotation period $P_{\rm rot}$ for solar-type stars in the IC 2391 (*filled triangles*), α Persei (*filled squares*), Pleiades (*filled diamonds*), and Hyades (*open circles*) clusters. *Source:* Patten, B. M., and Simon, T., *Astrophys. J. Suppl.*, **106**, 489, 1996.

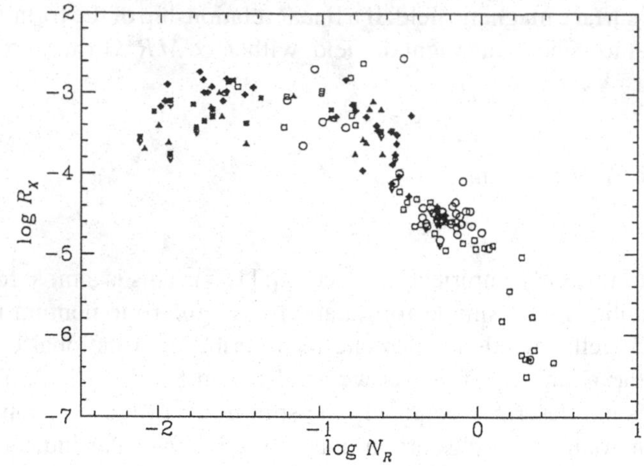

Fig. 7.2. The normalized X-ray luminosity $R_X = L_X/L_{\rm bol}$ as a function of the Rossby number N_R. Plotted are data for IC 2391 (*filled triangles*), α Persei (*filled squares*), the Pleiades (*filled diamonds*), the Hyades (*open circles*), and field main-sequence stars (*open squares*). *Source:* Patten, B. M., and Simon, T., *Astrophys. J. Suppl.*, **106**, 489, 1996.

stars in IC 2391 and, for comparison, for older stars from α Persei (age $\sim$ 50 Myr), the Pleiades (age $\sim$ 70 Myr), and the Hyades (age $\sim$ 600 Myr). One readily sees that there is an overall decline in the median rotation rate and X-ray luminosity with age. Note also that the older cluster stars trace out a definite correlation between L_X and $P_{\rm rot}$, whereas those in IC 2391 show at best a weak correlation between these two parameters. Following current practice, in Figure 7.2 we present an alternative representation of this activity–rotation plot, which greatly reduces the scatter when stars of different masses are combined together as in Figure 7.1. In Figure 7.2 we depict again the whole sample of stars, ranging in spectral type from late-F to M. The ordinate is the normalized X-ray luminosity, $R_X = L_X/L_{\rm bol}$, where $L_{\rm bol}$ is the bolometric luminosity; the abscissa is the *Rossby number*,

$$N_R = \frac{P_{\rm rot}}{\tau_{\rm conv}}, \tag{7.8}$$

where $\tau_{\rm conv}$ is the turnover time of turbulent convective motions in the outer convection zone. (Compare with Eq. [2.30], which is the standard definition of the Rossby number.) Note the clearly defined discontinuity near $\log_{10} N_R = -0.5$ and the *saturation plateau* at smaller values of the Rossby number. The existence of this plateau is often ascribed to a change in the nature of the stellar dynamo for the most rapid rotators.

7.3 Rotation of T Tauri and cluster stars

Clues to the initial angular momentum distribution of solar-type stars are mainly gathered from observation of much younger objects such as T Tauri stars, which are low-mass pre–main-sequence stars of age less than 10 Myr. Rotation periods and projected rotational velocities are available for more than one hundred of these stars. Figure 7.3 illustrates the histogram of rotation periods for stars belonging to the Orion Nebula cluster. One readily sees that this frequency distribution is distinctly bimodal, confirming the

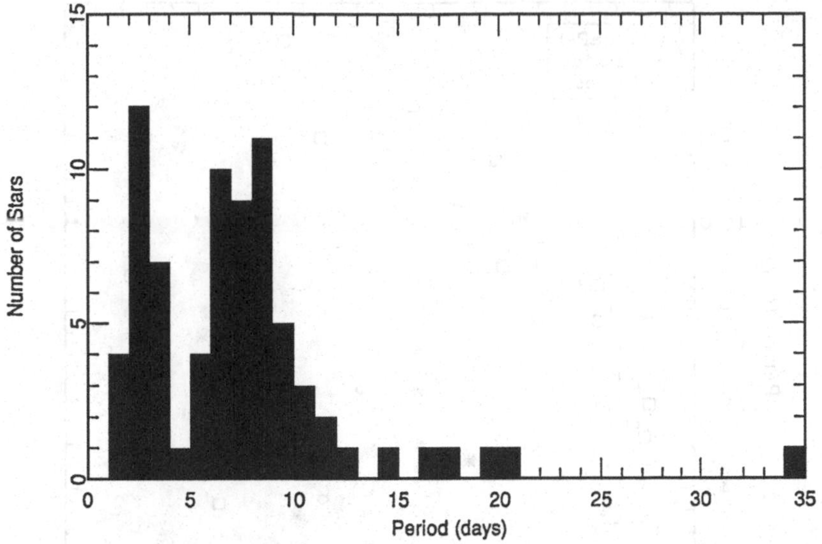

Fig. 7.3. Frequency distribution of rotation periods of T Tauri stars in the Orion Nebula cluster. *Source:* Choi, P. I., and Herbst, W., *Astron. J.*, **111**, 283, 1996.

discovery of Attridge and Herbst (1992). About one third of the stars are rapid rotators with a median period of 2.55 days and a dispersion of 0.7 days. The others are slow rotators with a median period of 8.30 days, a dispersion of 3.8 days, and a sparsely populated tail of very slow rotators extending to 34.5 days. It is important to note that this bimodal distribution of periods is not restricted to the Orion Nebula cluster since it is also apparent in Figure 1.9, which depicts the histogram of rotation periods for T Tauri stars in other clusters and associations. According to Choi and Herbst (1996), there is little doubt that 4–5 day periods are rare among T Tauri stars and so this bimodal period distribution is real.

Edwards et al. (1993) have also measured infrared color excesses for a sample of thirty-four T Tauri stars with photometrically derived rotation periods and spectral types later than K5. Their main conclusion is that the observed periods appear to be related to the presence or absence of a circumstellar accretion disk. Those stars that they infer to be surrounded by accretion disks (i.e., the *classical* T Tauri stars) are slow rotators with periods larger than 4 days, with a most probable period of 8.5 days, while those that lack accretion disk signatures (i.e., the *weak-line* T Tauri stars) cover a wide range of rotation periods, ranging from 1.5 to 16 days, including a significant number of objects with periods smaller than 4 days. This result was confirmed by Bouvier et al. (1993), who made a detailed study of T Tauri stars belonging to the Taurus–Auriga cloud. Their analysis shows that the mean rotation period is about 4 days for the weak-line T Tauri stars and about 8 days for the classical T Tauri stars. This apparent bimodality will be interpreted in Section 7.4.1.

Other clues to understanding the late pre–main-sequence/early main-sequence evolution of solar-type stars have been obtained from the study of late-type stars in the α Persei cluster (age $\sim$ 50 Myr), the Pleiades cluster (age $\sim$ 70 Myr), and the Hyades cluster (age $\sim$ 600 Myr). Figure 7.4 illustrates the rotation periods of low-mass stars

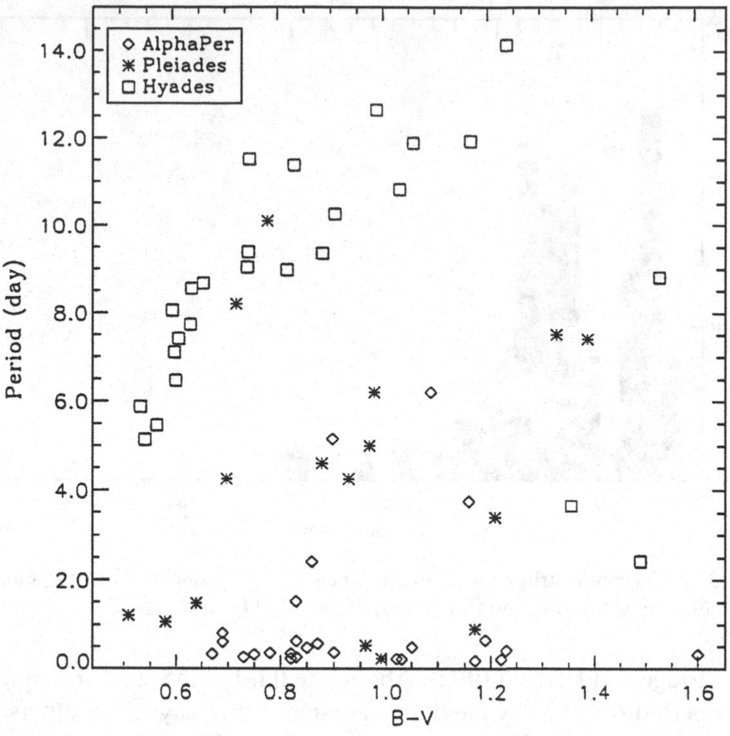

Fig. 7.4. Observations of stellar rotation periods in the open clusters α Persei, the Pleiades, and the Hyades. *Source:* Barnes, S., and Sofia, S., *Astrophys. J.*, **462**, 746, 1996.

belonging to these three clusters. It is immediately apparent that there is a significant increase in rotation period between the ages of α Persei and the Hyades (see also Figures 1.7 and 1.8). In Figure 7.5 we display the observed $v \sin i$ distributions of solar-type stars in the $(B - V)$ color range 0.55–0.85, corresponding to the mass range 0.8–1.0$M_\odot$. Again note the considerable spread in projected rotational velocities for the stars in α Persei.

The salient features of these observations have been summarized by Stauffer (1994). These are:

1. Very rapid rotators ($v \sin i > 100$ km s^{-1}) are present at all spectral types in α Persei.
2. Relatively rapid rotators ($v \sin i > 50$ km s^{-1}) are still present in the Pleiades among the K- and M-dwarfs but are nearly absent among the G dwarfs.
3. All of the G- and K-dwarfs in the Hyades are slow rotators ($v \sin i < 10$ km s^{-1}), although there are still one K8 dwarf and some M dwarfs with moderate rotation ($v \sin i \approx 15$–20 km s^{-1}) in the cluster.
4. In all three clusters, for all spectral types later than G0, more than half of the stars are slow rotators, with $v \sin i < 10$ km s^{-1}.

With the adopted ages for these clusters, the spin-down time during the early main-sequence evolution is a few 10 Myr for the G dwarfs, several 10 Myr for the K dwarfs, and a few 100 Myr for the M dwarfs.

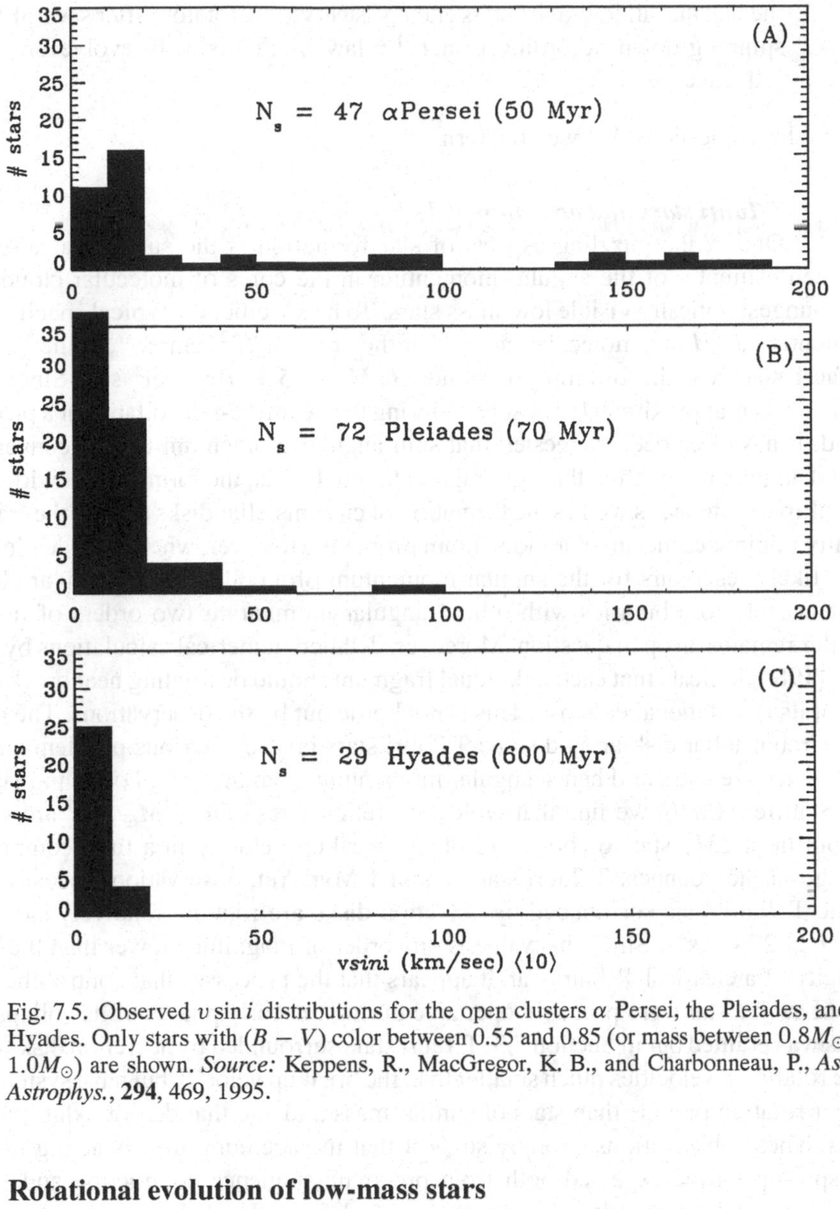

Fig. 7.5. Observed $v \sin i$ distributions for the open clusters α Persei, the Pleiades, and the Hyades. Only stars with $(B - V)$ color between 0.55 and 0.85 (or mass between $0.8M_\odot$ and $1.0M_\odot$) are shown. *Source:* Keppens, R., MacGregor, K. B., and Charbonneau, P., *Astron. Astrophys.*, **294**, 469, 1995.

7.4 Rotational evolution of low-mass stars

The challenge for any theoretical modeling of the rotational evolution of low-mass stars is to provide a convincing scenario that agrees with all of these observational facts. To be specific, the distribution of rotation rates that results from evolutionary calculations must account for the following constraints:

1. the apparent bimodal period distribution of the T Tauri stars,
2. the simultaneous presence of very slow rotators and a tail of very fast rotators in the vicinity of the zero-age main sequence,
3. the rapid spin down of the fastest rotators,
4. the apparent longer spin-down time for the lower mass stars,

5. the fact that all low-mass stars end up as very slow rotators after several 100 Myr, spinning down according to a $t^{-1/2}$ law as they slowly evolve on the main sequence.

It is to these questions that we now turn.

7.4.1 T Tauri stars and accretion disks

One of the puzzling aspects of star formation is the substantial discrepancy between estimates of the angular momentum in the cores of molecular clouds and in the youngest optically visible low-mass stars. To be specific, the typical specific angular momentum J/M in a molecular cloud is of the order of 10^{21} cm^2 s^{-1}, while in a typical T Tauri star near the birthline one finds $J/M \approx 5 \times 10^{17}$ cm^2 s^{-1}. Since angular momentum is approximately conserved during the near free-fall collapse of a protostellar cloud, it has often been suggested that spin angular momentum can be converted into orbital angular momentum through fragmentation. In fact, the formation of wide binaries or multiple systems, as well as the formation of circumstellar disks around the fragments, greatly minimizes the angular momentum problem. However, whereas wide binaries are most likely reservoirs for the angular momentum of a collapsing molecular cloud, the existence of close binaries with orbital angular momentum two orders of magnitude smaller remains an open question. Moreover, detailed numerical calculations by Durisen et al. (1984) indicate that each individual fragment should be rotating near break up at the end of its gravitational collapse. This is not borne out by the observations. The presence of a circumstellar disk around some T Tauri stars is also a serious problem since they tend to accrete mass and hence angular momentum from the disk. Following Hartmann and Stauffer (1989), we find that typical accretion rates of $10^{-7} M_\odot$ yr^{-1} are sufficient to spin up a $1 M_\odot$ star to about half of the breakup velocity in a time comparable to the age of the youngest T Tauri stars, about 1 Myr. Yet, observations reveal that most of the T Tauri stars surrounded by accretion disks are rotating relatively slowly, with $v \sin i \lesssim 20$ km s^{-1}. Since this value is one order of magnitude lower than the breakup velocity of a classical T Tauri star, it appears that the processes that control the rotation rate of such a star were probably operative during the early phases of its collapse.

As was pointed out in Section 7.3. T Tauri stars surrounded by accretion disks not only have rotational velocities much smaller than the breakup velocity but have systematically longer rotation periods than stars of similar mass and age that do not exhibit accretion disks. These observations strongly suggest that the accretion disk is acting to counter the spin-up torque expected both from pre–main-sequence contraction and from the deposition of high angular momentum material from the disk onto the star. Broadly speaking, two distinct angular momentum regulation mechanisms have been proposed, both of them relying on the interaction between the magnetosphere of a rotating star and a circumstellar accretion disk.

Königl (1991) has invoked the theory of Ghosh and Lamb (1979) for accreting magnetic neutron stars and white dwarfs to explain the slow rotation rate of T Tauri stars as the result of magnetic coupling to a truncated disk. In this model the poloidal magnetic field of the star has a closed global structure, modeled as an aligned dipole. Material in the disk that spirals slowly inward moves along the closed field lines and is channeled onto the star at high latitudes. That is to say, the dipolar field disrupts the inner parts of the circumstellar disk and the central star becomes effectively coupled to the disk several radii out. This possibility was investigated by Königl, who found that a kilogauss field could disrupt the disk at a distance of a few stellar radii from the center and that the spin-down torque transmitted by the field lines that thread the disk beyond the corotation radius could indeed balance the spin-up torque applied by accreting material. More recently, Cameron and Campbell (1993) have shown that a T Tauri star could evolve into a state of rotational equilibrium within the duration of the Hayashi phase, despite the rapid contraction of the star. The resulting rotation rates of their models, which have magnetic fields of a few hundred gauss and an accretion rate of a few $10^{-8} M_{\odot}\ \mathrm{yr}^{-1}$, are also found to be consistent with the observed rotation rates of classical T Tauri stars.

Alternatively, Shu et al. (1994) have proposed a model in which shielding currents in the surface layers of the disk are invoked to prevent penetration of the stellar field lines everywhere except near the corotation radius R_{corot}, where the Keplerian angular velocity of the disk matches the angular velocity of the star. Exterior to R_{corot}, matter diffuses onto field lines that bow outward, resulting in a magnetocentrifugally driven wind with a mass loss rate proportional to the disk accretion rate $\dot{M}_D$. Matter interior to R_{corot} diffuses onto field lines that bow inward and is funneled onto the star's surface. It is found that this flow actually results in a trailing-spiral configuration for the magnetic field and that it transfers angular momentum from the star to the disk as long as the corotation radius remains significantly greater than the star's equatorial radius R. As was shown by Ostriker and Shu (1995), for an aligned stellar dipole of strength $m = HR^3$ the corotation radius is given by

$$R_{\mathrm{corot}} = 0.923 \left(\frac{m^4}{GM\dot{M}_D^2} \right)^{1/7}, \tag{7.9}$$

where H is the field strength at the equator, G is the constant of gravitation, and M is the star's mass. Parenthetically note that the Ghosh–Lamb theory yields a similar relation, except for the value of the numerical constant, which is not exactly known but should be of order unity. Making use of Kepler's third law, one obtains the star's rotation period as

$$P_{\mathrm{rot}} = 5.57 \left(\frac{m^6}{G^5 M^5 \dot{M}_D^3} \right)^{1/7}. \tag{7.10}$$

Letting $P_{\mathrm{rot}} = 8$ days, $M = 0.5 M_{\odot}$, and $\dot{M}_D = 10^{-7} M_{\odot}\ \mathrm{yr}^{-1}$ in Eq. (7.10), one finds that $m = 7.32 \times 10^{36}$ in cgs units. This value corresponds to $H \approx 800$ gauss and $R_{\mathrm{corot}} \approx 4.4R$ for a star with $R = 3R_{\odot}$. Thus, given reasonable values for the stellar parameters, appropriately slow rotation rates are obtained for the classical T Tauri stars.

The hypothesis of disk-regulated angular momentum provides an attractive framework for understanding the rotational evolution of low-mass pre–main-sequence stars. No commonly accepted model exists at the present time, however, since the fine details of the disk–star interaction are still to be modeled quantitatively. Nonetheless, ample evidence now exists that an accretion disk may play a fundamental role in regulating the rotation rate of a classical T Tauri star, holding its angular velocity almost fixed during Hayashi track evolution. This locking results in net transfer of specific angular momentum from the central star to the disk, so that the total angular momentum of the star steadily decreases in time until its regulating accretion disk is fully dissipated. If so, then the observed bimodal period distribution for T Tauri stars clearly indicates that the fast rotators are stars that, for one reason or another, are not strongly locked to an accretion disk during Hayashi track evolution. Hence, because they remain free to spin up in response to changes in moment of inertia as they contract, they also cover a wider range of rotation periods than their disk-locked counterparts. As was noted by Choi and Herbst (1996), the gap in the histogram of T Tauri stars is evidence of the rapid evolution through which a star passes on its way to another mechanical equilibrium, once released from its disk induced rotational lock.

In Section 7.3 we have seen that the rotation distribution among main-sequence dwarfs of spectral type G and later in very young open clusters consists of a narrow peak at $v \sin i = 10$ km s^{-1} or less and an extended tail of rapid rotators, with $v \sin i > 100$ km s^{-1} (see Figure 7.5). As was noted by Cameron, Campbell, and Quaintrell (1995), the presence of fast and slow rotators in the same cluster suggests that this peak-and-tail distribution is already established when the cluster stars reach the zero-age main sequence. In order to check the validity of that assumption, they have thus expanded the work of Cameron and Campbell (1993) to determine how disk braking might affect the histogram of rotation periods for low-mass stars on the zero-age main sequence. For disk masses of a few hundredths of a solar mass or more, and dynamo-generated field strengths of a few hundred gauss, their numerical calculations indicate that the net (magnetic plus accretion) torque is sufficient to pull the star's rotation into quasi-static equilibrium before the end of the Hayashi phase, with the resulting rotation rate being one order of magnitude lower than the breakup rate. Thence, by the time this equilibrium breaks down due to the dwindling accretion rate, the star's rotation is effectively independent of both the disk mass and the initial angular momentum of the star. For lower disk masses, however, such an equilibrium is never established so that the star can retain a greater fraction of its initial angular momentum. The histogram of rotation rates that results from a reasonable choice for the distribution of disk masses has the form of a low-velocity peak and an extended high-velocity tail. The slow rotators are the stars that evolved into rotational equilibrium as classical T Tauri stars and gave away most of their initial angular momenta to their former disks; the stars in the tail are those with lower initial disk masses, in which rotational equilibrium was never established during the Hayashi phase. If this is the case, then their model provides a natural explanation for the histograms depicted in Figure 7.5.

7.4.2 Rotational evolution models

Since the early 1990s, much theoretical effort has been expended in trying to understand the rotational history of a low-mass star, both before and during the main-sequence phase. Notably, MacGregor and Brenner (1991) have developed a particularly simple description of the transport of angular momentum within the interior of a solar-type star. In this section I shall briefly describe their model, its use in conjunction with a suitable parameterization for the angular momentum loss resulting from magnetized stellar winds, and some of the numerical results obtained by Keppens, MacGregor, and Charbonneau (1995).

Broadly speaking, their approach to constructing an evolutionary sequence is to simplify matters by separating computation of the rotational evolution from that of the internal, structural evolution. Accordingly, we shall assume that the effects of rotation on internal structure are small, so that an evolutionary track for a spherical star of the same mass can be used to calculate the star's radius R, the radius of the convection zone base $R_{\rm conv}$, the mass of the radiative core $M_{\rm core}$, and the moments of inertia, $I_{\rm core}$, and $I_{\rm conv}$, of the core and envelope. We shall further assume that the radiative core and the convective envelope each rotate rigidly, although not necessarily at the same rate. If $\Omega_{\rm core}$ and $\Omega_{\rm conv}$ are the angular velocities of the core and envelope, then the angular momenta of these regions are $J_{\rm core} = I_{\rm core}\Omega_{\rm core}$ and $J_{\rm conv} = I_{\rm conv}\Omega_{\rm conv}$. With these assumptions, the equations governing the time evolution of these angular momenta can be derived by considering the processes by means of which angular momentum is redistributed and lost.

During pre–main-sequence contraction, angular momentum is reapportioned between the core and the envelope as a consequence of the gradual conversion of the stellar interior from a nearly fully convective state to one in which most of the mass is contained within the radiative core. Thus, if $dM_{\rm core}/dt$ denotes the rate of growth of the core mass, angular momentum exchange will occur at the rate $j\,dM_{\rm core}/dt$, where

$$j = \frac{2}{3}\,\Omega_{\rm conv}\,R^2_{\rm conv} \tag{7.11}$$

is the specific angular momentum of material in the thin spherical shell about the radius $r = R_{\rm conv}(t)$ that is undergoing assimilation at the core at time t.

We now assume that the torque exerted by the magnetically controlled wind extracts angular momentum only from the surface convection zone. The resulting deceleration of the convective envelope causes a shear to develop at the core–envelope interface. In a real star, this would lead to the creation of interfacial stresses that would act to redistribute angular momentum between the two regions. In the MacGregor–Brenner heuristic model, one simulates this transport process by assuming that an amount of angular momentum

$$\Delta J = \frac{I_{\rm core}I_{\rm conv}}{I_{\rm core} + I_{\rm conv}}\,(\Omega_{\rm core} - \Omega_{\rm conv}) \tag{7.12}$$

is transferred from the core to the envelope in a specified time $\tau_{\rm c}$. Note that an instantaneous exchange of angular momentum ΔJ would equilibrate $\Omega_{\rm core}$ and $\Omega_{\rm conv}$, thereby restoring an angular momentum distribution that satisfies the essential stability condition defined in Eq. (3.98).

In the absence of magnetic coupling with an accretion disk, the combination of the foregoing effects can be written down in the form

$$\frac{dJ_{\rm core}}{dt} = -\frac{\Delta J}{\tau_{\rm c}} + j\,\frac{dM_{\rm core}}{dt}, \tag{7.13}$$

for the core, and

$$\frac{dJ_{\rm conv}}{dt} = \frac{\Delta J}{\tau_{\rm c}} - j\,\frac{dM_{\rm core}}{dt} - \frac{J_{\rm conv}}{\tau_w}, \tag{7.14}$$

for the surface convection zone. In these equations, $\tau_{\rm c}$ is the prescribed core–envelope coupling time and τ_w is the e-folding time for wind-induced angular momentum loss from the convective envelope. (The time scale τ_w needs to be calculated from a reasonable model for the steady-state expansion of the stellar corona.) Once $R_{\rm conv}$, $dM_{\rm core}/dt$, $I_{\rm core}$,

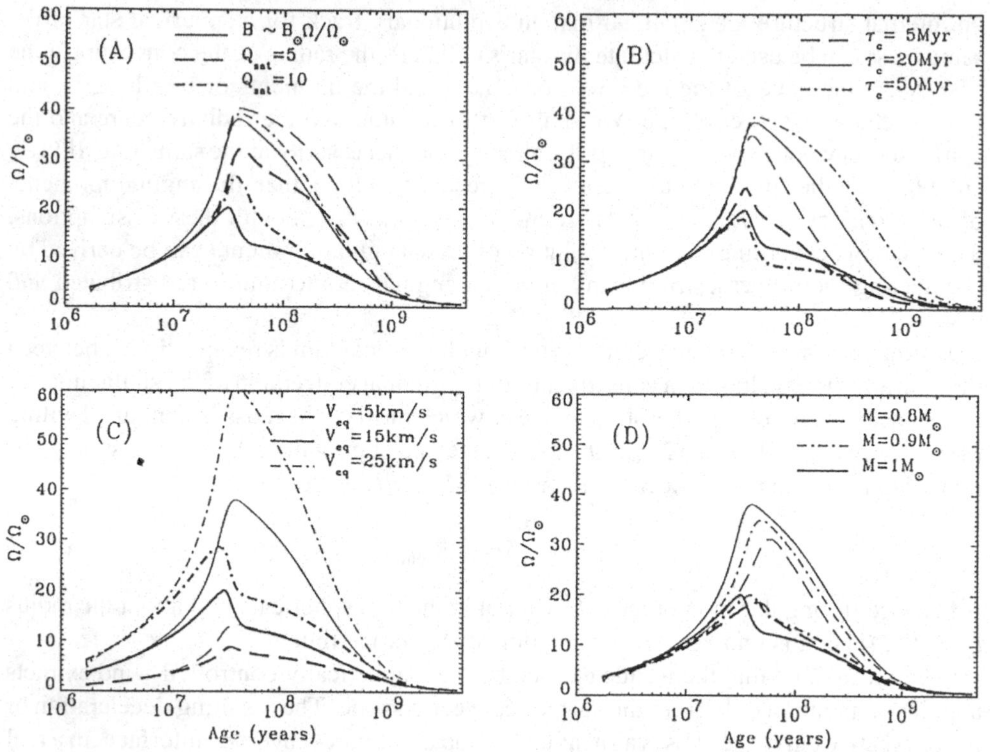

Fig. 7.6. The evolution of the rotation rate (in units of $\Omega_\odot = 3 \times 10^{-6}$ s^{-1}) of the core, $\Omega_{\rm core}$, and the convective envelope, $\Omega_{\rm conv}$ (*thicker lines*) for a single star. Panel A: For a $1M_\odot$ star, with initial equatorial velocity $v_{\rm eq} = 15$ km s^{-1} and coupling time scale $\tau_c = 20$ Myr, for three different dynamo prescriptions. The solid lines are for a linear dynamo; the dashed lines for a dynamo saturated at $\Omega_{\rm conv} \geq 5\Omega_\odot$; and the dash-dotted lines for a dynamo saturated at $\Omega_{\rm conv} \geq 10\Omega_\odot$. Panel B: The rotational histories for a $1M_\odot$ star having $v_{\rm eq} = 15$ km s^{-1} and a linear dynamo for $\tau_c = 5$ Myr (*dashed lines*), 20 Myr (*solid lines*), and 50 Myr (*dash-dotted lines*). Panel C: A $1M_\odot$ star, with $\tau_c = 20$ Myr and a linear dynamo, for $v_{\rm eq} = 5$ km s^{-1} (*dashed lines*), $v_{\rm eq} = 15$ km s^{-1} (*solid lines*), and $v_{\rm eq} = 25$ km s^{-1} (*dash-dotted lines*). Panel D: For a star of mass $0.8M_\odot$ (*dashed lines*), $0.9M_\odot$ (*dash-dotted lines*), and $1.0M_\odot$ (*solid lines*), with $\tau_c = 20$ Myr, a linear dynamo, and $v_{\rm eq} = 15$ cm s^{-1}. *Source:* Keppens, R., MacGregor, K. B., and Charbonneau, P., *Astron. Astrophys.*, **294**, 469, 1995.

and $I_{\rm conv}$ are known along an evolutionary track, Eqs. (7.13) and (7.14) can be integrated to yield the rotational evolution of the core and envelope of a low-mass star.

In Figure 7.6 we illustrate the influence of the model parameters on the rotational evolution of a single star. Panels A, B, and C are calculated for a $1M_\odot$ star; they depict the effect of varying the dynamo prescription, the coupling time scale τ_c, and the initial equatorial velocity $v_{\rm eq}$. Panel D illustrates the rotational evolution of stars of different mass. Obviously, an important feature of these solutions is the convergence of rotation rates after a time of the order of 1 Gyr. It is also apparent that *the rotational memory of a solar-type star is effectively lost at the age of the present-day Sun.* In fact, all models considered end up rotating at nearly the present-day solar rotation rate ($\Omega_\odot \approx 3 \times 10^{-6}$ s^{-1}), with essentially no internal differential rotation.

Panel A of Figure 7.6 shows how the phenomenological dynamo prescription influences the rotation evolution of a $1M_\odot$, with $v_{eq} = 15$ km s^{-1} and $\tau_c = 20$ Myr. The solid line corresponds to a *linear* dynamo, that is, a dynamo for which the strength of the mean coronal magnetic field increases linearly with rotation (see Eqs. [7.5] and [7.6]). The dashed lines and dash-dotted lines correspond to *saturated* dynamos, in which the mean coronal field saturates when the star rotates faster than, respectively, 5 and 10 times the present-day solar rotation rate. One readily sees that dynamo saturation reduces the angular momentum loss from the stellar wind since a lower field strength at the base of the corona causes less efficient magnetocentrifugal acceleration of the plasma. The angular momentum carried away by the stellar wind is therefore reduced, so that higher rotational velocities are achieved and sustained for a larger time. As was shown by Keppens and coworkers, a linear dynamo produces adequate spin-down early in the evolution but fails to produce sufficiently rapid rotators at the ages of α Persei and the Pleiades. Their analysis also shows that a saturated dynamo can explain the observed large spreads in rotation rates but the level of saturation is constrained by the requirement of achieving spin-down to slow rotation by the Hyades age (see Figure 7.5).

Making use of their parametric model for the rotational evolution of a single star, Keppens and coworkers have also investigated how the distribution of rotational velocities for late-type stars in the mass range 0.8–$1.0M_\odot$ evolves with age. Starting from an initial distribution compiled from observations of rotation among T Tauri stars, they found that reasonable agreement with the observationally inferred rotational evolution of solar-type stars is obtained for: (i) a linear dynamo that saturates beyond 20 times the present-day solar rotation rate, (ii) a coupling time scale τ_c of the order of 10 Myr, (iii) a mix of stellar masses consisting of roughly equal numbers of $0.8M_\odot$ and $1.0M_\odot$ stars, and (iv) disk regulation of the surface rotation up to an age of 6 Myr for stars with initial rotation periods larger than 5 days. The first requirement is in agreement with the observed saturation in chromospheric and coronal emission fluxes in the fastest rotators (see Section 7.2). As they noted, however, a number of discrepancies remain. In particular, their calculations fail to produce a sufficiently large proportion of slow rotators ($v_{eq} < 10$ km s^{-1}) on the zero-age main sequence.

At this juncture it is appropriate to compare these results with some of the model calculations made by Barnes and Sofia (1996). Following closely the method described in Section 5.4.1, these authors have computed the overall redistribution of angular momentum by making use of a simple diffusion equation and some ad hoc prescription for their coefficient of eddy viscosity (see Eq. [5.43]). As usual, the values of that coefficient were obtained by requiring that the present-day Sun rotates at the observed rate. A suitable parameterization was also used to describe the angular momentum loss through the action of a magnetically channeled stellar wind. An important conclusion of their work is that angular momentum loss without saturation is unable to account for the presence of the fastest rotators in young star clusters, regardless of the initial rotation periods. Moreover, calculations of evolutionary models in the mass range 0.6–$1.0M_\odot$ show that the saturation threshold is different for G, K, and M stars, with lower-mass stars saturating at lower angular velocities. Because lower-mass stars have deeper convective envelopes, this result seems to indicate that turbulent convection contributes significantly to the dynamo-generated magnetic fields of low-mass stars.

Insofar as comparison is possible, these results are quite similar to those obtained by Keppens and coworkers. In particular, both studies indicate that dynamo saturation is required to maintain a considerable spread in rotation rate at least until the age of the α Persei cluster (see Figure 7.5). Both studies also show that *the observed spin-down of the slow rotators in the young open clusters is in better agreement with differentially rotating models than with rigidly rotating models*. Since these investigations were carried out by means of models that make use of quite distinct parameterizations to treat angular momentum loss and redistribution, there is thus compelling evidence that saturated magnetized stellar winds, structural evolution, and core–envelope decoupling are the main agents determining the rotational history of a low-mass star. As was pointed out in Section 7.4.1, however, the effects of disk regulation during the pre–main-sequence phase should also be taken into account since disk–star magnetic coupling prevents, to some extent, spin-up associated with decreasing moment of inertia during that contraction phase.

7.5 Bibliographical notes

Sections 7.1 and 7.2. The following pioneering works are quoted in the text:

1. Schatzman, E., *Ann. Astrophys.*, **25**, 18, 1962.
2. Wilson, O. C., *Astrophys. J.*, **138**, 832, 1963; *ibid.*, **144**, 695, 1966.

See also References 36 and 37 of Chapter 1. The existence of rapidly rotating dwarfs in the Pleiades was originally reported in:

3. van Leeuwen, F., and Alphenaar, P., *The ESO Messenger*, No 28, p. 15, 1982.

Among the many papers on magnetic braking, reference may be made to:

4. Weber, E. J., and Davis, L., Jr., *Astrophys. J.*, **148**, 217, 1967.
5. Mestel, L., *Mon. Not. R. Astron. Soc.*, **138**, 359, 1968.
6. Durney, B. R., and Latour, J., *Geophys. Astrophys. Fluid Dyn.*, **9**, 241, 1978.
7. Mestel, L., in *Cool Stars, Stellar Systems, and the Sun* (Baliunas, S. L., and Hartmann, L., eds.), p. 49, Berlin: Springer-Verlag, 1984.
8. Mestel, L., and Spruit, H. C., *Mon. Not. R. Astron. Soc.*, **226**, 57, 1987.
9. Kawaler, S. D., *Astrophys. J.*, **333**, 236, 1988.
10. Cameron, A. C., Li, J., and Mestel, L., in *Angular Momentum Evolution of Young Stars* (Catalano, S., and Stauffer, J. R., eds.), p. 297, Dordrecht: Kluwer, 1991.

Discrete mass ejection is considered in:

11. Cameron, A. C., and Robinson, R. D., *Mon. Not. R. Astron. Soc.*, **236**, 57, 1989; *ibid.*, **238**, 657, 1989.
12. Jeffries, R. D., *Mon. Not. R. Astron. Soc.*, **262**, 369, 1993.

Other papers may be traced to Reference 12. Reference is also made to:

13. Vilhu, O., *Astron. Astrophys.*, **133**, 117, 1984.

A comprehensive discussion of the activity–rotation relationship will be found in:

14. Patten, B. M., and Simon, T., *Astrophys. J. Suppl.*, **106**, 489, 1996.

Other references may be traced to this paper. See also:

15. Wolff, S. C., and Simon, T., *Publ. Astron. Soc. Pacific*, **109**, 759, 1997

Section 7.3. See References 30–32 and 38–42 of Chapter 1. The following surveys are particularly worth noting:

16. Bouvier, J., in *Cool Stars, Stellar Systems, and the Sun* (Caillault, J. P., ed.), *A.S.P. Conference Series*, **64**, 151, 1994.
17. Stauffer, J. R., in *Cool Stars, Stellar Systems, and the Sun* (Caillault, J. P., ed.), *A.S.P. Conference Series*, **64**, 163, 1994.
18. Strom, S. E., in *Cool Stars, Stellar Systems, and the Sun* (Caillault, J. P., ed.), *A.S.P. Conference Series*, **64**, 211, 1994.

Accretion disks around T Tauri stars are also discussed in:

19. Edwards, S., Strom, S. E., Hartigan, P., Strom, K. M., Hillenbrand, L. A., Herbst, W., Attridge, J., Merrill, K. M., Probst, R., and Gatley, I., *Astron. J.*, **106**, 372, 1993.

Section 7.4.1. Comprehensive reviews of angular momentum evolution of young stars and accretion disks will be found in:

20. Bodenheimer, P., in *Angular Momentum Evolution of Young Stars* (Catalano, S., and Stauffer, J. R., eds.), p. 1, Dordrecht: Kluwer, 1991.
21. Bodenheimer, P., *Annu. Rev. Astron. Astrophys.*, **33**, 199, 1995.

See also:

22. Edwards, S., *Rev. Mexicana Astron. Astrofis.*, **29**, 35, 1994.

Detailed discussions are found in:

23. Ghosh, P., and Lamb, F. K., *Astrophys. J.*, **232**, 259, 1979; *ibid.*, **234**, 296, 1979.
24. Durisen, R. H., Yang, S., Cassen, P., and Stahler, S. W., *Astrophys. J.*, **345**, 959, 1989.
25. Hartmann, L., and Stauffer, J. R., *Astron. J.*, **97**, 873, 1989.
26. Königl, A., *Astrophys. J. Letters*, **370**, L39, 1991.
27. Shu, F., Najita, J., Ostriker, E., and Wilkin, F., *Astrophys. J.*, **429**, 781, 1994.
28. Ostriker, E. C., and Shu, F. H., *Astrophys. J.*, **447**, 813, 1995.

See also Reference 40 of Chapter 1. The rotational evolution of T Tauri stars with accretion disks is discussed in:

29. Cameron, A. C., and Campbell, C. G., *Astron. Astrophys.*, **274**, 309, 1993.
30. Cameron, A. C., Campbell, C. G., and Quaintrell, H., *Astron. Astrophys.*, **298**, 133, 1995.

See also:

31. Ghosh, P., *Mon. Not. R. Astron. Soc.*, **272**, 763, 1995.
32. Armitage, P. J., and Clarke, C. J., *Mon. Not. R. Astron. Soc.*, **280**, 458, 1996.

Section 7.4.2. The following review may be noted:

33. Charbonneau, P., Schrijver, C. J., and MacGregor, K. B., in *Cosmic Winds and the Heliosphere* (Jokipii, J. R., Sonett, C. P., and Giampapa, M. S., eds), p. 677, Tucson: University of Arizona Press, 1997.

See also:

34. Stauffer, J. R., and Hartmann, L. W., *Astrophys. J.*, **318**, 337, 1987.
35. MacGregor, K. B., and Brenner, M., *Astrophys. J.*, **376**, 204, 1991.
36. Soderblom, D. R., Stauffer, J. R., MacGregor, K. B., and Jones, B. F., *Astrophys. J.*, **409**, 624, 1993.
37. MacGregor, K. B., and Charbonneau, P., in *Cool Stars, Stellar Systems, and the Sun* (Caillault, J. P., ed.), *A.S.P. Conference Series*, **64**, 174, 1994.
38. Keppens, R., MacGregor, K. B., and Charbonneau, P., *Astron. Astrophys.*, **294**, 469, 1995.

See especially Reference 38. Another viewpoint will be found in:

39. Li, J., and Cameron, A. C., *Mon. Not. R. Astron. Soc.*, **261**, 766, 1993.
40. Cameron, A. C., and Li, J., *Mon. Not. R. Astron. Soc.*, **269**, 1099, 1994.

The Yale group has contributed the following papers:

41. Chaboyer, B., Demarque, P., and Pinsonneault, M. H., *Astrophys. J.*, **441**, 876, 1995.
42. Barnes, S., and Sofia, S., *Astrophys. J.*, **462**, 746, 1996.
43. Krishnamurthi, A., Pinsonneault, M. H., Barnes, S., and Sofia, S., *Astrophys. J.*, **480**, 303, 1997.

An illustration of their empirical coefficient of eddy viscosity will be found in Reference 40 (Fig. 16, p. 548) of Chapter 3. Further contributions have been made by:

44. Bouvier, J., Forestini, M., and Allain, S., *Astron. Astrophys.*, **326**, 1023, 1997.
45. Siess, L., and Livio, M., *Astrophys. J.*, **490**, 785, 1997.

Additional calculations based on the MacGregor–Brenner prescription and a useful review of the literature will be found in:

46. Allain, S., *Astron. Astrophys.*, **333**, 629, 1998.

8

Tidal interaction

8.1 Introduction

The main body of the book has been concerned with the effects of axial rotation upon the structure and evolution of single stars. As was pointed out in Section 1.4, further challenging problems arise from the study of double stars whose components are close enough to raise tides on the surface of each other. Indeed, tidal interaction in a detached close binary will continually change the spin and orbital parameters of the system (such as the orbital eccentricity e, mean orbital angular velocity Ω_0, inclination ω, and rotational angular velocity Ω of each component). Unless there are sizeable stellar winds emanating from the binary components, the total angular momentum will be conserved during these exchange processes. However, as a result of tidal dissipation of energy in the outer layers of the components, the total kinetic energy of a close binary system will decrease monotonically. Ultimately, this will lead to either a collision or an asymptotic approach toward a state of minimum kinetic energy. Such an equilibrium state is characterized by *circularity* ($e = 0$), *coplanarity* ($\omega = 0$), and *corotation* ($\Omega = \Omega_0$); that is to say, the orbital motion is circular, the rotation axes are perpendicular to the orbital plane, and the rotations are perfectly synchronized with the orbital revolution.

To be specific, unless the binary components rotate in perfect synchronism with a circular orbital motion, each star senses a variable external gravitational field – thus becoming liable to oscillatory motions that may be described as an "equilibrium tide" and a "dynamical tide." The former is just the instantaneous shape obtained by assuming that strict mechanical equilibrium prevails, even though the forcing potential depends on time, that is to say, it is assumed that the forced oscillations of the star are rapidly damped out and do not affect the "equilibrium distortion." The latter refers to the dynamical response of the star to the tidal forcing of its natural modes of oscillation. As we shall see in Section 8.2, the effects of turbulent viscosity retarding the equilibrium tide play an important role in binary components with a deep *convective* envelope; these stars experience a torque that tends to induce synchronization. However, because viscosity is much too small in stars having an outer *radiative* envelope, a different mechanism must be invoked to explain the high degree of synchronism and orbital circularization that is observed in the early-type binaries. In Section 8.3 we shall see that radiative damping can produce in part the required torque by retarding the dynamical tide in these stars.

In the late 1970s, the theoretical predictions based on these two distinct mechanisms were in agreement with the (then current) observations. Unfortunately, as will be shown in Sections 8.2.2 and 8.3.1, they are unable to explain all of the most recent observational data reported in Section 1.4. This is the reason why in Section 8.4 we shall consider another braking mechanism, which is much more efficient than the two classical ones but has hitherto escaped notice. In my opinion, this third mechanism was overlooked for so long because too much reliance had been placed on the deep-rooted tradition of celestial mechanics, with the hydrodynamical aspect of the problem being neglected altogether. As we shall see in Section 8.4.2, this mechanism is operative in the early-type and late-type binaries alike. It involves a large-scale meridional flow, superposed on the motion around the rotation axis of the tidally distorted star. These transient, mechanically driven currents are caused by the forced lack of *axial* symmetry in a binary component; they cease to exist as soon as synchronization has been achieved in the star. They are thus quite different from the steady, thermally driven currents presented in Section 4.6, which, as we recall, are caused by the forced lack of *spherical* symmetry in the radiative envelope of a tidally distorted binary component. They are also quite different from the large-scale atmospheric motions presented in Sections 2.5.1 and 2.5.2; as we shall see in Section 8.5, however, these *geostrophic* (or *astrostrophic*) currents are of direct relevance to the study of contact binaries.

8.2 The tidal-torque mechanism

The tidal-torque mechanism was originally discussed by Darwin (1879) with reference to a planet–satellite system. As was shown by Zahn (1966), it is also effective in binary-star components possessing an extended outer convection zone. In this model each component possesses tides lagging in phase behind the external field of force on account of eddy viscosity in its convective envelope. Accordingly, this misalignment of the tidal bulges with respect to the line joining the two centers of mass will introduce a net torque between the components. This torque will cause, in turn, a secular change in spin angular momentum of the individual components, the effects of which will be reflected in secular changes in the orbital elements of the binary.

8.2.1 Darwin's weak-friction model

For the sake of clarity, I shall first derive the synchronization time for a system of two rotating stars in circular orbits about their common center of mass; the rotation axes are assumed to be perpendicular to the orbital plane. We take as the origin of our system of coordinates the center of mass of the primary (of mass M and radius R). We shall also assume that the radii of the components are smaller than their mutual distance d, so that the secondary may be treated as a point mass when studying the tides raised on the primary. To lowest order in the ratio r/d, the tidal potential W due to the secondary (of mass M') is

$$W(r, \vartheta) = \frac{GM'r^2}{d^3} P_2(\cos\vartheta), \tag{8.1}$$

where r is the distance from the primary's center, ϑ is the angle between the direction to the field point and the line joining the two centers of mass, and P_2 is the Legendre polynomial of degree two. The dynamical tide will be neglected altogether.

If viscous dissipation is negligible, the equilibrium tide raised by the secondary can be described by an effective potential W_e whose value at the primary's surface is given by

$$W_e(R, \vartheta) = k \frac{GM'R^2}{d^3} P_2(\cos\vartheta), \tag{8.2}$$

where k is the apsidal-motion constant, which depends on the density stratification in the tidally distorted star. Outside the primary, the potential W_e will be the external solution of Laplace's equation, with Eq. (8.1) defining its boundary value at $r = R$. One obtains

$$W_e(r, \vartheta) = k \frac{GM'R^5}{d^3 r^3} P_2(\cos\vartheta), \tag{8.3}$$

whenever the tidal bulges are symmetrical about the line joining the two centers of mass.

Turbulent friction introduces a small time lag Δt, however; the tidal bulges lag (or lead) by a small angle δ if the rotational angular velocity Ω of the primary is smaller (or greater) than the orbital angular velocity Ω_0. This produces a torque component in the gravitational attraction of the two stars. The tidal torque Γ felt by the secondary is equal to $M' f_\theta d = -M' \partial W_e / \partial\vartheta$, where the angular derivative is evaluated at $r = d$ and $\vartheta = \delta$. One readily sees that

$$\Gamma = \frac{3}{2} k \frac{GM'^2}{R} \left(\frac{R}{d}\right)^6 \sin 2\delta \approx 3k \frac{GM'^2}{R} \left(\frac{R}{d}\right)^6 \delta. \tag{8.4}$$

The tidal torque acting upon the primary is exactly opposite of this torque.

Now, in the so-called weak-friction approximation one assumes that the small angle δ is linearly proportional to the departure from synchronism, with this angle being also proportional to the strength of viscous dissipation. We shall thus write

$$\delta = (\Omega - \Omega_0)\Delta t = (\Omega - \Omega_0)\frac{t_{\text{ff}}^2}{T}, \tag{8.5}$$

where $t_{\text{ff}} = (GM/R^3)^{-1/2}$ is the free-fall time and T is a typical time scale on which significant changes in the orbit take place through tidal evolution. Since the latter is inversely proportional to the efficiency of viscous dissipation, we shall further let $T = R^2/\nu_t$, where ν_t is the coefficient of eddy viscosity (see Section 2.4). We shall also assume that the primary goes through a succession of rigidly rotating states, during which the tidal torque causes a slow but inexorable change in the spatially uniform angular velocity Ω. One thus has $I\dot{\Omega} \approx -\Gamma$, where I is the moment of inertia of the primary about its rotation axis. (A dot designates a derivative with respect to time.) Thence, we can estimate the characteristic time for synchronization, t_{syn}, by

$$t_{\text{syn}} = -\frac{\Omega - \Omega_0}{\dot{\Omega}} \approx \frac{I(\Omega - \Omega_0)}{\Gamma} \approx \frac{r_g^2}{3k} \frac{T}{q^2} \left(\frac{d}{R}\right)^6, \tag{8.6}$$

where $r_g = (I/MR^2)^{1/2}$ is the fractional gyration radius and $q = M'/M$ is the mass ratio. Simultaneously, via the torque Γ, angular momentum is transferred from the primary's spin to the secondary's orbit. This results in a secular change in the distance ratio d/R and, hence, in the orbital angular velocity Ω_0.

The weak-friction model is ideally suited for a detailed study of tidal interaction in detached close binaries that have significant eccentricities. Following Hut (1981), we shall assume that the deviations from coplanarity are small enough to be treated linearly.

To simplify the discussion, the secondary is also assumed to be point like so that only on the primary tides will be raised. If so, then, it can be shown that the resulting tidal evolution equations for the primary are as follows:

$$\frac{da}{dt} = -6\frac{k}{T}q(1+q)\left(\frac{R}{a}\right)^8 \frac{a}{(1-e^2)^{15/2}}\left[f_1(e^2) - (1-e^2)^{3/2} f_2(e^2)\frac{\Omega}{\Omega_0}\right], \tag{8.7}$$

$$\frac{de}{dt} = -27\frac{k}{T}q(1+q)\left(\frac{R}{a}\right)^8 \frac{e}{(1-e^2)^{13/2}}\left[f_3(e^2) - \frac{11}{18}(1-e^2)^{3/2} f_4(e^2)\frac{\Omega}{\Omega_0}\right], \tag{8.8}$$

$$\frac{d\Omega}{dt} = 3\frac{k}{T}\frac{q^2}{r_g^2}\left(\frac{R}{a}\right)^6 \frac{\Omega_0}{(1-e^2)^6}\left[f_2(e^2) - (1-e^2)^{3/2} f_5(e^2)\frac{\Omega}{\Omega_0}\right], \tag{8.9}$$

$$\frac{d\omega}{dt} = -3\frac{k}{T}\frac{q^2}{r_g^2}\left(\frac{R}{a}\right)^6 \frac{\omega}{(1-e^2)^6}\frac{\Omega_0}{\Omega}\left[f_2(e^2) - \frac{1}{2}(1-\eta)(1-e^2)^{3/2} f_5(e^2)\frac{\Omega}{\Omega_0}\right], \tag{8.10}$$

where a is the semimajor axis, e is the eccentricity, Ω is the rotational angular velocity, and ω is the angle between the orbital plane and the equatorial plane of the primary. For brevity, we have defined the following quantities:

$$f_1(e^2) = 1 + \frac{31}{2}e^2 + \frac{255}{8}e^4 + \frac{185}{16}e^6 + \frac{25}{64}e^8, \tag{8.11}$$

$$f_2(e^2) = 1 + \frac{15}{2}e^2 + \frac{45}{8}e^4 + \frac{5}{16}e^6, \tag{8.12}$$

$$f_3(e^2) = 1 + \frac{15}{4}e^2 + \frac{15}{8}e^4 + \frac{5}{64}e^6, \tag{8.13}$$

$$f_4(e^2) = 1 + \frac{3}{2}e^2 + \frac{1}{8}e^4, \tag{8.14}$$

$$f_5(e^2) = 1 + 3e^2 + \frac{3}{8}e^4. \tag{8.15}$$

In Eq. (8.10) we have also let

$$\eta = \frac{r_g^2}{(1-e^2)^{1/2}}\frac{1+q}{q}\left(\frac{R}{a}\right)^2\frac{\Omega}{\Omega_0}, \tag{8.16}$$

which is the ratio of rotational to orbital angular momentum (see Eq. [8.18]). Finally, one can write Kepler's third law in the form

$$\Omega_0^2 = \frac{GM(1+q)}{a^3}, \tag{8.17}$$

where Ω_0 is the mean orbital angular velocity.

By making use of Eqs. (8.7)–(8.9), one easily verifies that the total angular momentum of the system is conserved. Here we have

$$\frac{d}{dt}\left\{I\Omega + \frac{MM'}{M+M'}[G(M+M')a(1-e^2)]^{1/2}\right\} = 0, \tag{8.18}$$

since the rotational contribution from a point-mass companion can be neglected. In the case of two extended deformable bodies, Eqs. (8.7) and (8.8) can be applied to each binary component, interchanging the role of primary and secondary and adding both contributions to the orbital parameters a and e.

The behavior of the solutions of Eqs. (8.7)–(8.10) around a state of equilibrium has been thoroughly investigated by Hut (1981). In particular, for moderately small eccentricities, he was able to derive the time scales for the exponential relaxation of the relevant parameters. In detached close binaries for which the orbital angular momentum is much larger than the sum of the rotational angular momenta, Hut found that the characteristic time for orbital circularization, $t_{\rm cir}$, is much larger than the other three, which are of comparable magnitude.* Making use of Eq. (8.9), one easily obtains the linearized equation

$$\frac{1}{\Omega - \Omega_0}\frac{d\Omega}{dt} = -\frac{1}{t_{\rm syn}}, \tag{8.19}$$

where

$$t_{\rm syn} = \frac{r_g^2}{3k}\frac{T}{q^2}\left(\frac{a}{R}\right)^6, \tag{8.20}$$

which confirms the order-of-magnitude estimate given in Eq. (8.6). For two extended bodies in almost circular orbits about their common center of mass, Eq. (8.8) further implies that

$$\frac{1}{e}\frac{de}{dt} = -\frac{1}{t_{\rm cir}(1)} - \frac{1}{t_{\rm cir}(2)}, \tag{8.21}$$

where the figures 1 and 2 refer to the primary and secondary, respectively. For the primary one has

$$t_{\rm cir} = \frac{2}{21k}\frac{T}{q(1+q)}\left(\frac{a}{R}\right)^8; \tag{8.22}$$

the secondary (of mass M' and radius R', say) makes a similar contribution to the effective circularization time, which is the *harmonic* mean of the circularization times obtained for the individual components.

One readily sees from Eqs. (8.20) and (8.22) that the ratio $t_{\rm syn}/t_{\rm cir}$ is of the order of the parameter η evaluated at equilibrium (see Eq. [8.16]). Since this quantity is much smaller than one in a detached close binary, we perceive at once that the synchronization of the components proceeds at a much faster pace than the circularization of the orbit. To the best of my knowledge, Hut (1981) was the first to point out that the rotation of each component in a detached close binary will synchronize with the instantaneous angular velocity at *periastron*, since during each revolution the tidal interaction will be the most important around that position (see Eq. [1.8]). Recall also that the inclination ω decreases rather quickly while at the same time rotation tends to synchronize with revolution, whereas the eccentricity of the orbit decreases at a much

* This ordering of the time scales was originally noticed by Alexander (1973, Figs. 7–10), who integrated numerically the tidal-friction equations for the close binary system AG Persei.

slower pace. This property of a detached close binary, which is probably independent of the exact nature of the underlying dissipative process, is a most likely explanation for the correlation between synchronism and coplanarity, as reported at the end of Section 1.4.

In deriving Eqs. (8.9) and (8.20) we have explicitly assumed that the tidally distorted star remains in a state of uniform rotation throughout its tidal evolution. As was shown by Scharlemann (1982), however, a tidally distorted star with an extended, differentially rotating convective envelope can be synchronized on the average, at a specific latitude on the surface of the star. Of course, because the tidal torque is applied mainly to the outer convective regions, the radiative core might rotate at a quite different speed – unless there is a strong coupling between the inner core and the outer envelope. In fact, even though such a coupling might exist in the late-type stars, once a star has evolved away from the main sequence it develops a helium-rich core whose rotation becomes decoupled from that of the envelope. This is particularly relevant to the case of a close binary star that has achieved synchronism and orbital circularity on the main sequence, since post–main-sequence expansion will desynchronize the components while maintaining a circular orbit as they move up to the giant branch. As we shall see in Section 8.4.1, in that case one must integrate Eqs. (8.19) and (8.21) along the evolutionary paths of the binary components, retaining the time dependence of the radii R and R' in the functions $t_{\rm syn}$ and $t_{\rm cir}$.

8.2.2 *Application to late-type binaries*

For tidally distorted stars possessing a deep convective envelope, it is generally believed that turbulent friction operating over the whole of that envelope is responsible for the tidal torque Γ, so that the characteristic time T must be a convective friction time scale derived from stellar envelope parameters (see Eqs. [8.4] and [8.5]). Lacking any better theory of turbulent convection in a star, we shall thus let $T = R^2/\nu_t$ and $\nu_t = L_c V_c$, where L_c is the typical size of the largest eddies and V_c is a typical convection velocity. Mixing-length theory provides crude estimates of these quantities. Following Zahn (1966), we have that the convective friction time scale T is of the order of $(MR^2/L)^{1/3}$, where L is the total luminosity of the star. (One has $T \approx 160$ days for a solar-type star.) By virtue of Eqs. (8.17) and (8.20), this particular prescription for the eddy viscosity implies that the synchronization time $t_{\rm syn}$ is proportional to the fourth power of the orbital period P $(= 2\pi/\Omega_0)$ – that is, $t_{\rm syn} \propto (a/R)^6$ or $t_{\rm syn} \propto P^4$. Similarly, by making use of Eqs. (8.17) and (8.22), one readily sees that $t_{\rm cir} \propto (a/R)^8$ or $t_{\rm cir} \propto P^{16/3}$. These rather high exponents make the characteristic times $t_{\rm syn}$ and $t_{\rm cir}$ strongly dependent on the separation of the components or, equivalently, on the orbital period.

The results obtained in Section 8.2.1 clearly show that the degree of synchronism and orbital circularization depends on how long the tidal torque has been operative on the binary components. Accordingly, because main-sequence stars evolve with almost constant radius, much information can be gained by merely comparing the evolutionary age of a main-sequence star with the corresponding time scales $t_{\rm syn}$ and $t_{\rm cir}$. The study of main-sequence close binaries belonging to open clusters is of particular interest, therefore, since stellar evolution theory provides an estimate of their ages from isochrone fitting. Here we shall mainly discuss the problem of orbital circularization, making use of the observational data summarized in Table 1.2 and Figure 1.12.

As was noted in Section 1.4, in a sample of low-mass main-sequence binaries belonging to the same cluster, all binaries with periods shorter than a cutoff period – $P_{\rm cut}$, say – have circular orbits, whereas binaries with longer periods have orbits with a distribution of eccentricities. From Table 1.2 it is also apparent that an older sample of binaries has a longer transition period, with $P_{\rm cut}$ increasing monotonically with the sample age t_a. This is in perfect agreement with the fact that in a coeval sample of binaries the cutoff period is time dependent, because the tidal interaction has more time to extend its influence in an old cluster than in a young one. The crucial test for Darwin's tidal-torque theory is to check whether the circulation time defined in Eq. (8.22) is indeed shorter than (or equal to) the age t_a at $P = P_{\rm cut}$.

To illustrate the problem, we shall assume that $M = M_\odot$, $R = R_\odot$, and $q = 1$. Although k probably lies within the range 0.01–0.02, I shall give an edge to the theory and let $k = 0.05$. By virtue of Kepler's third law, Eq. (8.22) can be recast in the practical form

$$t_{\rm cir}({\rm yr}) = 6 \times 10^5\, T({\rm yr})\, [P({\rm day})]^{16/3}. \tag{8.23}$$

Because we are considering similar binary components, the effective circularization time may differ from this by a factor of two, which is unimportant for our purpose. Since T is a free parameter in Eq. (8.23), let us prescribe that for each cluster the characteristic time $t_{\rm cir}$ is equal to its age t_a at $P = P_{\rm cut}$. Letting $P_{\rm cut} = 18.7$ days at $t_a = 17.6$ Gyr, one easily verifies that

$$t_a({\rm yr}) = 3 \times 10^3\, [P_{\rm cut}({\rm day})]^{16/3} \tag{8.24}$$

gives a moderately good fit for the other coeval samples listed in Table 1.2. Since Eqs. (8.23) and (8.24) must be equivalent at $P = P_{\rm cut}$, one readily sees that the mechanism is operative on the main sequence provided one has

$$T \lesssim 5 \times 10^{-3}\,{\rm yr} \approx 2\ {\rm days}, \tag{8.25}$$

which is much shorter than $T \approx 160$ days. Now, with $R \approx R_\odot$, $L_c \approx R_\odot/10$, and $T \approx 2$ days, the formula $T = R^2/v_t$ implies that the typical convection velocity V_c should be of the order of 40 km s^{-1}. Obviously, this independent evaluation of T is also too large by about two orders of magnitude.*

At this juncture it is appropriate to mention the work of Claret and Cunha (1997), who have integrated Eqs. (8.21) and (8.22) using a set of low-mass stellar models that are slowly evolving on the main sequence. Unless turbulent dissipation is artificially

* Zahn (1989) has argued that one has $v_t \propto P$ in the short-period binaries, thus implying that one should let $t_{\rm syn} \propto P^3$ and $t_{\rm cir} \propto P^{13/3}$ in these stars. According to Goldman and Mazeh (1991), however, one has $v_t \propto P^2$ in the short-period binaries, so that one should let $t_{\rm syn} \propto P^2$ and $t_{\rm cir} \propto P^{10/3}$. Unfortunately, although these modified versions of the standard theory can provide a somewhat better fit to the slope of the observed $\log t_a$–$\log P_{\rm cut}$ relation, they are still unable to resolve the basic weakness of the tidal-torque mechanism: Given a reasonable theoretical value for the convective friction time scale T, the circularization times are much too long at $P = P_{\rm cut}$ during the main-sequence phase. This inadequacy has been also confirmed by the independent analysis of Goodman and Oh (1997), who concluded that some mechanism other than turbulent convection circularizes solar-type binaries.

enhanced by a factor around 100–200, their calculations show that the tidal-torque mechanism is most ineffective in inducing orbit circularization on the lower main sequence. This result thus brings confirmation to the foregoing order-of-magnitude calculation.

Of direct relevance to the present discussion are the results of Zahn and Bouchet (1989), who have studied the pre–main-sequence evolution of solar-type binary stars. During this contraction phase, because a star undergoes great changes in size and structure, it is necessary to follow in time the dynamical state of the binary star along the evolutionary paths of its components (see Eqs. [8.21] and [8.22]). Their calculations strongly suggest that most of the orbital circularization takes place during the Hayashi phase, with the subsequent decrease in eccentricity on the main sequence being quite negligible. They found that the cutoff period of any sample should lie between 7.2 and 8.5 days, independent of the sample age. Unfortunately, this conclusion is not at all supported by the observational data reported in Table 1.2, which strongly suggest that the circularization mechanism is operative during the main-sequence lifetimes of the stars – pre–main-sequence tidal circularization is permitted but not required by present observations.

Let us also note that the tidal-torque mechanism, which is quite ineffective on the main sequence, may become operative again during the post–main-sequence phases. This is quite apparent from the work of Verbunt and Phinney (1995), who have shown that turbulent friction acting on the equilibrium tide can generate circular orbits up to $P \approx 200$ days in binaries containing giant stars. Yet, Figure 1.12 clearly shows that there exists a mixed population of circular and eccentric orbits in the whole period range 80–300 days. Obviously, independent calculations based on Eqs. (8.21) and (8.22) are needed to ascertain whether this mechanism can remain operative up to $P = 300$ days, or whether an additional circulation mechanism becomes operative during the expanding phases of stellar evolution.

For completeness, let us briefly discuss the problem of pseudo-synchronism in late-type main-sequence binaries. This is a much more difficult exercise, however, because the relevant observations are still very scarce. Letting $T \approx 160$ days in Eq. (8.22), one finds that the tidal-torque mechanism does contribute to the synchronization process up to $P \approx 20$ days in these binary stars. However, there exist a few binaries, with orbital periods in the range 40–50 days, that exhibit a definite tendency toward pseudo-synchronization in their solar-type components. Accordingly, we are led to conclude that this mechanism might not wholly account for the presence of pseudo-synchronous rotators in these binaries. A similar comment was made by Maceroni and van't Veer (1991) in their study of the dynamical evolution of G-type main-sequence binaries.

These results present quite a dilemma if one assumes that Darwin's tidal-torque theory alone can explain the whole set of observational data for the late-type binaries. As we shall see in Section 8.4.4, there is no longer any problem when one relaxes the assumption of strict uniform rotation, thus making allowance for tidally driven meridional currents in the asynchronously rotating components of a detached close binary.

8.3 The resonance mechanism

Cowling (1941) was the first to study the natural modes of oscillation in a centrally condensed star, suggesting that some of the gravity modes may enter into resonance with the periodic tidal potential in a close binary. As was originally noted by Zahn

(1975), in a binary component possessing a radiative envelope the resonances of these low frequency oscillations are heavily damped by radiative diffusion, which operates in a relatively thin layer below the star's surface. Owing to that dissipative process, the dynamical tide does not have the same symmetry properties as the forcing potential. Hence, a net torque is applied to the binary component, which tends to synchronize its axial rotation with the orbital motion.

To evaluate the characteristic time for synchronization in an early-type star, one must calculate the amplitude of the forced oscillation at the star's surface, taking radiative damping into account. Following Zahn's (1975) analysis, one can show that the bulge raised by the dynamical tide is much smaller than that produced by the equilibrium tide; however, unlike the latter, it can take any orientation with respect to the companion star, depending on the tidal frequency. Detailed calculations also show that the torque Γ_d resulting from the dynamical tide is proportional to the product $(R/a)^6(\Omega - \Omega_0)^{8/3}$, if the density gradient is assumed to be continuous across the core–envelope interface. (Compare with Eqs. [8.4] and [8.5].) It is therefore appropriate to introduce a new synchronization time $t_{\rm syn}$ defined as

$$\frac{1}{t_{\rm syn}} = \frac{d}{dt}\left|\frac{\Omega - \Omega_0}{\Omega_0}\right|^{-5/3}. \tag{8.26}$$

Again letting $I\dot{\Omega} \approx -\Gamma_d$ and making use of Eq. (8.17), one can estimate the time scale $t_{\rm syn}$ by

$$t_{\rm syn} = \frac{r_g^2}{5E_S}\frac{t_{\rm ff}}{q^2(1+q)^{5/6}}\left(\frac{a}{R}\right)^{17/2}, \tag{8.27}$$

where the structural constant E_S, which is the strict analog of the apsidal-motion constant k in Eq. (8.2), depends mainly on the size of the convective core. Along the main sequence, one has $E_S \approx 10^{-8}$ when $M = 2M_\odot$ and $E_S \approx 10^{-6}$ when $M = 10M_\odot$. For moderately small eccentricities, one also has

$$t_{\rm cir} = \frac{2}{21E_S}\frac{t_{\rm ff}}{q(1+q)^{11/6}}\left(\frac{a}{R}\right)^{21/2}, \tag{8.28}$$

with the secondary contributing a similar amount to the effective circularization time of the binary star. (Compare these two equations with Eqs. [8.20] and [8.22], respectively.)

8.3.1 Application to early-type binaries

Because the synchronization process involves a secular adjustment of the gaseous components in a binary star, Eqs. (8.17) and (8.27) clearly show that the degree of synchronism decreases as the distance ratio a/R and the orbital period P increase. Accordingly, if we consider a sample of binaries with a random distribution of ages, one may expect to find an increasingly mixed population of asynchronous and synchronous rotators as the orbital periods approach an upper period limit above which binaries are nearly all asynchronously rotating. Given any data set, it is therefore essential to check that the theory can indeed account for the whole period range for which there is still a significant tendency toward synchronization. The same remark can be made about orbital circularization.

Table 8.1. *The critical values a_c/R and P_c.*

	Synchronization		Circularization	
Mass ($M_\odot$)	a_c/R	P_c(day)	a_c/R	P_c(day)
1.6	6.11	1.21	4.44	0.75
2	7.05	1.59	4.99	0.95
3	6.81	1.92	4.85	1.10
5	6.52	2.19	4.68	1.33
7	6.72	2.69	4.80	1.62
10	6.67	3.30	4.77	2.00
15	7.04	3.98	4.99	2.38

Source: Zahn, J. P., *Astron. Astrophys.*, **57**, 383, 1977.

Following Zahn (1977), we shall define the limiting separations for synchronization and orbital circularization as the distance ratios a_c/R for which one has, respectively, $t_{\rm syn}/t_a = 0.25$ and $t_{\rm cir}/t_a = 0.25$. (The time t_a is the main-sequence lifetime of the binary star, which consists of two similar components.) Thence, by making use of Eqs. (8.17), (8.27), and (8.28), one can easily obtain the corresponding critical periods P_c. In Table 8.1 we list the numerical values of these limiting separations.

From Table 8.1 it is apparent that synchronism should be the rule up to $a/R \approx 6$–7 in the early-type, main-sequence stars. This is not in agreement with the observational results reported in Section 1.4, however, since they clearly show that the early-type (from O to F5) close binaries do exhibit a considerable tendency toward synchronization (or pseudo-synchronization) up to $a/R \approx 20$, with deviations from synchronism becoming the rule for $a/R \gtrsim 20$ only. In fact, this mechanism is also much too weak to account for the high degree of orbital circularization that is observed in the early-type, main-sequence stars. Indeed, whereas Figure 1.11 clearly shows that some binaries with A-type primary stars have circular orbits with periods as long as 10 days, Table 8.1 indicates that the mechanism is effective only up to $P \approx 1$–2 days in these stars.

A similar result was obtained by Claret and Cunha (1997), who have integrated Eqs. (8.21) and (8.28) using a set of early-type main-sequence models. Again, unless the effects of radiative damping acting on the dynamical tide are artificially increased by several orders of magnitude, it is found that the resonance mechanism is unable to explain the longest-period circular orbits shown in Figure 1.11. The same result was obtained by Pan (1997), who found that this mechanism does not explain the observed degree of synchronism in early-type binaries with orbital periods $P \approx 4$–8 days. In other words, unless the orbital periods are shorter than a few days only, it is a most ineffective tidal process.

Attempts to patch up Zahn's (1975, 1977) calculations have been made. In particular, Goldreich and Nicholson (1989) have pointed out that the synchronization process caused by the tidal forcing of the gravity modes proceeds from the outside toward the inside of an early-type star. Thence, assuming that the tides induce differential rotation by synchronizing the outer layers of the star while leaving its interior roughly unaltered, Savonije and Papaloizou (1997) have shown that rotational effects could significantly influence the tidal response in the surface layers of a $20M_\odot$ main-sequence star. In particular, in contrast to subsynchronous stars, which tend to spin up toward corotation as a result of resonances with damped g-modes, it is found that supersynchronous stars spin down toward corotation due to resonances with damped r-modes, analogous to Rossby waves in the Earth's atmosphere. In my opinion, although these rotational effects might also improve the efficiency of the resonance mechanism in the less massive stars, they do not change the inescapable fact that this mechanism is a short-range one, since the corresponding times t_{syn} and t_{cir} are proportional to $(d/R)^{8.5}$ and $(d/R)^{10.5}$, respectively. This is the reason why it is not likely to explain the largest circular orbits reported in Figure 1.11 and in p. 19n.

8.4 The hydrodynamical mechanism

In Section 8.2 we have assumed that an asynchronous binary component goes through a succession of rigidly rotating states, thus overlooking the ability of a gaseous body to develop large-scale currents in meridian planes passing through its rotation axis. The possibility that a secondary circulation controls, in part or in toto, the rotation rate of a tidally distorted star is discussed in this section. Admittedly, this is a comparatively new development that involves specialized concepts in hydrodynamics. In Section 2.5.3 we have already discussed the spin-down of a cyclonic vortex in the Earth's atmosphere, which is the archetype of many situations that are encountered in rotating fluids. For the sake of clarity, in Section 8.4.1 we shall also consider the spin-up and spin-down of an incompressible fluid confined between two parallel infinite plates, when these solid boundaries are subject to an impulsive change in the magnitude of their angular velocity. I shall then explain how to apply these results to the problem of an asynchronously rotating binary component. The transposition is far from being obvious, however, because one has to replace boundary conditions on a solid surface by boundary conditions on a free surface; conditions (2.17) and (2.18) are then replaced by conditions (2.20) and (2.21).

Unless the reader is already familiar with geophysical fluid dynamics, I recommend reading Sections 2.5.3 and 8.4.1, which are essential for the understanding of the double-star problem treated in Section 8.4.2. Sections 8.4.3 and 8.4.4 present practical applications to detached close binaries; they are almost self-contained and can be read without going through the mathematical derivations made in Sections 8.4.1 and 8.4.2.

8.4.1 *The spin-up and spin-down of a rotating fluid*

Consider the problem in which two parallel infinite plates and the fluid between them initially rotate with the constant angular velocity Ω_i. The angular velocity of the two plates is then impulsively changed to the new constant value Ω_0. We wish to describe the manner by which the fluid spins up (when $\Omega_i < \Omega_0$) or spins down (when $\Omega_i > \Omega_0$) to its new angular velocity Ω_0.

As was shown in Section 2.2.3, the equations describing the motion of a viscous incompressible fluid, in a frame rotating about the z axis with constant angular velocity Ω_0, are

$$\frac{\partial \mathbf{u}}{\partial \mathbf{t}} + \mathbf{u} \cdot \operatorname{grad} \mathbf{u} + 2\Omega_0 \mathbf{1}_z \times \mathbf{u} = \mathbf{g}_e - \frac{1}{\rho} \operatorname{grad} p + \nu \nabla^2 \mathbf{u} \tag{8.29}$$

and

$$\operatorname{div} \mathbf{u} = 0, \tag{8.30}$$

where **u** is the velocity in the rotating frame. Remaining symbols have their standard meanings (see Eqs. [2.9] and [2.27]). In cylindrical polar coordinates (ϖ, φ, z), the initial and boundary conditions corresponding to the impulsive change $(\Omega_i - \Omega_0)$ in the magnitude of the angular velocity are: $\mathbf{u} = (\Omega_i - \Omega_0)\varpi \mathbf{1}_\varphi$, for $t \leq 0$, and $\mathbf{u} = 0$, for $t \geq 0$, on the solid plates $z = +L$ and $z = -L$.

In order to discuss the relative importance of the terms $\mathbf{u} \cdot \operatorname{grad} \mathbf{u}$ and $2\Omega_0 \mathbf{1}_z \times \mathbf{u}$ in Eq. (8.29), it is convenient to define the dimensionless ratio

$$\epsilon = \frac{|\Omega_i - \Omega_0|}{\Omega}, \qquad \text{where} \quad \Omega = \max(\Omega_i, \Omega_0), \tag{8.31}$$

which may be described as a *Rossby number* varying between zero and one (see Eq. [2.30]). For the moment, we shall assume that the initial and final angular velocities differ by a small amount (i.e., $\epsilon \ll 1$), so that the nonlinear terms $\mathbf{u} \cdot \operatorname{grad} \mathbf{u}$ can be rightfully neglected in Eq. (8.29). Accordingly, by taking the curl of this equation, one finds that

$$\frac{\partial}{\partial t}(\operatorname{curl} \mathbf{u}) + 2\Omega_0 \operatorname{curl}(\mathbf{1}_z \times \mathbf{u}) = \nu \operatorname{curl}(\nabla^2 \mathbf{u}). \tag{8.32}$$

Equations (8.30) and (8.32) define three scalar equations for the three components of the velocity vector **u**. This is the so-called linear spin-up (or spin-down) problem. The corresponding nonlinear problem will be discussed *in fine*.

In the linear approximation, the unsteady solution is of the form

$$\mathbf{u} = \left[\frac{\partial \Psi}{\partial z} \mathbf{1}_\varpi + u\, \mathbf{1}_\varphi - \frac{1}{\varpi}\frac{\partial}{\partial \varpi}(\varpi \Psi)\, \mathbf{1}_z\right] \exp(-\Omega_0 \alpha t), \tag{8.33}$$

where Ψ is the stream function of the large-scale axisymmetric currents and α is a free parameter. As was shown by Greenspan and Howard (1963), boundary-layer theory can be used to solve this problem. Retaining only the highest order derivatives, we obtain

$$\left(\alpha + \delta^2 \frac{\partial^2}{\partial z^2}\right) u - 2\frac{\partial \Psi}{\partial z} = 0 \tag{8.34}$$

and

$$\frac{\partial^2}{\partial z^2}\left(\alpha + \delta^2 \frac{\partial^2}{\partial z^2}\right) \Psi + 2\frac{\partial u}{\partial z} = 0, \tag{8.35}$$

where $\delta = (\nu/\Omega_0)^{1/2}$ is the boundary-layer thickness. Parenthetically note that one has

$$\frac{\delta}{L} = \left(\frac{\nu}{\Omega_0 L^2}\right)^{1/2} = E^{1/2}, \tag{8.36}$$

where E is the *Ekman number* of the problem (see Eq. [2.32]).

Here we shall assume that $u = (\Omega_i - \Omega_0)\varpi$ for $t \leq 0$. The no-slip condition further implies that, at every instant ($t \geq 0$), one has

$$u = 0 \qquad \text{and} \qquad \frac{\partial \Psi}{\partial z}, \tag{8.37}$$

at $z = +L$ and $z = -L$ (see Eq. [2.18]). To ensure that the fluid does not penetrate into the solid walls, one must also let, at every instant ($t \geq 0$),

$$\Psi = 0, \tag{8.38}$$

at $z = +L$ and $z = -L$ (see Eq. [2.17]).

Making use of conditions (8.37), one can show that the appropriate boundary-layer solution of Eqs. (8.34) and (8.35) is

$$u = (\Omega_i - \Omega_0)\varpi(1 - e^{-\xi} \cos \xi) \tag{8.39}$$

and

$$\Psi = \frac{1}{2}(\Omega_i - \Omega_0)\,\varpi\,[\alpha z \mp \delta e^{-\xi} (\cos \xi + \sin \xi)]. \tag{8.40}$$

Here we have defined the stretched variables

$$\xi = \frac{L - z}{\delta} \quad \text{and} \quad \xi = \frac{L + z}{\delta}, \tag{8.41}$$

near the upper and lower plates, respectively. In Eq. (8.40) the minus sign refers to the boundary-layer solution near the upper plate $z = +L$ and the plus sign to that near the lower plate $z = -L$.

Conditions (8.38) further imply that we must let

$$\alpha = \frac{\delta}{L}. \tag{8.42}$$

Since Eq. (8.33) has a time dependence of the form $\exp(-\Omega_0 \alpha t)$ – or, equivalently, $\exp(-t/\tau)$ – it follows at once that the e-folding time of the velocity $\mathbf{u}$ in the rotating frame is equal to $(\Omega_0 \delta/L)^{-1}$. One thus has

$$\tau = \left(\frac{L^2}{\Omega_0 \nu}\right)^{1/2}. \tag{8.43}$$

(Compare with Eq. [2.101].)

The foregoing linearized problem has been also studied by Greenspan and Howard (1963) using Laplace transforms of Eqs. (8.34) and (8.35). As they showed, the detailed motion consists of three distinct phases: (i) the formation of thin Ekman layers near the two rotating plates, where viscous friction plays a dominant role, (ii) the formation of a large-scale meridional flow that spins up (or spins down) the fluid exponentially, with an e-folding time of the order of $t_d(\delta/L)^{-1}$, where t_d is the dynamical time scale and δ/L is the relative boundary-layer thickness, and (iii) a much slower decay of the small-amplitude residual motions over the characteristic time of viscous friction, which is of the order of $t_d(\delta/L)^{-2}$.

Figure 8.1 illustrates, at a given instant, the transient meridional flow between two parallel infinite plates that are impulsively spun down. Broadly speaking, the initial impulsive braking of the plates slows down the motion of the fluid, so that a radial inflow of matter takes place within the Ekman layers. By continuity, this radial inflow of matter requires motion along the rotation axis and a slow compensatory outward radial flow in the bulk of the fluid. Since viscous friction is negligible away from the solid walls, this slow outward motion approximately conserves the specific angular momentum $\Omega\varpi^2$.

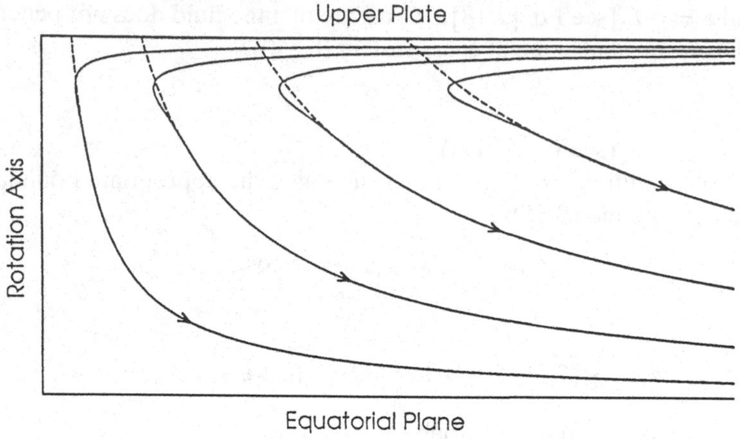

Fig. 8.1. Streamlines of the transient meridional flow in an incompressible fluid between two parallel infinite plates, with the solid walls being spun down impulsively at $t = 0$ (*solid lines*). Because the configuration is symmetric with respect to the equatorial plane, the lower half of the fluid is not represented. To illustrate the streamlines near the walls, we have let $\delta/L = 0.05$, which corresponds to a rather large viscosity. For comparison, the frictionless solution, which does not satisfy the boundary condition, is also illustrated (*dashed lines*). *Source:* Tassoul, J. L., and Tassoul, M., *Astrophys. J.*, **395**, 259, 1992.

Accordingly, by replacing high angular velocity fluid by low angular velocity fluid, the large-scale secondary flow serves to spin down the fluid far more rapidly than could mere viscous friction. An entirely analogous, but reverse, phenomenon occurs if the two plates are spun up slightly rather than spun down, but τ is then the spin-up time.

It is immediately apparent from these discussions that the spin-up and spin-down times are equal in the linear approximation (i.e., when $\epsilon \ll 1$). A quite different picture emerges when nonlinear effects are taken into account, that is to say, when the restriction to extremely small Rossby number is relaxed. Results for impulsive spin-up and spin-down between parallel infinite plates have been presented by Weidman (1976) for the complete range $0 \leq \epsilon \leq 1$. Of practical interest is the time it takes for the bulk of the fluid to spin up or spin down. In Figure 8.2 we present these two characteristic times – τ_{99}, say – in units of the e-folding time τ, as functions of the Rossby number. (By definition, they are the elapsed times for which the fluid locally reaches 99% of the change in angular velocity imposed on the solid walls.) Figure 8.2 obviously shows two important features of the problem: (1) The nonlinearity monotonically increases the spin-up and spin-down times and (2) a nonlinear spin-up is achieved somewhat faster than a nonlinear spin-down. Note also that the effects of nonlinearity become of paramount importance as the Rossby number approaches unity, with both characteristic times becoming then much larger than their common value obtained in the linear approximation (see Eq. [8.43]).

8.4.2 *Ekman pumping in a tidally distorted star*

Following closely the assumptions made in Sections 8.2.1, we have a gaseous star (of mass M, radius R, and luminosity L) acted on by tidal forces originating from a point-mass companion (of mass M'). The primary is assumed to move in a circular orbit

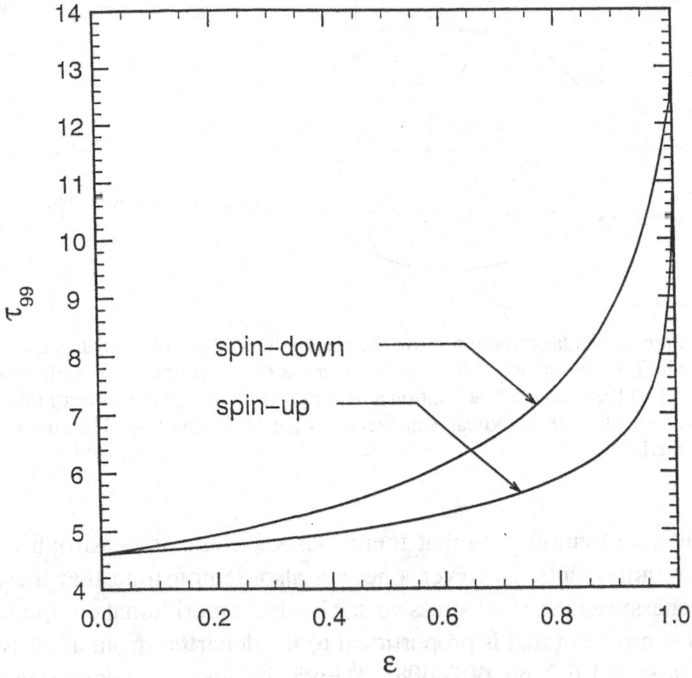

Fig. 8.2. Characteristic times for spin-up and spin-down between two parallel infinite plates as functions of the Rossby number (see Eq. [8.31]). *Source:* Weidman, P. D., *J. Fluid Mech.*, **77**, 685, 1976.

about their common center of mass, with its rotation axis perpendicular to the orbital plane. Let the center of mass of the primary be taken as the origin of our system of spherical polar coordinates (r, θ, φ). As usual, the x axis points toward the point-mass companion, and the z axis is parallel to the rotation axis.

If synchronization has not yet been achieved, it is evident that the primary is not at rest with respect to the frame corotating with the orbital angular velocity Ω_0. To be specific, if Ω_i is a typical value of the initial rotational angular velocity, then the rotational velocity in the corotating axes is $\mathbf{u} = (\Omega_i - \Omega_0)\varpi \mathbf{1}_\varphi$. (In this particular frame, thus, a state of perfect synchronism corresponds to $\mathbf{u} \equiv \mathbf{0}$.) Such a purely azimuthal motion can only be approximate, however, because the primary is always elongated in the direction of the line joining the two centers of mass. This lack of axial symmetry around the rotation axis is illustrated in Figure 8.3, where the four arrows indicate the tidal attraction corrected for the gravitational attraction at the center of mass of the primary. (The small tidal lag is not represented because it plays a negligible role in the Ekman-pumping process. This is an approximation, of course, since it is this tidal lag that will eventually permit a secular exchange of energy and of rotational and orbital angular momenta.) Evidently, if there were no tidal bulges, the motion would remain forever axisymmetric in the corotating frame. Because of the presence of these tidal distortions, however, each fluid parcel in the surface layers is forced to move along an ellipse, in a plane parallel to the equator, with slight accelerations and decelerations along its trajectory.

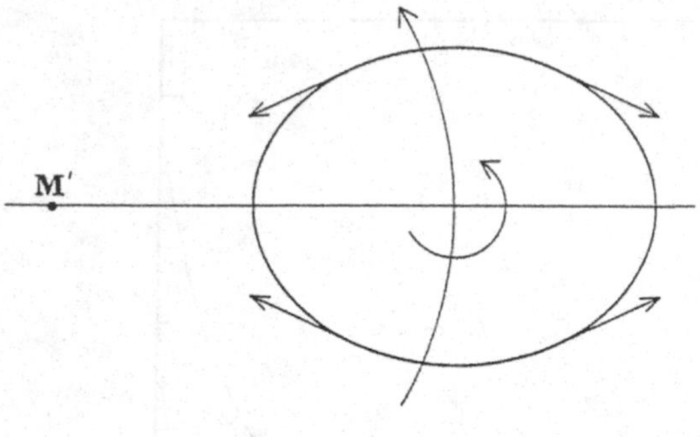

Fig. 8.3. *Differential* tidal attraction due to the mass point M' (i.e., the tidal attraction corrected for the gravitational attraction at the center of mass of the primary), at four places in the equatorial belt of the primary. The rotation axis is perpendicular to the plane of this schematic drawing. The vertical arrow indicates the sense of the orbital motion. The small tidal lag is not represented.

It is a simple matter to demonstrate that there exists a class of geostrophic motions that satisfy these requirements. However, one can also demonstrate that these purely azimuthal flows generate a tangential stress vector having a meridional component and a smaller azimuthal component that is proportional to the departure from axial symmetry (see Tassoul and Tassoul 1992, pp. 607–608). Thence, because both components of the stress vector must vanish at a free boundary, one can show that slow but inexorable meridional currents are needed to cancel out its azimuthal component at the surface of an asynchronous rotator. We may anticipate, therefore, that *a large-scale meridional flow will always be required to satisfy all the basic equations and all the boundary conditions, when synchronization has not yet been achieved.* As we shall see, these transient currents play a role that is quite similar to that of the secondary flows described in Sections 2.5.3 and 8.4.1.

To formulate the problem in its most general terms, one should solve simultaneously the hydrodynamical equations and the equations that describe the tidal interaction between two deformable bodies. In particular, because the magnitude of the orbital angular velocity Ω_0 is slowly varying in time, the correct form of the equations of motion is

$$\frac{\partial \mathbf{u}}{\partial \mathbf{t}} + \mathbf{u} \cdot \operatorname{grad} \mathbf{u} + 2\Omega_0 \mathbf{1}_z \times \mathbf{u} + \dot{\Omega}_0 \mathbf{1}_z \times \mathbf{r} = \mathbf{A} - \operatorname{grad}\left(V - \frac{1}{2}\Omega_0^2 \varpi^2 - W\right) - \frac{1}{\rho}\operatorname{grad} p + \frac{1}{\rho}\mathbf{F}(\mathbf{u}), \tag{8.44}$$

where $\mathbf{A}(t)$ describes the acceleration of the common center of mass with respect to our frame of reference, W is the tidal potential, and $\mathbf{F}$ is the (turbulent) viscous force per unit volume.* Remaining symbols have their standard meanings (see Eqs. [2.27]). A dot designates a derivative with respect to time.

To make the problem tractable, we shall neglect the nonlinear terms $\mathbf{u} \cdot \operatorname{grad} \mathbf{u}$ in Eq. (8.44). This implies that one has $|\mathbf{u} \cdot \operatorname{grad} \mathbf{u}| \ll |2\Omega_0 \mathbf{1}_z \times \mathbf{u}|_z$ and $|\mathbf{u} \cdot \operatorname{grad} \mathbf{u}| \ll |\operatorname{grad} W|$ (i.e., $|\Omega_i - \Omega_0| \ll \Omega_0$ and $|\Omega_i - \Omega_0|^2 \ll GM'/d^3$, where d is the mutual

* See, e.g., Landau, L. D., and Lifshitz, E. M., *Mechanics*, Section 39, Oxford: Pergamon Press, 1959.

separation of the two components). For the sake of simplicity, we shall neglect the secular variations in time of the orbital angular velocity, and we shall assume also that the star is a barotrope (see Section 3.2.1). Given these simplifying approximations, by taking the curl of Eq. (8.44), we obtain

$$\frac{\partial}{\partial t}(\text{curl}\,\mathbf{u}) + 2\Omega_0\,\text{curl}(\mathbf{1}_z \times \mathbf{u}) = \text{curl}\left[\frac{1}{\rho}\,\mathbf{F}(\mathbf{u})\right], \tag{8.45}$$

which is quite similar to Eq. (8.32). The function **F** can be neglected in the bulk of the star; in the surface boundary layer, however, one has

$$\mathbf{F} = \frac{\partial}{\partial r}\left(\mu_V \frac{\partial \mathbf{u}}{\partial r}\right) + \cdots, \tag{8.46}$$

where μ_V is the vertical (i.e., in the direction of gravity) coefficient of eddy viscosity. To ensure mass conservation we must also prescribe that

$$\text{div}(\rho\mathbf{u}) = 0. \tag{8.47}$$

Boundary conditions (2.20) and (2.21) further imply that

$$\mathbf{n}\cdot\mathbf{u} = 0 \qquad \text{and} \qquad \mathbf{n}\times[\mathbf{n}\cdot\mathbf{T}(\mathbf{u})] = \mathbf{0} \tag{8.48}$$

on the free surface of the tidally and rotationally distorted primary. As usual, **n** is the unit outer normal to the free surface, and **T** are the Reynolds stresses, which depend linearly on μ_V and the first-order derivatives of **u** (see Section 3.6). Equations (8.45)–(8.48) specify the vector **u** completely.

Boundary-layer theory can be used to describe the general features of these time-dependent motions in the corotating frame. To be specific, one writes

$$\mathbf{u} = \sum_k \mathbf{u}_k(r, \theta, \varphi)\exp(-2\Omega_0\beta_k t), \tag{8.49}$$

where the $\mathbf{u}_k$ can be expanded in terms of radial functions and spherical harmonics ($k = 0, 1, 2, \ldots$). Thence, performing a boundary-layer analysis of Eqs. (8.45)–(8.47) and applying boundary conditions (8.48), one can obtain the permissible values for the β_k in Eq. (8.49). Obviously, the lowest eigenvalue β_0 is the most important one since it defines the *e*-folding time τ^* of the transient motions, which is equal to $(2\Omega_0\beta_0)^{-1}$.

Detailed mathematical calculations show that there always exists a thin Ekman-type *suction* layer that induces a large-scale flow of matter within the almost frictionless interior of an asynchronous binary component. Figure 8.4 illustrates the streamlines of the tidally induced meridional flow in a model with constant density and constant eddy viscosity, when the mass ratio is equal to unity ($M = M'$). In the case of a spin-down ($\Omega_i > \Omega_0$), these motions correspond to a *quadrupolar* circulation pattern that is weakly dependent on the longitude φ, with the fluid entering the boundary layer in the equatorial belt and returning with decreased angular momentum to the poles. (The reverse phenomenon occurs in the case of a spin-up, when $\Omega_i < \Omega_0$.) Given our simplifying approximations, the typical speed of the meridional flow is of the order of $\epsilon_T(\delta/R)(\Omega_i - \Omega_0)\varpi$ while the *e*-folding time τ^* is approximately equal to $(2\Omega_0\epsilon_T\delta/R)^{-1}$, where δ/R is the relative boundary-layer thickness – which is of the order of $(\mu_V/2\rho\Omega_0 R^2)^{1/2}$ – and ϵ_T is the ratio of the tidal attraction to gravity at the equator. For moderately small

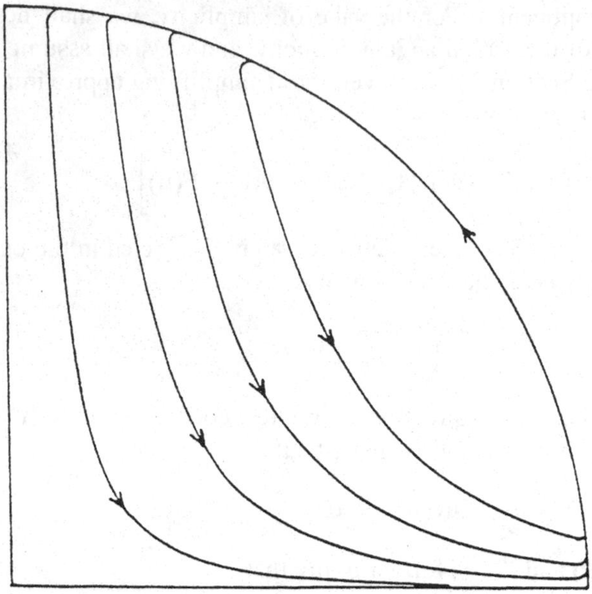

Fig. 8.4. Streamlines of the transient, tidally induced meridional circulation in an asynchronously rotating model with constant density and constant eddy viscosity, when $\Omega_i > \Omega_0$. The rotation axis is vertical. The streamlines do not penetrate into the free boundary, with matter flowing from the equator to the poles in the outermost surface layers. *Source:* Tassoul, J. L., and Tassoul, M., *Astrophys. J.*, **359**, 155, 1990.

eccentricities and masses of comparable magnitude, it follows that

$$\tau^* = \frac{P}{4\pi}\left(\frac{\delta}{R}\right)^{-1}\frac{M}{M'}\left(\frac{a}{R}\right)^3, \tag{8.50}$$

since, for unequal masses, one has

$$\epsilon_T = \frac{M'}{M}\left(\frac{R}{a}\right)^3. \tag{8.51}$$

As usual, P $(= 2\pi/\Omega_0)$ is the orbital period and a is the semimajor axis.

As was already noted, in obtaining these results we have made use of several simplifying approximations. Since I want to avoid any misunderstanding about the applicability of the time scale τ^* to a sample of binary stars, I shall conclude this section by making a few practical remarks.

First, in view of application to actual binary stars, it is much more realistic to evaluate the ratio δ/R for a model in which $\rho \propto (R-r)^3$ in the surface layers. To simulate the case of an eddy viscosity that is larger than the microscopic viscosity, we shall also assume that $\mu_V = 10^N \mu_{\rm rad}$, where $\mu_{\rm rad}$ is the radiative viscosity and N is a constant (see Eq. [4.62]). To be specific, if we let $\rho = \rho_b(R-r)^3$ and $\mu_{\rm rad} = \mu_b(R-r)$, a straightforward dimensional analysis shows that $\delta/R = 10^{N/4}(\mu_b/\rho_b\Omega_0 R^4)^{1/4}$. For a

Cowling point-source model with electron-scattering opacity, this relation takes the form

$$\frac{\delta}{R} = 3 \times 10^{-5+N/4}[P(\mathrm{day})]^{1/4} \left(\frac{L/L_\odot}{M/M_\odot}\right)^{1/4} . \qquad (8.52)$$

Parenthetically note that the small exponent $1/4$ considerably reduces the uncertainties on the coefficient of eddy viscosity in the surface layers.

Second, in deriving Eq. (8.50) we have neglected the secular variations in time of the orbital angular velocity Ω_0. For the sake of simplicity, we have also assumed that the free boundary of the primary is nearly coincident with the steady surface corresponding to synchronism. It is evident that these approximations impose a severe restriction on the mass ratio M'/M. Indeed, if the mass M was much larger than M', a small change in the rotational angular velocity Ω would lead to large variations in time of the quantities Ω_0 and a/R. *Because such large changes were not permitted in our model calculations, it follows at once that Eq. (8.50) does not apply to binary systems having extreme mass ratios.* Such a restriction is unimportant for binary-star systems, since the masses M and M' are in general of comparable magnitude. It is of paramount importance for planetary-satellite systems, however, for they generally have very small mass ratios.

Third, because we have neglected the nonlinear terms $\mathbf{u} \cdot \operatorname{grad} \mathbf{u}$ in our analysis, the e-folding time defined in Eq. (8.50) is no more than a lower limit to the actual synchronization time, t_{syn} (say), in a real binary star. Indeed, as was properly shown in Section 8.4.1, *nonlinearity increases the spin-up and spin-down times of the flow between parallel infinite plates* (see Figure 8.2). Accordingly, because these laboratory problems are quite similar to ours, the synchronization time t_{syn} should also be much larger than the e-folding time τ^*. Finally, recall that we have considered barotropic models only; that is, we have explicitly assumed that Eq. (8.45) contains no term proportional to the vector $\operatorname{grad} p \times \operatorname{grad} \rho$. General considerations in geophysics indicate that *baroclinicity effects inhibit large-scale circulations*. Hence, given the great similarities between the geophysical and astronomical problems, we conclude that in a more realistic stellar model the inherent departures from barotropy should also inhibit the tidally driven currents. Since the effects of nonlinearity and baroclinicity cannot be ascertained at this time, hereafter we shall make the reasonable assumption that $t_{\mathrm{syn}} = 10^{\sigma}\tau^*$, where σ is a constant of order unity (see Section 8.4.3).

Fourth, in an early-type binary component, the tidally driven currents are most probably confined to its radiative envelope, since the core–envelope interface can act as an effective barrier. This fact is of little concern to us, because it is only the surface rotation rates that can be measured. On theoretical grounds, a concomitant braking of the convective core by turbulent diffusion of linear momentum is quite plausible, however.

Fifth, since the tidally driven currents do not depend on eddy viscosity in the bulk of an asynchronous binary component, it follows at once that the Ekman-pumping process is also operative in stars possessing a deep convective envelope. In fact, the hydrodynamical mechanism should be more effective in late-type binary components than in early-type ones, because the ratio δ/R takes larger values in stars that have a larger eddy viscosity in their outermost surface layers.

8.4.3 *The characteristic times*

In Sections 8.2.2 and 8.3.1 we have expounded the main shortcomings of the two well-known mechanisms that are usually invoked to explain synchronism in the close binary stars. In Section 8.4.2 we have considered a third mechanism that tends to synchronize the axial and orbital motions in the components of a detached close binary. This spin-down (or spin-up) process involves large-scale meridional currents superposed on the azimuthal motion around the rotation axis of an asynchronous rotator. Figure 8.4 illustrates these *transient* motions, which vanish altogether when the tidally distorted body has reached a state of hydrostatic equilibrium in the frame corotating with the orbital angular velocity Ω_0, that is, when synchronization has been attained. They are quite similar, therefore, to the transient meridional currents that are responsible for the decay of various rotational motions near solid boundaries (see Figures 2.2, 2.3, and 8.1).

Although problems with solid boundaries have some obvious features in common with the double-star problem, it is evident that they differ in the manner by which the secondary meridional flow comes into existence. For example, in the case of a midlatitude cyclonic vortex in the Earth's atmosphere, turbulent friction acting on the ground slows down the azimuthal motion, thus producing a radial inflow of matter toward the rotation axis. By continuity, this horizontal transport is balanced by a small vertical flux of matter into the free atmosphere above the surface boundary layer. It is this upward motion that eventually produces the secondary flow illustrated in Figure 2.2 (as well as those illustrated in Figures 2.3 and 8.1). In contrast, *in the double-star problem there are no solid boundaries that may spin down (or spin up) the azimuthal motion in an asynchronously rotating binary component.* But then, as was explained in Section 8.4.2, it is the self-gravitational attraction of the star that acts as the "container," forcing the tidal bulges to remain almost aligned with the line joining the two centers of mass. Given this severe constraint on the free surface of a tidally distorted star, one can show that it is this forced lack of axial symmetry that prevents the azimuthal motion from being wholly one of pure rotation in an asynchronous rotator, thus leading to the formation of transient meridional currents, as illustrated in Figure 8.4.

Note also that the problems with solid and free boundaries differ in their respective time scales. In the former case, the spin-down time is proportional to the dynamical time scale (i.e., the final period) divided by the relative thickness of the Ekman pumping layer (see Eqs. [2.101] and [8.43]). In the case of a tidally distorted star, however, the e-folding time τ^* is proportional to the dynamical time scale (i.e., the orbital period) divided by the *product* $\epsilon_T(\delta/R)$, where ϵ_T is a measure of the small departure from axial symmetry (see Eq. [8.51]) and δ/R is the relative boundary-layer thickness (see Eq. [8.52]). Obviously, the presence of a free (rather than solid) boundary reduces the efficiency of the Ekman-pumping process. As we shall see in Section 8.4.4, however, even though δ/R lies in the range 10^{-5} to 10^{-3} and despite the fact that ϵ_T is also a small parameter, this is quite sufficient to provide the general trend of the observational data.

Making use of the results presented in Section 8.4.2, one finds that

$$t_{\rm syn}({\rm yr}) = \frac{1.44 \times 10^{\sigma - N/4}}{q(1+q)^{3/8}} \left(\frac{L_\odot}{L}\right)^{1/4} \left(\frac{M_\odot}{M}\right)^{1/8} \left(\frac{R}{R_\odot}\right)^{9/8} \left(\frac{a}{R}\right)^{33/8} \tag{8.53}$$

or, by virtue of Kepler's third law,

$$t_{\rm syn}({\rm yr}) = 5.35 \times 10^{2+\sigma-N/4}\,\frac{1+q}{q}\left(\frac{L_\odot}{L}\right)^{1/4}\left(\frac{M}{M_\odot}\right)^{5/4}\left(\frac{R_\odot}{R}\right)^{3}[P({\rm day})]^{11/4}, \tag{8.54}$$

where q is the mass ratio. As explained in Section 8.4.2, these formulae do not apply to binary systems having extreme mass ratios. The meaning of the adjustable factor $10^{\sigma-N/4}$ is properly explained at the end of Section 8.4.2 also. Following Claret, Giménez, and Cunha (1995), the most plausible values are $\sigma \approx 1.6$, with $N \approx 0$ in a radiative envelope and $N \approx 10$ in a convective envelope. However, future discussions based on a larger sample of binaries could well lead to a smaller value for σ and more refined values for N.

Simultaneously, because viscous dissipation retards the equilibrium tide, angular momentum is exchanged between the orbit and the rotation of each component, thus modifying the orbital eccentricity of the binary star. To a good degree of approximation, the ratio $t_{\rm syn}/t_{\rm cir}$ is of the order of the ratio of rotational and orbital angular momentum. Hence, for moderately small eccentricities and masses of comparable magnitude, we can estimate the time to circularize the orbit by

$$t_{\rm cir}({\rm yr}) = \frac{1.44 \times 10^{\sigma-N/4}}{(1+q)^{11/8} r_g^2}\left(\frac{L_\odot}{L}\right)^{1/4}\left(\frac{M_\odot}{M}\right)^{1/8}\left(\frac{R}{R_\odot}\right)^{9/8}\left(\frac{a}{R}\right)^{49/8} \tag{8.55}$$

or

$$t_{\rm cir}({\rm yr}) = 9.4 \times 10^{3+\sigma-N/4}\,\frac{(1+q)^{2/3}}{r_g^2}\left(\frac{L_\odot}{L}\right)^{1/4}\left(\frac{M}{M_\odot}\right)^{23/12}\left(\frac{R_\odot}{R}\right)^{5}[P({\rm day})]^{49/12}, \tag{8.56}$$

where r_g is the fractional gyration radius. As usual, the secondary makes a similar contribution to the effective circularization time of the binary (see Eq. [8.21]).

At this juncture, it is worth noting that the two mechanisms presented in Sections 8.2 and 8.3 are mutually exclusive, in the sense that one of them applies to stars having a deep convective envelope whereas the other one applies to stars having a radiative envelope. Because the third mechanism can be operative in both groups of stars, however, the resonance mechanism and the hydrodynamical mechanism both produce secular changes in the spin and orbital parameters of the early-type stars. Similarly, because the tidal-torque mechanism and the hydrodynamical mechanism are not mutually exclusive, both of them can be operative in the late-type binaries. As we shall see in Section 8.4.4, in the early-type binaries it is always the hydrodynamical mechanism that is the most effective; Eqs. (8.53) and (8.55) thus provide the dominant contributions to the times $t_{\rm syn}$ and $t_{\rm cir}$. In contrast, *in stars having a deep convective envelope both the tidal-torque mechanism and the hydrodynamical mechanism can be operative, albeit for different values of the parameters R, L, and P*. Accordingly, for these binaries Eqs. (8.19) and (8.21) can be used to discuss jointly the synchronization and orbital circularization caused by both mechanisms, provided that one inserts

$$\frac{1}{t_{\rm syn}} = \frac{1}{t_{\rm syn}({\rm Eq.~[8.20]})} + \frac{1}{t_{\rm syn}({\rm Eq.~[8.53]})} \tag{8.57}$$

and, for each component,

$$\frac{1}{t_{\text{cir}}} = \frac{1}{t_{\text{cir}}(\text{Eq. [8.22]})} + \frac{1}{t_{\text{cir}}(\text{Eq. [8.55]})} \tag{8.58}$$

in the corresponding equations.

To conclude this section, it may not be inappropriate to stress again the importance of solving Eqs. (8.19) and (8.21) rather than making comparison between time scales. As was already noted in Section 8.2.2, the latter approach is probably adequate, in most cases, for binary components evolving without mass loss on the main sequence. For rapidly contracting or expanding components, however, it might prove quite misleading. It is also worth noting that *any meaningful comparison between theory and observation can be made on a statistical basis only, preferably with a large sample of binaries.* In other words, one should not evaluate the merits of a theory by making use of a few short-period binaries that display a very eccentric orbit and/or a large degree of asynchronism. These binaries may have had a large initial eccentricity, with a large initial departure from synchronism. Other evolutionary processes, which were not included in the theory, might also significantly modify both axial rotation and the strength of tidal interaction. Recall also that the tidal-torque and hydrodynamical mechanisms both depend on two adjustable parameters: the viscous time scale T in Eqs. (8.20) and (8.22) and the factor $10^{\sigma - N/4}$ in Eqs. (8.53)–(8.56). One should not expect these two quantities to be universal constants, however. As a matter of fact, the mathematical difficulties of the problem are such that there is little or no hope of calculating their values from first principles alone. It would thus seem that progress can be made only through an efficient cooperation between theory and observation, using – as I said – a very large sample of binaries.

8.4.4 Pseudo-synchronization and orbital circularization

Let us first discuss the early-type main-sequence binaries. As was pointed out in Section 8.3.1, the resonance mechanism is unable to account for the observed degree of orbital circularization among these stars. However, by making use of Eqs. (8.55) and (8.56), one can show that the hydrodynamical mechanism is quite effective in inducing orbital circularization during the main-sequence lifetime t_a of these stars. Yet, if one considers a model consisting of two A-type components with little or no turbulence in their outer layers (i.e., with $N = 0$), a comparison between the times t_{cir} and t_a shows that this mechanism can explain circular orbits up to $P \approx 6$ days only. This is somewhat shorter than the 10-day cutoff shown in Figure 1.11. This finding strongly suggests either that turbulence in the outer layers might play a somewhat greater role in these stars (with $N \approx 4$, say, instead of $N = 0$) or that their observed eccentricity distribution might result from main-sequence and pre–main-sequence circularization. Accordingly, a mere comparison of the time scales t_{cir} and t_a is probably insufficient in this case; the problem requires a direct integration of Eqs. (8.21) and (8.55).

As far as synchronization is concerned, the hydrodynamical mechanism is also a much more efficient process than the resonance mechanism, since the time t_{syn} depends primarily on the factor $(a/R)^{4.125}$ instead of $(a/R)^{8.5}$ (see Eqs. [8.27] and [8.53]). The presence of a smaller exponent makes it a long-range mechanism that can enforce synchronization up to $a/R \approx 20$ in the early-type main-sequence binaries. This is in agreement with the observations reported in Section 1.4. Note also that the effects of the hydrodynamical mechanism can still be felt for larger separations, up to $P \approx 100$ days (say) – without bringing complete synchronization beyond $P \approx 15$–25 days, however. This is in agreement with the finding that most late A-type dwarfs in binaries with $P \leq 100$ days are Am stars, with rotational velocities smaller than 100 km s^{-1} (see Section 6.3.5). There is thus no need to invoke pre–main-sequence braking to explain the paucity of normal A-type dwarfs from binaries with orbital periods smaller than 100 days.

Now, in Section 8.2.2 we have shown that the tidal-torque mechanism is quite ineffective in inducing orbital circularization in the late-type main-sequence binaries (see Eqs. [8.23]–[8.25]). To show that the hydrodynamical mechanism can be operative in these stars, let us apply Eq. (8.56) to a typical solar-type binary component. For reasonable values of the parameters, one finds that

$$t_{\rm cir}({\rm yr}) = 3 \times 10^{7-N/4}[P({\rm day})]^{49/12} , \qquad (8.59)$$

where N is a fitting constant. In fact, because we have let $\mu_V = 10^N \mu_{\rm rad}$, the factor 10^N is some mean value of the Reynolds number Re in the surface layers (see Eq. [2.51]). Since N is a free quantity in Eq. (8.59), we shall thus prescribe that for each cluster listed in Table 1.2 one has $t_{\rm cir} = t_a$ at $P = P_{\rm cut}$. For the three oldest clusters one obtains $N \approx 9.3$–9.7 or $Re \approx 10^9$–10^{10}. For the Pleiades one must let $N \approx 12$, whereas the pre–main-sequence cluster requires the value $N \approx 14$. These two values may not be quite reliable, however, because Eq. (8.59) is directly applicable only to *static* stars. Anyhow, these crude evaluations of N are quite reasonable for late-type main-sequence binaries because, owing to the extreme smallness of the microscopic viscosity, the outer layers of a convective envelope can easily sustain Reynolds numbers of the order 10^9–10^{10}. Hence, we conclude that the hydrodynamical mechanism can be responsible for orbital circularization on the main sequence, even though it may not be equally efficient during the pre–main-sequence contraction. Since the two relevant mechanisms are not mutually exclusive, this result strongly suggests that the tidal-torque mechanism is the dominant one during the pre–main-sequence phase up to 8 days (as reported in Section 8.2.2), whereas the hydrodynamical mechanism becomes fully responsible for orbital circularization during the main-sequence phase, at a much slower pace, beyond $t_a = 1$ Gyr (say). If so, then, the inefficiency of the tidal-torque mechanism on the main sequence is no longer an issue. Obviously, an integration of Eqs. (8.21) and (8.58) would be most welcome.

In Section 8.2.2 we also pointed out that the tidal-torque mechanism does not wholly account for pseudo-synchronization in the late-type main-sequence binaries. Making use of Eq. (8.54), one can show that the hydrodynamical mechanism is operative in these stars, although it is difficult to quantify with any certainty the counter-effects of magnetically driven winds on the synchronization process. In this connection, let us mention the work of van't Veer and Maceroni (1992), who have shown that the hydrodynamical mechanism is much more effective than the tidal-torque mechanism in the angular-momentum losing, G-type binaries belonging to the main-sequence group.

A very interesting case of asynchronism is that of the double-lined eclipsing binary TZ Fornacis. It is a system with an orbital period of 75.7 days and a circular orbit. Its components have nearly equal masses ($M = 2.05 M_\odot$ and $M' = 1.95 M_\odot$) but unequal radii ($R = 8.32 R_\odot$ and $R' = 3.96 R_\odot$). The more massive component is rotating synchronously with the orbital motion while the companion is spinning 16 times faster than the orbital period rate. This puzzling binary star has been recently investigated by Claret and coworkers, who integrated Eqs. (8.19) and (8.21) along the evolutionary path of each binary component. Within the theoretical and observational error bars, these calculations describe how the binary components may pass through a stage characterized by a synchronous primary and a supersynchronous companion in circular orbits about their common center of mass. However, two independent sets of calculations strongly suggest that the case of TZ Fornacis can be explained either by the hydrodynamical mechanism alone or by a combination of the tidal-torque and resonance mechanisms. Since all three mechanisms can operate in a tidally distorted star, one may therefore argue that they become almost equally efficient during some periods of post–main-sequence evolution, so that all of them must be taken into account during this expanding phase.

8.5 Contact binaries: The astrostrophic balance

In Sections 4.6 and 8.2–8.4 we have considered detached close binaries, in which the tidal distortions are relatively small and where components display physical characteristics that are similar to those of single stars. When the two components are separated by a few radii only, these tidal distortions may become, however, quite large. This is well illustrated by eclipsing binaries that exhibit sinusoidal-type light curves, and for which the first-order scheme of approximation adopted in the above sections becomes utterly inadequate. In what follows I shall thus consider the *Roche model*, in which practically all the mass of each component is concentrated in a central point surrounded by a tenuous envelope of vanishingly small density.* The importance of this model stems from the fact that it provides a good approximation for binary components that are in physical contact and share a common envelope.

Let M_I and M_{II} denote the masses of the two components, and let D be their mutual separation. We choose a rotating frame of reference with the origin at the center of gravity of the mass M_I. The x axis points toward the center of gravity of the mass M_{II}, and the z axis is perpendicular to the orbital plane. The effective gravity at any point P can be described as the gradient of a potential Ψ, where

$$\Psi = G\frac{M_I}{r_I} + G\frac{M_{II}}{r_{II}} + \frac{1}{2}\Omega^2\left[\left(x - \frac{M_{II}}{M_I + M_{II}}D\right)^2 + y^2\right], \tag{8.60}$$

in which r_I and r_{II} are the distances from P to the centers of gravity of the two masses. Let us further assume that the rotational angular velocity occurring in Eq. (8.60) is equal to the Keplerian orbital angular velocity. We thus let

$$\Omega^2 = \frac{G\,(M_I + M_{II})}{D^3}. \tag{8.61}$$

If we adopt D as the unit of length and GM_I/D as the unit of potential, we can then write (except for an additive constant)

$$\Psi = \frac{1}{r_I} + q\left(\frac{1}{r_{II}} - x\right) + \frac{1}{2}(1+q)(x^2 + y^2), \tag{8.62}$$

where $q = M_{II}/M_I$ is the mass ratio ($q \leq 1$).

* The geometry of the equipotentials that surround a rotating gravitational dipole was originally investigated by the French mathematician Edouard Roche (1820–1883) in 1873. For a detailed historical account the reader should consult Kopal's (1989) book.

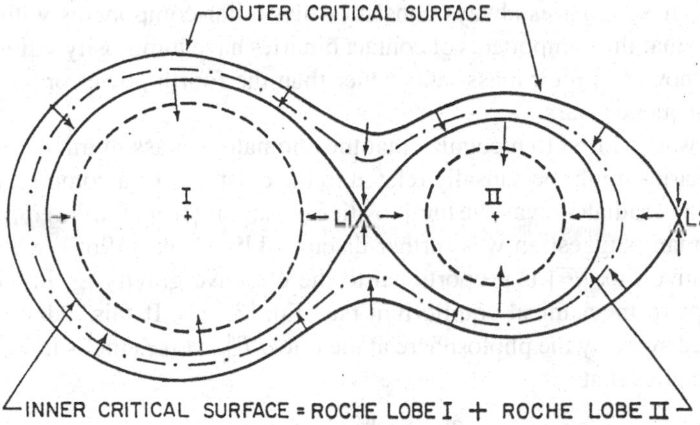

Fig. 8.5. The inner and outer critical surfaces of the binary Roche model plotted in the equatorial plane. The arrows indicate the direction of the effective gravity. For a detached system, the two stellar surfaces (*dashed curves*) both lie beneath the inner critical surface; for a contact binary, the common stellar surface (*dash-dotted curve*) lies between the inner and outer critical surfaces. *Source:* Shu, F. H., Lubow, S. H., and Anderson, L., *Astrophys. J.*, **209**, 536, 1976.

Figure 8.5 represents a section of the equipotentials $\Psi =$ *constant* cut by the orbital plane $z = 0$. Quite generally, level surfaces corresponding to high values of Ψ form separate lobes enclosing each one of the two centers of gravity and differ little from spheres. With diminishing values of Ψ, the two lobes become increasingly elongated in the direction of their common center of gravity until, for a certain critical value $\Psi = \Psi_{\text{in}}$ characteristic of each mass ratio, both lobes will come into contact to form a dumbbell-like configuration. It will henceforth be called the *inner critical surface*, and its two lobes will be called the *Roche lobes*. Note that the Roche lobes unite at a point where the effective gravity vanishes (i.e., at the Lagrangian point L_1). For even smaller values of Ψ, the connecting part of the dumbbell will open up so that single level surfaces enclose both bodies, thus providing us with a convenient representation of a contact binary. Below a critical value $\Psi = \Psi_{\text{out}}$ ($< \Psi_{\text{in}}$) characteristic of each mass ratio, however, gravitational confinement of a binary against the expansive tendency of its internal pressure is no longer possible. An inspection of Figure 8.5 shows that this *outer critical surface* also contains a point where the effective gravity vanishes (i.e., the Lagrangian point L_2). For a contact binary, the common stellar surface thus lies between the inner and outer critical surfaces corresponding to the equipotentials $\Psi = \Psi_{\text{in}}$ and $\Psi = \Psi_{\text{out}}$.

By definition, contact binary stars have both components filling or overfilling their Roche lobes. Practically all known contact systems are eclipsing binaries. The light curves of these extremely close systems have a sinusoidal appearance, which is due to the severe tidal distortion of the components. They also have eclipse minima of almost equal depth, implying very similar effective temperatures for both components. In fact, this property of the contact binaries seems to be continuous over a wide range of spectral types, from stars as early as O type to stars as late as K type. (They range in orbital period from 5.6 days to 0.22 days.) The similarity of effective temperatures would not be surprising if contact binaries consisted of identical stars. However, for some as yet unknown reason, these binaries always consist of dissimilar components with unequal masses. Note also that the components of contact binaries have luminosity ratios roughly equal to the first power of their mass ratio rather than the fourth power or so observed for single main-sequence stars.

Struve (1948) was the first to recognize that the anomalous mass–luminosity relation of the contact binaries might be causally related to the existence of a common envelope that redistributes and radiates away the luminosities emanating from the two independent cores. This important suggestion was further discussed by Osaki (1965), who pointed out that the radiative flux $|\mathcal{F}|$ is proportional to the effective gravity $|\mathbf{g}|$ in a common *radiative* envelope in mechanical equilibrium (see Eq. [3.41]). If this radiative flux is ultimately radiated away by the photosphere at the rate $\sigma T_{\rm eff}^4$, von Zeipel's law of gravity darkening thus implies that

$$T_{\rm eff} \propto g^{0.25}, \tag{8.63}$$

where $T_{\rm eff}$ is the effective temperature and g is the local surface gravity. Now, the condition that the free surface of a contact binary must be an equipotential implies a relation between the radii and masses of the components. For the binary Roche model, this relation may be approximated by

$$\frac{R_{II}}{R_I} = \left(\frac{M_{II}}{M_I}\right)^{0.46}. \tag{8.64}$$

It follows at once that the average surface gravities ($\approx GM/R^2$) of the two components are nearly equal. Hence, by virtue of Eq. (8.63), their effective temperatures should be also nearly equal.

The case of a common convective envelope in mechanical equilibrium was subsequently discussed by Lucy (1967), who found that the variation of effective temperature with local surface gravity is of the form

$$T_{\rm eff} \propto g^{0.08}. \tag{8.65}$$

Again, because the average surface gravities of the two components are closely equal, this gravity-darkening law predicts little variation of effective temperature over the free surface of a late-type contact binary.

Following Osaki (1965) and Lucy (1968), we can now derive a theoretical mass–luminosity relation that is valid for both the early-type and late-type contact binaries. It follows at once from Eq. (8.64) that the ratio of surface areas ($\approx R^2$) of the two Roche lobes is closely equal to the mass ratio. Hence, because we have shown that the components of a contact binary have similar surface brightnesses ($\approx L/4\pi R^2$), we obtain the approximate relation

$$\frac{L_{II}}{L_I} = \frac{M_{II}}{M_I}. \tag{8.66}$$

This relation closely agrees with the observational data. We therefore conclude that the anomalous mass–luminosity relation of the contact binaries merely reflects the ratio of surface areas for components having similar effective temperatures.

It is generally accepted that the main features of the photosphere of a contact binary star are to be understood in terms of energy transport within a common (radiative or convective) envelope. The foregoing discussion clearly shows that the top layers of the common envelope are barotropic (with equal pressures, densities, and temperatures over the equipotentials). Yet, because the two underlying radiating cores have unequal masses, we know that the temperature distribution cannot be uniform over the Roche lobes. By continuity, temperature differences over each equipotential above the inner critical surface do exist, therefore implying that the bottom layers of the common envelope are baroclinic (see Section 3.2.1). We are thus faced at once with the following two questions: First, what is the exact nature of the energy flow that brings nearly equal effective temperatures in the two components of a contact binary? And, second, is it possible to build a common-envelope model that is barotropic in its outermost surface layers while being baroclinic near the two dissimilar Roche lobes?

As was pointed out by Lucy (1968), the paradox of overluminous secondaries in the late-type contact binaries can be resolved by assuming some lateral energy transfer in a common convective envelope. The existence of early-type contact binaries makes it clear that this energy transfer can occur even in contact binaries with radiative envelopes. This fact strongly suggests that there exists a transfer mechanism common to both the late-type and early-type contact binaries that is quite independent of the underlying envelope structure. This is the reason why it has often been conjectured that the required interchange of heat and mass is directly attributable to a lateral temperature or entropy gradient, in a direction roughly parallel to the equipotentials, near the base of the inner critical surface. For the sake of simplicity, I shall consider the case of a common radiative envelope.

In the frame rotating with the Keplerian orbital angular velocity Ω, the basic equations governing the motion in an early-type contact binary are

$$\frac{\partial \rho}{\partial t} + \operatorname{div}(\rho \mathbf{u}) = 0, \tag{8.67}$$

$$\frac{\partial \mathbf{u}}{\partial t} + \mathbf{u} \cdot \operatorname{grad} \mathbf{u} + 2\boldsymbol{\Omega} \times \mathbf{u} = -\frac{1}{\rho} \operatorname{grad} p + \operatorname{grad} \Psi + \frac{1}{\rho} \mathbf{F}(\mathbf{u}), \tag{8.68}$$

$$\rho T \left(\frac{\partial S}{\partial t} + \mathbf{u} \cdot \operatorname{grad} S \right) = \operatorname{div}(\chi \operatorname{grad} T) + \rho \epsilon_{\rm Nuc}, \tag{8.69}$$

$$p = \frac{\mathcal{R}}{\bar{\mu}} \rho T + \frac{1}{3} a T^4, \tag{8.70}$$

where $\mathbf{u}$ is the velocity relative to the rotating axis, $\mathbf{F}$ is the turbulent viscous force per unit volume, and Ψ is the Roche potential defined in Eq. (8.62). Remaining symbols have their standard meanings (see Section 3.2). These six scalar equations are to be solved subject to appropriate initial and boundary conditions at the two stellar centers and at the top of the common envelope.

Numerous attempts have been made to build a contact-binary model consisting of two stars having different masses but equal effective temperatures. Yet, as is well known, the internal structure of a contact binary remains a puzzle. It is not my intention in this section to review all the conflicting models that can be found in the literature. Rather, I shall briefly comment on one important but often forgotten ingredient of the problem, namely, the *geostrophic* (or *astrostrophic*) flow that is required to prevent the appearance of unwanted discontinuities in the solutions. Since *all* proposed models do exhibit discontinuities at the base of the common envelope, astrostrophy could well provide the solution for the existing impasse.

We begin by describing the barotropic models originally proposed by Shu, Lubow, and Anderson (1976) because they may be viewed as the zeroth-order solution for unevolved main-sequence contact binaries. Following these authors, we make the a priori assumption that, apart for the slow thermally driven currents discussed in Chapter 4, the system is at rest in the corotating frame (i.e., $\mathbf{u} \equiv \mathbf{0}$). Since the condition of mechanical equilibrium is in general incompatible with the energy equation in a circulation-free barotrope, we must therefore assume that radiative equilibrium holds on average on each equipotential $\Psi =$ *constant* (see Sections 3.3.1 and 6.2). To derive the zeroth-order equations, we shall also introduce a system of curvilinear coordinates (ξ, η, ζ) with $\xi = \Psi$ and with η and ζ defining the "horizontal" position on a level surface. Assuming further that the chemical composition is uniform over each equipotential, one finds that $p = p_0(\Psi)$, $\rho = \rho_0(\Psi)$, and $T = T_0(\Psi)$. Hence, letting $\mathbf{u} \equiv \mathbf{0}$ in Eqs. (8.67)–(8.69), we obtain

$$\frac{dp_0}{d\Psi} = \rho_0 \tag{8.71}$$

and

$$\chi \frac{dT_0}{d\Psi} = \frac{L}{\bar{g}A(\Psi)}, \tag{8.72}$$

in the radiative regions. As usual, L is the total luminosity, $\bar{g}$ is the effective gravity averaged over an equipotential, and A is the area of that closed surface. Similar ordinary differential equations can be written down for the two convective cores in which nuclear burning is taking place.

Since Eq. (8.62) gives a complete specification of the effective gravitational field, detailed solutions of these ordinary differential equations can be obtained using the standard boundary conditions at the two centers and at the shared surface. Unfortunately, as was correctly pointed out by Shu and coworkers, there are too many boundary conditions to satisfy for all thermodynamic variables to be continuous across the inner critical surface. Because mechanical requirements imply that the pressure $p_0(\Psi)$ must be continuous across the Roche lobes, it was therefore concluded that no barotropic solutions with unequal stellar components exist unless one makes allowance for discontinuous changes in the density $\rho_0(\Psi)$ and the temperature $T_0(\Psi)$ at one of the Roche lobes. This is the *contact-discontinuity hypothesis*. Although the models constructed according to this idea look very much like observed contact binaries, they have been widely criticized on the ground that a contact discontinuity should disappear on a thermal time scale.* I shall not go into the disputes because, unsatisfactory as these zeroth-order barotropic models might be, they could provide the foundation for a more satisfactory solution of the basic equations.

* Lucy (1976) and others have suggested that a newly formed contact binary will evolve on a thermal time scale toward a state of marginal contact and that, if contact is then broken, the system will evolve back into contact, again on a thermal time scale. This is the *thermal-relaxation-oscillation hypothesis*. As was pointed out by Shu (1980), however, *a contact discontinuity will also naturally arise in these oscillatory models*, with the zeroth-order barotropic models constituting the equilibrium states about which Lucy's (1976) models might undergo thermal relaxation oscillations.

Obviously, if the two stellar components had the same mass, their common envelope could be treated as a barotrope, all the way from the Roche lobes to the photosphere. Because the mass ratio is not in general equal to one, however, the luminosities generated in the two separate cores become necessarily unequal at the inner critical surface. This uneven distribution of the sources of heat generates a lateral interchange of heat and mass in the bottom layers of the common envelope. As we shall demonstrate, this inescapable fact implies the existence of a large-scale astrostrophic flow along the equipotentials that lie above the inner critical surface.

In Section 2.2.3 we have seen that the relative importance of the inertial and Coriolis effects is measured by the Rossby number Ro $(= U/\Omega D)$, where U characterizes the scale of the horizontal velocity. In the present case, the Rossby number is of the order of the ratio of the orbital period $(= 2\pi/\Omega)$ to the characteristic time of the flow $(\approx D/U)$. Since this ratio is undoubtedly much smaller than one, the inertia of the relative motion can be neglected in Eq. (8.68). Hence, restricting attention to steady motions in the corotating frame, we can rewrite that equation in the form

$$2\boldsymbol{\Omega} \times \mathbf{u} = -\frac{1}{\rho} \operatorname{grad} p + \operatorname{grad} \Psi. \tag{8.73}$$

Note that we have omitted the viscous force because it plays a negligible role away from the boundaries (see, however, below).

To present a self-consistent formulation of the problem, we shall first write each thermodynamic variable as the sum of the (known) zeroth-order solution and a baroclinic "correction." We thus let

$$p = p_0(\Psi) + p_1(\Psi; \eta, \zeta), \tag{8.74}$$

and we write similar expressions for the density and the temperature in the common envelope. Making use of Eq. (8.71), we can thus rewrite the "vertical" component of Eq. (8.73) in the form

$$2\rho\,(\boldsymbol{\Omega} \times \mathbf{u})_V = -\frac{\partial p_1}{\partial \Psi} + \rho_1, \tag{8.75}$$

where the subscript "V" designates a component along the effective gravity. Similarly, the two "horizontal" components of that equation are

$$2\rho\,(\boldsymbol{\Omega} \times \mathbf{u})_H = -\operatorname{grad}_H p_1, \tag{8.76}$$

where the subscript "H" designates a component parallel to the equipotentials.

The key point to our discussion is that the thickness d of the common envelope is much smaller than the typical horizontal length D, which is the mutual separation. Since the horizontal scale of variation of p_1 is $\mathcal{O}(D)$, it readily follows from Eq. (8.76) that $p_1 = \mathcal{O}(\rho\Omega U D)$. Accordingly, because in Eq. (8.75) the vertical pressure gradient $\partial p_1/\partial\Psi$ is $\mathcal{O}(p_1/d)$, one finds that $\partial p_1/\partial\Psi$ is $\mathcal{O}(\rho\Omega U D/d)$. Taking into account that the vertical component of the Coriolis force is $\mathcal{O}(\rho\Omega U)$, we obtain

$$\frac{|\rho\,(\boldsymbol{\Omega} \times \mathbf{u})_V|}{|\partial p_1/\partial\Psi|} = \mathcal{O}\left(\frac{d}{D}\right) \ll 1. \tag{8.77}$$

We therefore conclude that the Coriolis acceleration can rightfully be neglected in Eq. (8.75).

Following Pedlosky (1987), we can also derive an estimate of the density ratio ρ_1/ρ_0 since, by virtue of Eqs. (8.75) and (8.77), we know that the quantity $\rho_1 g$ is of the same order of magnitude as the vertical component of grad p_1. Excepting perhaps the point L_1, we can thus write

$$\rho_1 = \mathcal{O}\left(\frac{p_1}{gd}\right) = \mathcal{O}\left(\frac{\rho\Omega U D}{gd}\right), \tag{8.78}$$

so that

$$\frac{\rho_1}{\rho} = \mathcal{O}\left(\frac{U}{\Omega D}\right)\mathcal{O}\left(\frac{\Omega^2 D^2}{gd}\right). \tag{8.79}$$

Making use of Eq. (8.61) and letting $g \approx GM/R^2$, we obtain

$$\frac{\rho_1}{\rho} = \mathcal{O}\left(\frac{U}{\Omega D}\right)\mathcal{O}\left(\frac{R^2}{Dd}\right). \tag{8.80}$$

Because the ratio R^2/Dd is of order one in a contact binary, we therefore conclude that, as long as the Rossby number remains much smaller than one, we can let $\rho_1 \ll \rho_0$ in the common envelope.

Thus, within the framework of our approximations, Eqs. (8.75) and (8.76) become

$$\frac{\partial p_1}{\partial \Psi} = \rho_1 \tag{8.81}$$

and

$$2\rho_0 (\mathbf{\Omega} \times \mathbf{u})_H = -\text{grad}_H\, p_1. \tag{8.82}$$

Similarly, because the flow produces only slight density changes as long as the Rossby number is small, Eq. (8.67) can be approximated by

$$\text{div}\, \mathbf{u} = 0, \tag{8.83}$$

from which it follows that the ratio of the vertical to horizontal speeds is $\mathcal{O}(d/D)$ and, hence, much smaller than one.

Now, adding Eqs. (8.71) and (8.81), we obtain

$$\frac{\partial p}{\partial \Psi} = \rho, \tag{8.84}$$

which describes an approximate balance in the vertical direction between the vertical pressure force and the effective gravity. This is the *hydrostatic approximation.* It is quite different from the approximation made in Eq. (8.82), in which the horizontal Coriolis force is made to balance the horizontal pressure force that is permanently maintained near the base of the common envelope. This is known as the *astrostrophic approximation*, and it is the strict analog of the geostrophic approximation discussed in Section 2.5.1.

In order to specify the astrostrophic velocity **u** in the common radiative envelope of an early-type contact binary, it is necessary to make explicit use of a relation between the horizontal pressure and temperature gradients. In the case of a simple ideal gas, this relation is quite straightforward since Eq. (8.70) then reduces to the linear relation

$$p = \frac{\mathcal{R}}{\bar{\mu}}\rho_0 T, \tag{8.85}$$

so that Eq. (8.82) becomes

$$2(\boldsymbol{\Omega} \times \mathbf{u})_H = -\frac{\mathcal{R}}{\bar{\mu}} \operatorname{grad}_H T. \tag{8.86}$$

The necessity for large-scale astrostrophic currents also requires that we solve the nonlinear equation (8.69) for the temperature field in the common envelope. Here we have

$$\rho_0 T \mathbf{u} \cdot \operatorname{grad} S = \operatorname{div}(\chi \operatorname{grad} T). \tag{8.87}$$

By making use of Eqs. (2.11) and (8.83), we can also rewrite Eq. (8.87) in the more convenient form

$$c_V \rho_0 \mathbf{u} \cdot \operatorname{grad} T = \operatorname{div}(\chi \operatorname{grad} T), \tag{8.88}$$

where c_V is the specific heat at constant volume.

Equations (8.83), (8.86), and (8.88) are the fundamental equations of the problem. They provide a simple but adequate description of the astrostrophic flow and the lateral energy transfer in a common radiative envelope. As usual, appropriate boundary conditions must be prescribed at the inner critical surface and at the outer boundary. Moreover, since we do not expect the astrostrophic flow to penetrate into the Roche lobes, we must ensure that conditions (2.20) and (2.21) are properly satisfied at the inner critical surface. Not unexpectedly, these requirements lead to the formation of a thin viscous boundary layer immediately above the Roche lobes. Since the outer part of the common envelope has to be closely barotropic, we must also require that the baroclinic corrections (i.e., p_1, ρ_1, and T_1) and the astrostrophic velocity $\mathbf{u}$ vanish at a distance from the inner critical surface.

Although a detailed solution of this hydrodynamical problem still lies in the distant future, I hope that the above discussion has made clear the need for a consistent treatment of the astrostrophic balance at the base of the common envelope in a contact-binary star (see Eq. [8.86]). Indeed, there can be little doubt that large-scale currents flowing along (and not across) the equipotentials play an essential role in the problem since, without them, it would be impossible to obtain a solution that satisfies all the basic equations and all the boundary conditions, while being continuous across the inner critical surface.

Admittedly, I have so far considered the astrostrophic balance in static models only, that is, binary systems in which there is no net flux of matter from one stellar component to the other. However, one can show that the necessity of having large-scale astrostrophic currents in the lower part of the common envelope also applies to evolving binary configurations in which one stellar component is losing matter to its companion, so that the shape of their equipotentials is gradually changing in time. In other words, *the concept of astrostrophy is equally relevant to both the static and evolving systems*. A quantitative study based on Eqs. (8.83)–(8.87) would be most useful, therefore, since these theoretical results could provide considerable insight into the nature of the lateral energy transfer in the common (radiative or convective) envelope of an evolving contact binary.

8.6 Discussion

Chapter 4 and some parts of Chapter 5 have been devoted to the study of large-scale circulations generated by nonspherical perturbations to the structure of a star. More specifically, Section 4.6 was concerned with the steady, *thermally driven* meridional flows that result from tidal distortion and mutual heating of the stellar components in a detached close binary. The necessity for *mechanically driven* currents in an asynchronously rotating binary component was further discussed in Section 8.4.2. As was noted, the importance of these transient motions lies in the fact that they serve to synchronize the axial and orbital motions far more rapidly than could turbulent diffusion of momentum. For completeness, in Section 8.5 we have also presented a qualitative study of the *astrostrophic* currents that arise from the nonuniform heating at the base of the common envelope in a contact binary. Since each of these four independent flows exhibits quite distinct features, I shall briefly discuss their differences and similarities.

In all studies of meridional circulation in stars, the assumption is made that turbulent friction can be neglected altogether in the bulk of the configuration. It is generally accepted that the flow calculated on the base of a simple frictionless model does provide an adequate representation of the motion at some distance from the boundaries. However, as I have several times mentioned in the book, a frictionless solution does not satisfy the kinematic boundary condition

$$\mathbf{n} \cdot \mathbf{u} = 0\,, \qquad \text{with} \quad |\mathbf{u}| \text{ finite}, \tag{8.89}$$

neither at the free surface nor at the $\bar{\mu}$-barrier defining a core–envelope interface. This is the basic reason why one *must* retain turbulent friction in a very thin layer of fluid adjacent to each boundary. In this thin boundary layer the normal component of the velocity is diminished continuously from its interior value to a limiting value of zero at the boundary. It is thus the turbulent viscous force, which contains second-order derivatives in the velocity $\mathbf{u}$, that allows both the radial component $\mathbf{n} \cdot \mathbf{u}$ to vanish and the tangential component $\mathbf{n} \times \mathbf{u}$ to remain finite at the boundaries. In other words, the presence of viscous friction increases the order of the equations in the boundary layers so that it is possible to satisfy as many boundary conditions as the basic equations demand (see Section 2.2.2). For some reason, however, this well-known fact has been (and still is) frequently ignored in the astronomical literature.

A good example of the importance of boundary layers in stars is provided by the reflection effect in detached close binaries. Indeed, as was shown in Section 4.6.2, the presence of a "hot spot" on the photosphere of a binary component generates large-scale superficial currents. It is immediately apparent from Figure 4.9 that this axially symmetric circulation remains confined to a thin thermo-viscous boundary layer, with the speed of the flow decreasing exponentially with optical depth. Of course, to this boundary-layer flow we must add the steady circulations generated by rotation and tidal attraction alone, and which are illustrated in Figures 4.3 and 4.8. In each case, the time scale of the thermally driven currents is equal to the Kelvin–Helmholtz time, GM^2/RL, divided by a small number that measures the corresponding departure from spherical symmetry (see Eqs. [4.9], [4.127], and [4.136]). As was explained in Section 4.8, these currents are quite different from those encountered in geophysics and laboratory hydrodynamics.

By contrast, there is an evident similarity between the problems treated in Sections 8.4.1 and 8.4.2. (Compare Figures 8.1 and 8.4.) In each case, the motion consists of three distinct phases: the formation of an Ekman-type suction layer near the boundaries, the establishment of a large-scale meridional flow that spins down (or spins up) the frictionless interior, and finally the viscous decay of small residual oscillations.

In spite of obvious differences, these two problems have one essential feature in common, namely, a change in the azimuthal motion near the boundaries. In the laboratory problem, this change is due to the frictional force acting near the solid walls; in the double-star problem, it is caused by the tidal attraction that forces the fluid particles in the surface layers to move nonuniformly along noncircular orbits (see Section 8.4.3). No matter whether the boundaries are solid or free, however, the spin-down results mainly from the conservation of specific angular momentum in the frictionless interior, with the large-scale advection of angular momentum being regulated by the Ekman layer near the boundaries. Actually, the difference between the time scales defined in Eqs. (8.43) and (8.50) can be ascribed to the nature of the pumping mechanism itself: either an impulsive change in the rotation rate of the two parallel infinite plates or the forced lack of axial symmetry in the azimuthal motion of an asynchronously rotating binary component.

To illustrate this point, let us calculate the spin-down times by means of a simple qualitative argument. In each case, the specific angular momentum is essentially preserved in the frictionless interior. Hence, in the linear approximation, a fluid particle with specific angular momentum $\Omega_i \varpi^2$ will acquire the lower angular velocity Ω_0 by moving radially outward the distance

$$l = \frac{1}{2} \frac{\Omega_i - \Omega_0}{\Omega_0} \varpi. \tag{8.90}$$

The spin-down time, which is the time required for the fluid to cover this length at the speed of the meridional currents, is therefore equal to

$$\frac{l}{|\mathbf{u}|} = \frac{(\Omega_i - \Omega_0)\varpi}{2\Omega_0 |\mathbf{u}|}, \tag{8.91}$$

where $\mathbf{u}$ is the typical speed of the meridional flow.

In the laboratory problem, by virtue of Eqs. (8.40) and (8.42), one has $|\mathbf{u}| \approx (\delta/2L)(\Omega_i - \Omega_0)\varpi$. Hence, making use of Eq. (8.91), one readily sees that the spin-down time is of the order of $(\Omega_0 \delta/L)^{-1}$, which is the result obtained in Eq. (8.43). In contrast, in the double-star problem, we have shown that

$$|\mathbf{u}| \approx \epsilon_T \frac{\delta}{R} (\Omega_i - \Omega_0)\varpi. \tag{8.92}$$

Equations (8.91) and (8.92) imply at once that the spin-down time is of the order of $(2\Omega_0 \epsilon_T \delta/R)^{-1}$, which is the result obtained in Eq. (8.50).

For small Rossby numbers, there is also a great similitude between a geostrophic wind in the Earth's atmosphere and the large-scale circulatory currents in the common envelope of a contact binary having dissimilar components (see Sections 2.5.1, 2.5.2, and 8.5). As is well known, the geostrophic balance is a good approximation for the velocity field in the free atmosphere, at some distance above the Earth's surface. It provides for the wind to follow the direction of the surfaces of constant pressure, and for the geostrophic velocity to vary with height according to the thermal wind equation (see Eq. [2.84]). The situation is quite similar in a contact binary with unequal masses in the sense that nonuniform heating at the inner critical surface generates a lateral temperature gradient and, hence, an astrostrophic flow in the bottom layers of the common envelope. In this case, however, because it is impossible to observe the interior of a contact binary, one must solve simultaneously the coupled equations for the large-scale motion and the temperature field in the common envelope (see Eqs. [8.83], [8.86], and [8.88]). This is not expected to be a straightforward task, for the Roche geometry is awkward, to say the least. As far as I can see, the problem can be made more tractable by using the triply orthogonal system of Roche coordinates that is associated with purely tidal distortions (see, e.g., Kopal 1989, pp. 41–44). Obviously, the removal of the centrifugal potential from Eq. (8.62) is a minor approximation because the flattening caused by the centrifugal force can hardly affect the energy transfer between the components in their common envelope. Although the assumed steadiness of the flow is perhaps a more questionable approximation, it should be of no serious concern at this stage, however, since – as was noted in Section 8.5 – static models could prove indispensable to the development of a rational theory of evolving contact-binary stars.

8.7 Bibliographical notes

The presentation in Sections 8.2–8.4 largely follows:

1. Tassoul, J. L., and Tassoul, M., *Fund. Cosmic Physics*, **16**, 377, 1996.

This review paper contains many additional references as well as a more detailed comparison between theory and observation. See also:

2. Claret, A., and Giménez, A., *Astron. Astrophys.*, **296**, 180, 1995.
3. Claret, A., Giménez, A., and Cunha, N. C. S., *Astron. Astrophys.*, **299**, 724, 1995.
4. Claret, A., and Cunha, N. C. S., *Astron. Astrophys.*, **318**, 187, 1997.

Reference 35 is a useful addendum to these four papers.

Section 8.2. The tidal-torque theory was originally considered in:

5. Darwin, G. H., *Phil. Trans. Roy. Soc. London*, Part II, **170**, 447, 1879 (reprinted in *Scientific Papers*, II, p. 36, Cambridge: Cambridge University Press, 1908).

Application to stars with an outer convective envelope was first made by:

6. Zahn, J. P., *Ann. Astrophys.*, **29**, 489, 1966.

Tidal evolution in close binary systems for high eccentricities is discussed in:

7. Alexander, M. E., *Astrophys. Space Science*, **23**, 459, 1973.
8. Mignard, F., *The Moon and the Planets*, **20**, 301, 1979; *ibid.*, **23**, 185, 1980.
9. Hut, P., *Astron. Astrophys.*, **99**, 126, 1981; *ibid.*, **110**, 37, 1982.

Other contributions are by:

10. Zahn, J.P., *Astron. Astrophys.*, **57**, 383, 1977; *ibid.*, **67**, 162, 1978; *ibid.*, **220**, 112, 1989.
11. Scharlemann, E. T., *Astrophys. J.*, **246**, 292, 1981; *ibid.*, **253**, 298, 1982.
12. Goldman, I., and Mazeh, T., *Astrophys. J.*, **376**, 260, 1991.
13. Goodman, J., and Oh, S. P., *Astrophys. J.*, **486**, 403, 1997.

Detailed comparisons between theory and observation will be found in References 1, 4, and 13. See also:

14. Zahn, J. P., and Bouchet, L., *Astron. Astrophys.*, **223**, 112, 1989.
15. Maceroni, C., and van't Veer, F., *Astron. Astrophys.*, **246**, 91, 1991.
16. Verbunt, F., and Phinney, E. S., *Astron. Astrophys.*, **296**, 709, 1995.

Section 8.3. The classical references on the subject are those of:

17. Cowling, T. G., *Mon. Not. R. Astron. Soc.*, **101**, 367, 1941.
18. Zahn, J. P., *Astron. Astrophys.*, **41**, 329, 1975; *ibid.*, **57**, 383, 1977.

Additional calculations are due to:

19. Rocca, A., *Astron. Astrophys.*, **111**, 252, 1982; *ibid.*, **175**, 81, 1987; *ibid.*, **213**, 114, 1989.

See especially her third paper. Related discussions are those of:

20. Goldreich, P., and Nicholson, P. D., *Astrophys. J.*, **342**, 1079, 1989.
21. Ruymaekers, E., *Astron. Astrophys.*, **259**, 349, 1992.
22. Savonije, G. J., and Papaloizou, J. C. B., *Mon. Not. R. Astron. Soc.*, **291**, 633, 1997.

Detailed comparisons between theory and observation will be found in References 1 and 4; see also:

23. Pan, K. K., *Astron. Astrophys.*, **321**, 202, 1997.

Section 8.4.1. See especially Reference 14 of Chapter 2. The following review paper is particularly worth noting:

24. Benton, E. R., and Clark, A., Jr., *Annu. Rev. Fluid Mech.*, **6**, 257, 1974.

The reference to Weidman is to his paper:

25. Weidman, P. D., *J. Fluid Mech.*, **77**, 685, 1976.

Sections 8.4.2–8.4.4. The hydrodynamical mechanism was originally discussed in:

26. Tassoul, J. L., *Astrophys. J.*, **322**, 856, 1987; *ibid.*, **358**, 196, 1990.
27. Tassoul, J. L., *Astrophys. J. Letters*, **324**, L71, 1988.
28. Tassoul, J. L., and Tassoul, M., *Astrophys. J.*, **359**, 155, 1990; *ibid.*, **395**, 259, 1992.
29. Tassoul, M., and Tassoul, J. L., *Astrophys. J.*, **395**, 604, 1992.

See also:

30. van't Veer, F., and Maceroni, C., in *Binaries as Tracers of Stellar Formation* (Duquennoy, A., and Mayor, M., eds.), p. 237, Cambridge: Cambridge University Press, 1992.

The combined effects of the tidal-torque and hydrodynamical mechanisms are treated in:

31. Tassoul, J. L., *Astrophys. J.*, **444**, 338, 1995.
32. Keppens, R., *Astron. Astrophys.*, **318**, 275, 1997.

Reference 1 (pp. 408–412) presents a detailed comparison between theory and observation; see also References 2 and 3, as well as Budaj's contribution (Reference 46 of Chapter 6).

The hydrodynamical mechanism has been criticized by:

33. Rieutord, M., *Astron. Astrophys.*, **259**, 581, 1992.
34. Rieutord, M., and Zahn, J. P., *Astrophys. J.*, **474**, 760, 1997.

In Reference 33 the claim is made that Ekman pumping is not efficient enough to reduce the synchronization time, which should remain of the order of the viscous time. For some reason, the same argument was repeated in Reference 34.

To be specific, in both papers they describe the internal motion by means of a series in the powers of the small parameter δ/R, which is the relative thickness of the Ekman layer at the free surface. By making use of this one-parameter expansion in the powers of δ/R, they conclude that the meridional currents described in Section 8.4.2, which are proportional to the *product* $\epsilon_T(\delta/R)$, should not exist. It is a simple matter to show that their analysis of the second-order terms is inadequate because they have failed to prescribe that these terms, which are intricately coupled to the first-order terms, must satisfy the vorticity equation as well as the boundary conditions on the tensions (see Eqs. [8.45] and [8.48]). *Their analysis of the meridional flow in a tidally distorted configuration is incomplete, therefore, and so cannot be presented as a proof that these currents should be of order* $(\delta/R)^2$. (Note also that their one-parameter expansion is quite inadequate to describe tidally driven currents, since it defines meridional motions, of order $(\delta/R)^2$, that do not vanish in the limiting case $\epsilon_T \to 0$.) Accordingly, there is no reason to claim that tidally driven currents of order $\epsilon_T(\delta/R)$ do not exist in a nonsynchronous binary component.

In both papers, the claim is also made that the synchronization time should be equal to the ratio of the available kinetic energy to the power dissipated by friction in the surface boundary layer. This may be true in the case of a laboratory fluid with fixed solid boundaries in the rotating frame. It is certainly incorrect in the double-star problem, however, because the outer surface of a binary component is not perfectly fixed in the corotating frame, so that the tidally distorted body is liable to exchange kinetic energy and angular momentum with the orbital motion. This is another fundamental difference between the well-known problems with solid boundaries and the double-star problem for which the time-dependent torque caused by the small tidal lag plays an essential role.

This and other misapprehensions presented in References 33 and 34 are discussed in:

35. Tassoul, M., and Tassoul, J. L., *Astrophys. J.*, **481**, 363, 1997.

In Reference 35 (p. 367) it is also explained why planetary systems, such as Io–Jupiter or recently discovered planet–star systems, do not fulfill the stringent conditions under which the time scale defined in Eq. (8.50) has been obtained.

Sections 8.5 and 8.6. An exhaustive discussion of the binary Roche model, its associated coordinates, and associated harmonics will be found in:

36. Kopal, Z., *The Roche Problem*, Dordrecht: Kluwer, 1989.

See also Reference 51. The difficulties of constructing contact systems composed of two unequal stellar components were originally noted by:

37. Kuiper, G. P., *Astrophys. J.*, **93**, 133, 1941.

The following key references are discussed in the text:

38. Struve, O., *Ann. Astrophys.*, **11**, 117, 1948.
39. Osaki, Y., *Publ. Astron. Soc. Japan*, **17**, 97, 1965.
40. Lucy, L. B., *Zeit. Astrophys.*, **65**, 89, 1967.
41. Lucy, L. B., *Astrophys. J.*, **151**, 1123, 1968; *ibid.*, **205**, 208, 1976.
42. Shu, F. H., Lubow, S. H., and Anderson, L., *Astrophys. J.*, **209**, 536, 1976; *ibid.*, **229**, 223, 1979; *ibid.*, **239**, 937, 1980.
43. Lubow, S. H., and Shu, F. H., *Astrophys. J.*, **216**, 517, 1977; *ibid.*, **229**, 657, 1979.

Good critical reviews of these and other theoretical contributions have been given by:

44. Shu, F. H., in *Close Binary Stars: Observations and Interpretation* (Plavec, M. J., Popper, D. M., and Ulrich, R. K., eds.), p. 477, Dordrecht: Reidel, 1980.
45. Smith, R. C., *Quart. J. R. Astron. Soc.*, **25**, 405, 1984.
46. Rucinski, S. M., in *Interacting Binary Stars* (Pringle, J. E., and Wade, R. A., eds.), p. 113, Cambridge: Cambridge University Press, 1985.

See also:

47. Wang, J. M., *Astron. J.*, **110**, 782, 1995.

The difficulties of modeling large-scale thermally driven currents in Roche geometry are well illustrated in the following papers:

48. Webbink, R. F., *Astrophys. J.*, **215**, 851, 1977.
49. Smith, R. C., and Smith, D. H., *Mon. Not. R. Astron. Soc.*, **194**, 583, 1981.
50. Zhou, D. Q., and Leung, K. C., *Astrophys. J.*, **355**, 271, 1990.

Application of the astrostrophic balance to contact binaries was first explicitly made in:

51. Tassoul, J. L., *Astrophys. J.*, **389**, 375, 1992.

See also Reference 7 (pp. 45–56) of Chapter 2.

Epilogue

Although stellar rotation has aroused the interest of many distinguished astronomers and mathematicians for almost four hundred years, the theoretical study of the basic physical processes is largely a development of the twentieth century, indeed of the past thirty years or so. In this book I have attempted to present the theory of rotating stars as a branch of classical hydrodynamics, pointing out the differences and similarities between stars and other systems in which rotation is an essential ingredient, such as the Earth's atmosphere and the oceans. Throughout this volume I have thus assumed that the laws governing the internal dynamics of a rotating star are the usual principles of classical mechanics – basically mass conservation, Newton's second law of action, and the laws of thermodynamics. As is well known, one of the reasons why fluid motions in huge natural systems are so complex derives from the fact that the Navier–Stokes equation of motion is inherently nonlinear, so that the superposition of two solutions of a given problem is not necessarily a solution of that problem. In physical terms, this means that it is not possible in general to describe only the largest scale motions in a rotating star, since these flows will almost certainly interact with a whole spectrum of smaller-scale motions. The necessity of incorporating these small-scale, eddylike and/or wavelike motions into the large-scale flows remains as one of the important problems to be solved in astrophysical fluid dynamics.

With very few exceptions, geophysical and astrophysical problems involve motions of such complexity that progress can be made only through a cooperation between formal theory and observation. Since the late 1940s, together with the observations there have been great advances in our theoretical understanding of large-scale phenomena in the Earth's atmosphere and the oceans. By contrast, until the mid-1980s astronomers always had to make use of analytic or numerical models that could not be adequately verified with the available data base. There is little doubt that this lack of direct measurements can explain, at least in part, why the theory of rotating stars is lagging somewhat behind the Earth sciences. It cannot be presented as a complete explanation, however, since prior to the 1970s the oceanographers too had great difficulties in observing the deep interior of the oceans. As a matter of fact, in the first half of the twentieth century, while the geophysicists were assembling the fundamental mechanisms governing the large-scale atmospheric and oceanic flows, the astronomers still had to explain, among other problems, the origin of the energy radiated by the Sun and the stars. Research on stellar interiors thus became primarily a branch of modern physics, with great emphasis being laid on the atomic and nuclear processes. Actually, spherically symmetric stellar models in hydrostatic equilibrium were so successful in accounting for the major observed properties of stars that the most challenging problems of stellar hydrodynamics received comparatively much less attention.

Meridional currents in rotating stars provide a good example of the slow maturation of ideas in stellar hydrodynamics. Indeed, already in 1925, Eddington and Vogt reasoned that the transport of radiation in a rotationally distorted star should cause large-scale motions in meridian planes passing through the rotation axis. Eddington even went a step further, noting that "(these) currents will be deflected east and west by the star's rotation, just as similar currents in our own atmosphere are deflected by the earth's rotation." More importantly, he also wrote: "Presumably when the current has attained a moderate speed a steady state will be reached because the viscosity of the stellar material is considerable and the fundamental equations of equilibrium will be modified by the addition of viscous stresses."* Important advances in our knowledge of the internal dynamics of a rotating star were thus made in the 1920s. Yet, despite some far-reaching but incomplete contributions made over the next three decades, the building blocks that explain how the meridional currents and concomitant differential rotation are sustained in an early-type star were not properly assembled until the 1980s – essentially during the period 1982–1995. Chapter 4 presents an overall picture of these techniques, which have been successfully applied to the development of analytic models in which circulation and rotation are represented explicitly but the smaller-scale motions parametrically. Undoubtedly, the most exciting prospects for the future are associated with possibilities of incorporating into numerical models the transfers achieved by these small-scale motions, resolving individual eddy events in sufficient detail to reproduce their transfer properties adequately rather than making use of ad hoc coefficients of eddy viscosity.

Because asteroseismology is still in its infancy, the interior of an upper-main-sequence star has remained so far *terra incognita*.† Hence, there has been as yet little contact between observation and the theoretical studies of large-scale flows in the early-type stars (see, however, Section 6.4). Yet, the accumulation of relatively recent observations has made it clear that, while we understand the fundamentals of stellar evolution, the so-called standard models are in error in a number of details. This is particularly true for early-type stars, and stellar rotation is currently the favorite candidate to explain the discrepancies. Indeed, Herrero et al. (1992) have found that all fast rotators among O-type stars show large surface helium abundances correlated with the rotation rate, which indicates that there is probably a link between rotation and turbulent mixing in these

* Eddington, A. S., *The Internal Constitution of the Stars*, p. 285, Cambridge: Cambridge University Press, 1926 (New York: Dover Publications, 1959).

† The number of oscillation modes detected in main-sequence stars and white dwarfs is by many orders of magnitude smaller than that in the Sun. This number is much too small to determine the radial structure of a star directly from measured frequencies, as it is done in helioseismology (see Section 1.2.2). Even in the most favorable cases (such as the white dwarf PG 1159-035, in which about one hundred frequencies have been identified with high-order g-modes), the observed splitting of the modes only provides global information about the star's rotation: the value of its rotation period, $P_{\rm rot} = 1.35$ d, and the evidence that it is rotating nearly uniformly. For a recent survey of asteroseismology, see W. A. Dziembowski, in *Sounding Solar and Stellar Interiors* (Provost, J., and Schmider, F. X., eds.), I.A.U. Symposium No 181, p. 317, Dordrecht: Kluwer, 1997.

stars.* Unfortunately, as was repeatedly pointed out in Sections 3.6 and 5.4.1, it is not possible at this writing to calculate unequivocally the coefficients of eddy diffusivity in the radiative interior of rotating stars. Accordingly, since the choice of these coefficients is far from being unique, this will necessarily bring about a certain amount of uncertainty in the numerical treatment of stellar evolution. That is to say, despite the fact that rotation has long been known to be capable of inducing turbulent mixing in stellar radiative zones, we are not yet in a position to provide a fully quantitative explanation for the data. Moreover, because the practical evaluation of the eddy diffusivities of matter and momentum in the radiative interior of a rotating star is at least partly an art, not just a science, there is so far no clear expectation for the large-scale flow deep inside an upper-main-sequence star. This could hardly be more different than the situation encountered in late-type star studies, since new observational techniques have recently provided a great deal of information about the internal rotation of the Sun and the rotational evolution of low-mass stars.

Till the late 1980s, theoretical models invariably predicted that the angular velocity in the solar convection zone was constant on cylinders concentric to the rotation axis; moreover, there were then some indications that the Sun's radiative core might be rotating much more rapidly than the surface. According to the most recent helioseismological data, however, it is now generally thought that the rotation rate in the solar convection zone is similar to that at the surface, with the outer parts of the Sun's radiative core rotating uniformly at a rate somewhat lower than the surface equatorial rate. (The rotation rate in the inner core is more uncertain, but recent measurements indicate that these regions might indeed rotate rigidly down to the center.) The 1980s have thus seen our knowledge of the Sun's internal rotation go from the level of mere speculation to that of a field in which the interplay between theory and observation has become indispensable. Yet, it is clear that we are still a long way from an understanding of the interaction between rotation and turbulent convection. Furthermore, because we cannot infer the internal motions of the Sun in a purely deductive manner from the basic equations, our present understanding of its rotational history remains at best phenomenological. As was pointed out in Chapter 5, refined measurements of the Sun's angular velocity in its most central regions will be needed to identify unequivocally the mechanisms that are continuously redistributing the internal angular momentum in response to the rotational deceleration of the solar convection zone.

The 1990s have also witnessed rapid progress in the theoretical study of low-mass stars, both before and during the main-sequence phase. Again, numerical models have provided the opportunity to delve into the component mechanisms responsible for the rotational evolution of these stars, namely, disk–star magnetic coupling during the early phases and internal angular momentum redistribution and saturated magnetized stellar winds during the later phases. It may not be inappropriate to recall, however, that the numerical simulations presented in Chapters 5 and 7 do not "explain" the current observations but rather provide new insights into processes that are not easily explored with the available

* Herrero, A., Kudritzki, R. P., Vilchez, J. M., Kunze, D., Butler, K., and Haser, S., *Astron. Astrophys.*, **261**, 209, 1992. For a comprehensive review of these and related matters, see Marc Pinsonneault, "Mixing in Stars," *Annu. Rev. Astron. Astrophys.*, **35**, 557, 1997.

instrumentation. This is all the more true in the case of those evolutionary sequences of rotating models that can reproduce the abnormal abundances of the light chemical elements in the Sun and solar-type stars. Indeed, because several adjustable parameters are usually needed to describe the turbulent diffusion processes in the radiative interior of a star, it is quite clear that these models cannot provide the kind of understanding that one would develop from a theory based on first principles alone. Yet, these parameterized models serve a useful purpose because they can be constrained by requiring that the present-day Sun depletes the light elements in the observed proportions, and so they can be used to estimate the gross amount of turbulence present in stellar radiative interiors. Not unexpectedly, the more we progress the more we uncover new, unresolved problems.

In conclusion, it is well to recall that throughout this book I have made use of concepts and methods that were originally introduced in the Earth sciences – barotropy and baroclinicity, geostrophy, eddy–mean flow interaction, boundary-layer theory, etc. In particular, following the example set by the meteorologists and the oceanographers, I have attempted to present consistent solutions that satisfy all the basic equations and all the boundary conditions. This is the reason why we have found that the large-scale motions in the radiative or convective regions of a rotating star always consist of an overall motion around the rotation axis together with much slower but inexorable meridional currents – a situation not unlike those encountered in the Earth's atmosphere and the oceans. In most cases, these secondary flows are dynamically unimportant in the sense that they have little or no effect on the global structure of a rotating star. There is, however, an important exception: the transient meridional currents that advect angular momentum throughout the interior of an asynchronously rotating binary component. As was shown in Chapter 8, this mechanism is closely related to *Ekman pumping*, and so it is of direct relevance to the study of synchronization and orbital circularization in the close (and not so close) binary stars. This is a fairly new concept in astronomy that was essentially developed between 1987 and 1997; hence, unlike other approaches based on celestial mechanics or resonant interactions with natural modes of oscillation, it has not yet become a part of the astronomical tradition. In this, as in many other debatable issues, it is the accumulation of new observational data that will eventually resolve the controversy. In the present problem, it is essential to improve, observationally, the upper period limits above which detached binaries are asynchronously rotating or have noncircular orbits.

Subject index

Author index